KU-996-862

Population and Community Biology

INVASIVE SPECIES AND BIODIVERSITY MANAGEMENT

WITHDRAWN
FROM STOCK
QMUL LIBRARY

Population and Community Biology Series

VOLUME 24

The study of both populations and communities is central to the science of ecology. Th series of books explores many facets of population biology and the processes that determi the structure and dynamics of communities. Although individual authors are given freedo to develop their subjects in their own way, these books are scientifically rigourous and quantitative approach to analysing population and community phenomena is often used.

The titles published in this series are listed at the end of this volume.

INVASIVE SPECIES AND BIODIVERSITY MANAGEMENT

Edited by

Odd Terje Sandlund

Norwegian Institute for Nature Research (Nina),
Trondheim, Norway

Peter Johan Schei

Directorate for Nature Management (DN),
Trondheim, Norway

and

Åslaug Viken

Zoological Institute (NTNU),
Norwegian University of Science and Technology,
Dragvoll, Norway

KLUWER ACADEMIC PUBLISHERS

DORDRECHT / BOSTON / LONDON

A C.I.P. Catalogue record for this book is available from the Library of Congress.

ISBN 0-412-84080-4

Published by Kluwer Academic Publishers,
P.O. Box 17, 3300 AA Dordrecht, The Netherlands.

Sold and distributed in North, Central and South America
by Kluwer Academic Publishers,
101 Philip Drive, Norwell, MA 02061, U.S.A.

In all other countries, sold and distributed
by Kluwer Academic Publishers,
P.O. Box 322, 3300 AH Dordrecht, The Netherlands.

Printed on acid-free paper

Printed in the Netherlands.

Contents

Part 6 Where do we go from here?

Contributors

Michael Bean
Environmental Defense Fund,
1875 Connecticut Avenue, Suite
1016, Washington, DC 20009, USA
phone: 1 202 387 3500.
fax: 1 202 234 6049
e-mail: mb@edf.org

Charles F. Boudouresque
CNRS – University Research Unit
DIMAR, University of the
Mediterranean, Endoume Marine
Station, Rue de Batterie des Lions,
F-13007 Marseilles, France
phone: 33 91269130
fax: 33 91411265
e-mail: boudour@com.univ.mrs.fr

Ralph T. Bryan
Division of Quarantine, National
Center for Infectious Diseases, IHS-
HQW, Epidemiology Branch
5300 Homestead Drive, N.E.,
Albuquerque, NM 87110, USA
phone: 1 505 248 4226
fax: 1 505 248 4393
e-mail: rrb@cdc.gov

James T. Carlton
Williams College – Mystic Seaport,
PO Box 6000, 75 Greenmanville
Ave., Mystic, CO 06355, USA
phone: 1 860 572 5359
fax: 1 860 572 5329
e-mail: james.t.carlton@williams.edu

Michael N. Clout
Centre for Conservation Biology,
School of Biological Sciences,
University of Auckland
Tamaki Campus, PO Box 92019,
Auckland, New Zealand
phone: 64 9 3737599
fax: 64 9 3737001
e-mail: m.clout@auckland.ac.nz

Jeffrey A. Crooks
Marine Life Research Group,
Scripps Institution of Oceanography,
La Jolla, CA 92093-0218, USA
phone: 1 619 534 3579
fax: 1 619 822 0562
e-mail: jcrooks@ucsd.edu

Cyrille de Klemm
IUCN Commission on Environmental
Law,
21 Rue de Dantzig,
F-75015 Paris, France
phone: 33 1 4532 2672
fax: 33 1 4533 4884

Robert E. Eplee
U.S. Department of Agriculture,
Animal and Plant Health Inspection
Service, Plant Protection and
Quarantine, PO Box 279, Whiteville,
NC 28472, USA
phone: 1 910 642 3991
fax: 1 910 642 0757
e-mail: eplee@weblnk.net

Lyle Glowka
IUCN Environmental Law Centre,
Adenauerallee 214,
D-53113 Bonn, Germany
phone: 49 228 2692 231
fax: 49 228 2692 250
e-mail: IUCN-ELC@wunsch.com

Rob Hengeveld
Institute of Forest and Nature
Research, PO Box 23, 6700 AA
Wageningen, The Netherlands
phone: 31 26 35 46 868
fax: 31 26 22 44 175
e-mail: r.hengeveld@ibn.dlo.nl

Kjetil Hindar
Norwegian Institute for Nature
Research (NINA), Tungasletta 2,
N-7005 Trondheim, Norway
phone: 47 73 80 14 00
fax: 47 73 80 14 01
e-mail:
kjetil.hindar@ninatrd.ninaniku.no

Annika Hofgaard
Climate Impact Research Centre
(CIRC), ANS, PO Box 62,
S-98107 Abisko, Sweden
phone: 46 980 40006
fax: 46 980 40171
e-mail:
annika.hofgaard@ ans.kiruna.se

Alan Holt
The Nature Conservancy of Hawaii,
1116 Smith St., Suite 201, Honolulu,
Hawaii 96817, USA
phone: 1 808 537 4508
fax: 1 808 545 2019
e-mail: aholt@tnc.org

Brian J. Huntley
National Botanical Institute,
Private Bag X7,
7735 Claremont, South Africa
phone: 27 21 762 1166
fax: 27 21 761 4687
e-mail: huntley@nbict.nbi.ac.za

Peter Jenkins
Biopolicy Consulting, PO Box 772,
Placitas, NM 87043, USA
phone: 1 505 867 0641
fax: 1 505 771 0737
e-mail: jenkinsbiopolicy@msn.com

Jeffrey A. McNeely
IUCN – The World Conservation
Union, Rue Mauverney 28,
CH-1196 Gland, Switzerland
phone: 41 22 999 0284
fax: 41 22 999 00 15
e-mail: jam@hq.iucn.ch

Harold A. Mooney
Stanford University, Department of
Biological Sciences,
Stanford, CA 94305-5020, USA
phone: 1 415 723 1179
fax: 1 415 723 9253
e-mail:
hmooney@jasper.stanford.edu

Peter B. Moyle
University of California, Department
of Wildlife,
Davis, CA 95616, USA
phone: 1 916 752 6353
fax: 1 916 752 9154
e-mail: pbmoyle@ucdavis.edu

George I. Oduor
CAB International, Africa Regional Centre, PO Box 633, Village Market, Nairobi, Kenya
phone: 254 2 521450
fax: 254 2 522150
e-mail: CABI-IIBC-KENYA@cabi.org

Richard Ogutu-Ohwayo
Fisheries Research Institute, PO Box 343, Jinja, Uganda
phone: 256 43 20484
fax: 256 43 21727
e-mail: firi@mukla.ac.ug

Roger P. Pech
CSIRO Wildlife and Ecology, GPO Box 284, Canberra, ACT 2601, Australia
phone: 61 6 242 1657
fax: 61 6 241 3343
e-mail: r.pech@dwe.csiro.au

Marcel Rejmánek
University of California, Division of Biological Sciences, Section of Evolution and Ecology, Davis, CA 95616, USA
phone: 1 916 752 0617
fax: 1 916 752 5410

David M. Richardson
University of Cape Town, Institute for Plant Conservation, Botany Department, Rondebosch 7700, South Africa
phone: 27 021 650 2440
fax: 27 021 650 4046
e-mail: rich@botzoo.uct.ac.za

Michael J. Samways
Invertebrate Conservation Research Centre, Department of Zoology and Entomology, University of Natal, Private Bag X01, Scottsville 3209, South Africa
phone: 27 331 2605328
fax: 27 331 2605105
e-mail: samways@zoology.unp.ac.za

Odd Terje Sandlund
Norwegian Institute for Nature Research (NINA), Tungasletta 2, N-7005 Trondheim, Norway
phone: 47 73 80 15 48
fax: 47 73 80 14 01
e-mail: odd.t.sandlund@ninatrd.ninaniku.no

Peter Johan Schei
Directorate for Nature Management (DN), Tungasletta 2, N-7005 Trondheim, Norway
phone: 47 73 58 05 00
fax: 47 73 91 54 33
e-mail: peter-johan.schei@dn.dep.no

Vandana Shiva
Research Foundation for Science and Technology, A-60 Hauz Khas, New Dehli, 110016, India
phone: 911 1 69 68 077
fax: 911 1 68 56 795
e-mail: TWN@Unv.ernet.In

Eystein Skjerve
Norwegian College of Veterinary Medicine, Department of Pharmacology, Microbiology and Food, P.O. Box 8146 Dep., N-0033 Oslo, Norway
phone: 47 22 96 48 44
fax: 47 22 96 48 50
e-mail: eystein.skjerve@veths.no

Michael E. Soulé
University of California, Santa Cruz, CA 95064, USA
phone: 1 408 459 4837
fax: 1 408 429 5427
e-mail: soule@co.tds.net

Wendy Strahm
IUCN – The World Conservation Union, 28 Rue Mauverney, CH-1196 Gland, Switzerland
phone: 41 22 999 0157
fax: 41 22 999 0015
e-mail: WAS@HQ.iucn.ch

Åslaug Viken
Zoological Institute, Norwegian University of Science and Technology (NTNU), N-7036 Dragvoll, Norway
phone: 47 73 59 62 78
fax: 47 73 59 13 09
e-mail: aslaug.viken@chembio.ntnu.no

Yngvild Wasteson
Norwegian College of Veterinary Medicine, Department of Pharmacology, Microbiology and Food, P.O. Box 8146 Dep., N-0033 Oslo, Norway
phone: 47 22 96 48 01
fax: 47 22 96 48 50
e-mail: yngvild.wasteson@veths.no

Randy Westbrooks
U.S. Department of the Interior – U.S. Department of Agriculture, 233 Border Belt Drive, P.O. Box 279, Whiteville, NC 28472, USA
phone: 1 910 648 6762
fax: 1 910 648 6763
e-mail : rwestbrooks@weblnk.net

Preface

This volume is based on papers presented at "The Norway/United Nations (UN) Conference on Alien Species" which was hosted by the Norwegian Ministry of the Environment in collaboration with the United Nations Environment Programme (UNEP), the Secretariat of the Convention on Biological Diversity, the United Nations Educational, Scientific and Cultural Organisation (UNESCO), the International Conservation Union (IUCN) and the Scientific Committee on Problems of the Environment (SCOPE) of the International Council of Scientific Unions (ICSU). The organisation and sponsoring of the conference was also a joint venture between the Norwegian Ministry of Environment, the Ministry of Agriculture, the Ministry of Fisheries and the Ministry of Foreign Affairs. The Conference was held at Royal Garden Hotel, Trondheim, Norway, 1-5 July 1996. This was the second Trondheim Conference on Biodiversity, the first being the Norway/UNEP Expert Conference on Biodiversity, 24-28 May 1993.

The conference was organised by the Norwegian Directorate of Nature Management, the Norwegian Institute for Nature Research (NINA) and the Norwegian University for Science and Technology (NTNU) and its Centre for Environment and Development (SMU), all based in Trondheim. These institutions are all active in the fields of management, education and research related to biological diversity.

We are extremely grateful for the support we have received from all these national and international institutions.

A total of approx. 180 scientists, managers and policymakers from developing and developed countries, as well as representatives from international organisations and NGOs, attended the 1996 Trondheim Conference. Among the participants were representatives from the sectors of environment, agriculture, forestry and fisheries with experiences and responsibilities related to prevention and management of invasive species. This broad constituency, which is engaged in the implementation of the CBD, was updated on present scientific knowledge on this issue, through the lectures and discussions at the Conference.

We hope that this peer reviewed volume will further contribute to the understanding of the problem and the effective implementation of the Biodiversity Convention on this challenging issue.

Trondheim, 1998-03-23

Peter J. Schei
Conference Chair

1 Introduction: the many aspects of the invasive alien species problem

ODD T. SANDLUND, PETER J. SCHEI and ÅSLAUG VIKEN
Norwegian Institute for Nature Research (NINA), Directorate for Nature Management (DN), Norwegian University for Science and Technology (NTNU), Trondheim, Norway

The problem

"Near the Guardia we find the southern limit of two European plants, now become excessively common. The fennel in great profusion covers the ditch banks in the neighbourhood of Buenos Ayres, Montevideo, and other towns. But the cardoon (Cynara cardunculus) *has a far wider range: it now occurs in these latitudes on both sides of the Cordillera, across the continent. I saw it in unfrequented spots in Chile, Entre Rios, and Banda Oriental. In the latter country alone, very many (probably several hundred) square miles are covered with one mass of these prickly plants, and are impenetrable by man or beast. Over the undulating plains, where these great beds occur, nothing else can live. Before their introduction, however, I apprehend the surface supported as in other parts a rank herbage. I doubt whether any case is on record, of an invasion of so grand a scale of one plant over the aborigines."*

(C. Darwin 1839. ***Voyage of The Beagle****).*

When Charles Darwin in the 1830s observed two European plant species that had established and dominated in seminatural and urbanized habitats over large areas in South America, he considered it quite remarkable. Today, this is a general phenomenon in many parts of the world. In most countries, the number and proportion of alien species in the flora and fauna are frightfully high, so that seminatural ecosystems may be dominated by non-native species. This has devastating effects on native biodiversity, and introductions of alien invasive species is one of the four members of Jared Diamond's (1985) "evil quartet" of major threats to native biodiversity. Moreover, with the present development in international trade and travel, the transport of species, and thereby the risk of introduction into new areas is bound to increase (Jenkins, Ch. 15). Thus, we are in an urgent need for

O. T. Sandlund et al. (eds.), Invasive Species and Biodiversity Management, 1–7.

tools and methods to reduce this threat to natural biodiversity. The tools must be based on understanding the invasive species and the invasible ecosystem, and the processes involved in invasions. We need to develop predictive models for invasions, and practical ways and means of stopping or reducing invasions.

During the negotiations of the Convention on Biological Diversity (CBD) the threats to indigenous biodiversity posed by alien invasive species were highly recognized. Most biologists consider this the second most important threat factor after habitat destruction. Therefore the obligations laid down in the Convention are quite extensive and demanding. Article 8 h of the Convention states: "Prevent the introduction of, control or eradicate those alien species which threaten ecosystems, habitats or species".

The complex, difficult and global nature of this task indicates the need for a thoroughly planned worldwide strategy and action plan for the prevention of invasions and the management of invasive species. An important element is the cooperative development and use of new, innovative tools in this struggle. A global strategy is now underway, partly as a follow-up to the conference in Trondheim in 1996 (Mooney, Ch. 27). The third Conference of the Parties (COP-3) of CBD in Buenos Aires in 1996 gave its full support to this initiative and the support has later been reinforced by CBD's Subsidiary Body for Scientific, Technical and Technological Advice (SBSTTA) at its third meeting in Montreal in 1997. The expert group on marine and coastal diversity which was established under SBSTTA in 1995 has also emphazised the importance of a global alien species strategy.

It may be noted that the CBD applies the expression "alien species" to denote any species which is introduced into new habitats by human intervention; i.e. a process considered different from the natural process of species migrating and establishing in new areas. In the biological literature we find several different words for this phenomenon. Introduced species, non-native species, and exogeneous species are more or less synonymous and means any species which has been established after human-moderated introduction to the habitat. Alien species as a threat to native biological diversity implies a more restricted concept, and is commonly called invasive or aggressive species. In this book we have accepted the terminology preferred by the individual authors, as the meaning emerges clearly regardless of words. In any case, it is probably a futile effort to try to develop a standard terminology on this subject. "Species" in CBD's article 8h must also be understood to include any alien stock, population or subspecies.

The human dimension of alien species: across disciplines and sectors

The problems related to the invasion and spreading of unwanted organisms are obviously not restricted to the seminatural and urbanized habitats as observed by Charles Darwin in South America more than 160 years ago. Organisms which employ humans as their habitat, or which inhabit domesticated food producing species or food products are on the same bandwagon of increased transport across oceans and continents, as demonstrated by, e.g., Skjerve and Wasteson (Ch. 17), Jenkins (Ch. 15), and Bryan (Ch. 11) in this volume. This impose a major threat to human and animal health, and agricultural production. Disease organisms like the tuberculosis bacterium, which were considered close to extinction, are apparently increasing in frequency and area of occurrence. This is a combined result of resistant strains of bacteria and an increased frequency and distance of migration by people who may carry the disease organism. Diseases and parasites of domesticated food-producing animals and cultivated plants are in many cases increasing, mainly due to liberalized trade and related reduction in quarantine restrictions.

The spreading of alien invasive species which may do harm to human, animal and plant health as well as to the native biodiversity in our environments is a problem handled by agencies responsible for many of society's sectors. An important aim of this volume is to demonstrate that the various sectors of society, e.g., environmental management, human health, agriculture, and veterinary medicine may benefit greatly from sharing their experiences and learning from each other. Holt (Ch. 5) provides an excellent example of cross-sectorial collaboration on alien species management in Hawaii.

The numerous dilemmas inherent in alien species management is amply demonstrated by the fact that most of the food production of the world is based on a great reshuffling of species through agriculture (McNeely, Ch. 2). We are not able to imagine the present level of global food production without this reshuffling. More than 90 % of the world's agricultural production is based on less than 20 plant and 6 animal species. These few species presently have a nearly global distribution, while their areas of origin are restricted to a few "Vavilov centers" (Smith, 1995). Fortunately, most of the dominating domesticates do not invade natural or seminatural habitats. However, the transport of seed material and products from agriculture pose a serious and continuing threat by bringing unintentionally weeds, parasites and disease organisms.

Intentional introductions of alien species most often happens because management institutions or individuals expect to gain some short or long term benefit (McNeely, Ch. 2). The expected benefits are commonly economic, but individual or public demand for an enjoyable environment is also a common motivation for introductions. Thus, the objectives might be to, e.g. increase biomass production in fisheries, agriculture or forestry (Huntley Ch. 24; Richardson, Ch. 16) or to create nice garden or park sceneries by planting alien species of ornamental and garden plants. However, the efforts to reach laudable objectives have often resulted in serious side effects (Oguto-Ohwayo, Ch. 4) or outright disasters (Huntley, Ch. 24). Both Oguto-Ohwayo, and Shiva (Ch. 3) also points to the cultural, socio-economic and social impacts often experienced from introductions with otherwise laudable intentions.

All these cases highlight the dilemma that faces the informed manager in most, if not all, cases where an introduction is the proposed action. Your action may give some benefit, but will at the same time cause environmental damage. It may be said, however, that the problems we face today in many instances are the results of the action of uninformed managers or the general public, e.g. in the shape of tourists bringing ornamental plants back to their garden, or families emptying their aquarium into the nearest pond or river before leaving for a holiday.

The ecology of introductions: do we understand it?

Although much scientific work has been performed on the ecology of invasions since the subject was highlighted by Elton (1958) fourty years ago (cf. Drake *et al.*, 1989), our understanding is still quite restricted. What is clear is that there is a series of steps in the process from an organism is introduced into a non-native habitat until it is established in numbers to give major effects on the new habitat. The trajectory taken by this process depends on characteristics of the introduced species as well as the conditions of the receiving ecosystem (Rejmanek, Ch. 6; Hengeveld, Ch. 8). Empirical evidence shows clearly that many introduced species may stay passive for a long time. Many generations may be spent in a non-expanding state, until suddenly the species becomes invasive and "aggressive", and expands to dominate the environment (Crooks and Soulé, Ch. 7). The reasons for this phenomenon are poorly understood. There is obviously a great need for predictive models to describe the invasive process, but the models presently available mainly enable us to see what factors might be important (Hengeveld, Ch. 8).

The most devastating effects of introduced species have been observed on islands (Strahm, Ch. 22; Clout, Ch. 23). While freshwater systems have been subject to extensive introductions, with a larger number of introduced than native species in some regions, the ecosystem and species diversity effects often appears small (Moyle, Ch. 12). However, in lakes with a high number of endemic species, the effects of introductions appear similar to what is observed on islands with high endemism (Oguto-Ohwayo, Ch. 4; Clout, Ch. 23; Huntley, Ch. 24).

Ecosystems modified by man generally appears more receptive to alien species than undisturbed ecosystems. Although this may also be related to the higher number of anthropogenic introductions into these habitats (cf. Williamson 1996), it appears natural that global change processes, e.g., climate change, will seriously influence the impact of alien species on ecosystems (Mooney and Hofgaard, Ch. 9).

The process where man seeks to cultivate natural populations or ecosystems to increase the yield of some desired product commonly involves introduction of genetically different individuals to strengthen populations. This has been common practice in forestry and inland fisheries. One fairly well studied example is the domesticated or semi-domesticated escapees from salmon culture which pose a serious threat to natural Atlantic salmon stocks (Hindar, Ch. 10). Research on this problem may also bring important understanding of the potential impacts from introduction of genetically modified or engineered organisms (GMOs, GEOs) into the environment.

As Bryan (Ch. 11) demonstrates, the spreading and re-emergence of human diseases may be better understood if analysed as ecological processes, whereas the often well documented epidemiology and migration of various human diseases may teach environmental researchers and managers many lessons regarding pathways, mechanisms and population variations in invasives.

International pathways

The globalization trends in the world's trade, tourism and migration, offer unrestricted opportunities for unintentional introductions of invasive alien species. This includes, but also adds to sea transport, which has been an important pathway for both aquatic and terrestrial species for several hundred or even thousands of years. Sea transport continues to be a dominant agent in what might be called the internationalisation of coastal marine fauna (Carlton, Ch. 13). Sea transport is also an important indirect cause for the spreading of catastrophic human disease organisms; witness for example the *Yersinia*-infected Norway rat (*Rattus norvegicus*) brought by ships to the American west coast in 1899 (Bryan, Ch. 11). Significant changes in the marine biota are also brought about through technical development schemes. The Suez Canal is one eloquent example (Boudouresque, Ch. 14).

Sea transport is but one agent in the global trade systems of today. As pointed out by Peter Jenkins (Ch. 15), the galloping liberalization of world trade is accompanied by transport and introduction of an ever increasing number of alien species among all the world's regions. The reduction in quarantine regulations resulting from trade liberalization also negatively impacts human and animal health (Skjerve and Wasteson, Ch. 16). The important economic activity of forestry has been a dominant agent in reshaping many of the world's terrestrial biota. Several temperate and subtropical tree species have been extensively planted over large areas in many biogeographic regions (Richardson, Ch. 17). Considering the major effect of forest trees on both biotic and abiotic environmental parameters, it is disturbing that this activity is still going on at a large scale, while our understanding of the effects on biodiversity, soil chemistry etc. from the planting of alien tree species is poorly understood (Huntley, Ch. 24).

Management tools: can we learn from experience?

Considering the impacts of alien invasive species, and the difficulties in predicting effects of any new introduction, it appears that the only solution would be to put a stop to introductions of aliens. As this is clearly not a feasible solution, management institutions in most of society's sectors are in dire need of effective management tools. Legislation and regulations to restrict the movement of unwanted organisms are clearly very important (Bean, Ch. 18). Experiences from the trade on agricultural products like live animals and animal products before the present trade liberalization seems to indicate that application of strict regulations is a very efficient tool (Skjerve and Wasteson, Ch. 17). Prevention of introductions is clearly more cost efficient than eradication or control of an invasive once it is established (Westbrooks and Eplee, Ch. 19). This is clearly the case both in plants, insects (Samways, Ch. 20), and disease organisms (Skjerve and Wasteson, Ch. 17). Once an invasive species is established, introducing some biological control agent may be the only possible way out (Oduor, Ch. 21; Pech, Ch. 25). This solution often, however, involves its own risks of unforeseen negative effects.

Where do we go from here?

With international agents being instrumental in the steadily increasing movement of species around the world, international agreements and protocols are obviously needed to reduce the pressure of invasives on natural ecosystems, cultivated lands and human and animal health (Glowka and de Klemm, Ch. 26). At the same time, a concerted effort is needed in a Global Strategy for dealing with Alien Invasive Species (Mooney, Ch. 27). The development of new and innovative preventive tools is badly needed, and the sharing of experiences and knowledge across borders and disciplines will be an important element in this global effort.

Acknowledgements

We thank the Conference Committee and the International Advisory Group to the second Trondheim Conference on Biodiversity for their essential contributions to a successful conference in 1996. Thanks are also extended to all the contributors to this volume for their efforts to enable us to keep to schedule, and to the peer reviewers for their assistance in the development of the chapters. Ms. Synnøve Vanvik and Mr. Knut Kringstad, at NINA, have helped us with the word processing and graphics, respectively. Dr. Bob Carling, formerly of Thomson Science, helped develop this into a book project.

References

Diamond, J.M. 1985. Introductions, extinctions, exterminations, and invasions, in *Community Ecology*, (eds T.J. Case and J.M. Diamond), Harper & Row, New York, pp. 65-79.

Drake, J.A., Mooney, H.A., di Castri, F., Groves, R.H., Kruger, F.J., Rejmanek, M. and Williamson, M., eds (1989) *Biological Invasions. A Global Perspective*. John Wiley, Chichester, UK.

Elton, C.S. (1958) *The Ecology of Invasions by Animals and Plants*. Methuen, London, UK.

Smith, B.D. (1995) *The Emergence of Agriculture*. Scientific American Library, W.H. Freeman & Co, New York.

Williamson, M. (1996) *Biological Invasions*. Population and Community Biology Series 15, Chapman & Hall, London, UK.

Part 1

Human dimensions

2 The great reshuffling: how alien species help feed the global economy

JEFFREY A. McNEELY
IUCN – The World Conservation Union, Gland, Switzerland

Abstract

The global economy fosters the spread of alien species, one of the negative by-products of the globalization of the world economy. The issues are complex and do not lend themselves to clear-cut solutions under current conditions. One element is that introduced species often seem to do better in their new home than in their place of origin, perhaps because of a paucity of natural enemies or competitors. For example, eucalyptus from Australia is widespread in Southeast Asia, India, California, and various parts of Africa. Second, "natural" is becoming an increasingly elusive concept, as virtually all ecosystems have a strong and increasing anthropogenic component. People are designing – either on purpose or by accident – the kinds of ecosystems they find congenial. Third, the growing trade in ornamental species has also meant a great increase in the introduction of aliens, often leading to a net increase in species richness in their destination. It is quite likely, for example, that many parts of the temperate world have far more species now than ever before, and this may also be the case in at least certain parts of the tropics (such as Sri Lanka and most other island nations), though this great increase of species numbers is usually at the expense of indigenous species (and thus reduces global species diversity). Fourth, moving goods around the world quickly provides ideal opportunities for the accidental introduction of species ranging from zebra mussels to disease-carrying mosquitoes to bacteria and viruses. And fifth, most purposeful introductions have been done for economic reasons, but usually without a careful consideration of the full costs involved. When the costs have become apparent, they usually must be paid by someone other than those who sponsored or promoted the introduction – often the general public. Decision-makers need to invest more in assessing the potential impacts before allowing introductions and to incorporate more biosafety measures once the species has been introduced. Accidental introductions by definition are not exposed to a prior cost-benefit assessment, but assessments of the costs of such introductions can justify increased budgets to control and limit such accidental introductions. Whereas purposeful introductions might be controlled by legislation or regulation, accidents may be far more important in the spread of introduced species and much more difficult to control. The Convention on Biological Diversity offers an important opportunity for addressing global problems of introduced species, a threat to biodiversity that is far more immediately significant than the introduction of living modified organisms (LMOs), which to date has received far more attention under the guise of biosafety. A biosafety protocol which also addresses the issues of alien species and international trade would be far more useful for achieving all the objectives of the CBD.

O. T. Sandlund et al. (eds.), Invasive Species and Biodiversity Management, 11–31.

Introduction

Many people warmly welcome globalization of trade, and growing incomes in many parts of the world are leading to increased demand for imported products. A generally unrecognized side effect of this globalization is the introduction of exotic species, at least some of which may be harmful. These species are of interest to the Conference of the Parties of the Convention on Biological Diversity (CBD), which calls on the Parties to "prevent the introduction of, control or eradicate those alien species which threaten ecosystems, habitats, or species" (Article 8h).

This paper will examine the history and ecology of the global trade in species of plants and animals, explore the concept of "naturalness" and what it means at a time when human influence on ecosystems is pervasive, compare purposeful and accidental introductions, suggest how climate change relates to the global economy and alien species, introduce some economic concepts relevant to the issue of global trade and alien species, and recommend steps that could be taken by the global community to deal more effectively with the issue of harmful alien species.

Species are introduced into new habitats by people for a number of reasons. Levin (1989) identified 3 major categories: (i) accidental introductions; (ii) species imported for a limited purpose which then escape; and (iii) deliberate introductions. Many of the introductions relate to the human interest in providing species that are helpful to people. This is particularly true of agricultural species; indeed, in most parts of the world, the great bulk of human dietary needs are met by species that have been introduced from elsewhere (Hoyt, 1992). Species introductions in this sense, therefore, are an essential part of human welfare in virtually all parts of the world. Further, maintaining the health of these introduced species of undoubted benefit to humans may require the introduction of additional species for use in biological control programmes which import natural enemies of, for example, agricultural pests (Waage, 1991).

Most harmful exotics are not the result of intentional releases or contraband brought in by international travellers, but rather due to unintentional "hitchhiking" through international trade, with exotics stowing away in ships, planes, trucks, shipping containers, and packing materials, or arriving on nursery stock, unprocessed logs, fruits, seeds, and vegetables (OTA, 1993). Jenkins (1996) concludes, "Increased international trade has the potential to cause more harmful exotic species introductions. More proactive, more comprehensive, and better funded international efforts are needed to ensure that widely adapted invasive exotics do not further homogenize biological systems on a global scale."

Indeed, the biggest hidden danger from introduced species may well be their contribution to global homogenization, a seemingly inevitable process that involves multiple factors ranging from communications technology to consumer mentality. People have always been on the move, which helps explain how our species spread over the planet. Australian aborigines brought in the dingo, Polynesians sailed with pigs, taro, yams, and at least 30 other species of plants (and rats as stowaways), and the Asians who first peopled the Americas also brought dogs with them. The impact of these earliest colonists was devastating on the local species, leading to numerous extinctions (see, for example, Martin and Klein, 1984) and numerous introductions, at least by the later colonists who already had developed agriculture (Cuddihy and Stone, 1990). The period of European colonialism which began with the voyage of Columbus in 1492 (Crosby, 1972) (leaving aside the earlier unsuccessful attempts by Vikings) ushered in a new era of species introductions, as the Europeans sought to recreate the familiar conditions of home (Crosby, 1986). They took with them species such as wheat, barley, rye, cattle, pigs, horses, sheep, and goats, but in the early years were limited by the available means of transport. Once steam-powered ships came into common use, the floodgates opened and over 50 million Europeans emigrated between 1820 and 1930, taking numerous plants and animals with them and often overwhelming the native flora and fauna.

More recently, global trade has greatly increased: the growth in global economic output during the 1980s was greater than that during the several thousand years from the beginning of civilization until 1950. The value of total imports increased from US$ 192 146 000 000 in 1965 to US$ 3 316 672 000 000 in 1990, a 17-fold increase in 25 years (WRI, 1994). Imports of agricultural products and industrial raw materials – those which have the greatest potential to contribute to the problem of invasive species – amounted to US$ 482 308 000 000 in 1990, up from US$ 55 132 000 000 in 1965. This tremendous economic performance has been built on an increasingly homogenized foundation of information, finance, culture, and ecosystems.

This homogenization – which has been termed "biological pollution" (Luken and Thieret, 1996) – reduces the diversity of crops and livestock and can increase their vulnerability to both native and exotic pests, often leading to the increased use of pesticides which may have broad negative impacts on ecosystems. Thus introductions may lead to "cascades" of effects that were not part of the decisions that led to the introduction. Species introductions may thus be considered part of the class of phenomena that economists call "externalities", impacts of an activity that affect others outside the activity; the interests of those others are usually ignored by those undertaking the activity (see below).

As a biodiversity issue it is not always possible to identify invasions as inherently "bad"; di Castri (1989) asserts that overall, the central European flora has undergone an enrichment of diversity over historical time as a result of human-induced plant invasions. Britain's mammalian fauna totaling about 49 species includes some 21 introduced species, including eight large mammals (wild goat *Capra hircus*, fallow deer *Dama dama*, Sika deer *Cervus nippon*, Indian muntjak *Muntiacus muntjak*, Chinese muntjak *Muntiacus reevesi*, Chinese water deer *Hydropetes inermis*, Bennett's wallaby *Macropus rufogriseus bennetti*, reindeer *Rangifer tarandus*). It is thus highly likely that, due to human influence, the mammalian fauna of Britain is more species-rich now than at any time since the Neolithic (though with few large carnivores). The genetic diversity of this fauna is also very high, including species originating from Asia, North America, Europe, and Australia (Jarvis, 1979). Jacobs (1975) describes the transformation of the saline Lake Nakuru from an ecosystem of very low diversity (a large population of flamingoes, two species of algae and a few invertebrate species) to one of much higher diversity (including 30 species of fish-eating birds) after the introduction of a fish, *Tilapia grahami*, to control mosquitoes in 1961. More generally, cities – where the majority of the world's people will live by the turn of the century – are greatly enriched by invasive species of plants. Many invasive species seem to do best in urban and urban-fringe environments where long histories of human disturbance have created vacant niches and abundant bare ground. Cities tend to be the focal points of the global economy and the entry points for many invasives. They also tend to be largely anthropogenic habitats, often of very great species richness. For example, London has some 2 100 species of flowering plants and ferns growing wild while the rest of Britain has no more than 1500 species, and Berlin has 839 native species of plants and 593 invasives (Kowarik, 1990; McNeely, 1995).

Despite some arguably positive effects on biodiversity at the local level, however, overwhelming evidence indicates the profoundly negative effects of introductions on species and genetic diversity at both the local and global level. Such introductions can lead to severe disruption of ecological communities (Smith, 1972; Zaret and Paine, 1973; Mooney and Drake, 1986; Drake, 1989; Carlton and Geller, 1993), and heavily influence the genetic diversity of indigenous species. Some protected areas established to conserve native species have been profoundly affected by introduced species (Bratton, 1982), and on some islands introduced species closely match or even outnumber native ones (Table 2.1). If one judges biodiversity only by species richness, then those islands are now twice as valuable as they were when they were "natural". However, most known extinctions – at least of birds – have taken place on islands, so while the individual islands may have more species, the world as a whole has lost diversity. The unique has been replaced by the commonplace.

Table 2.1 Known numbers of invasive and native species in various countries/areas.

Country/Area	Number of native species	Number of invasive species	Source
New Zealand (plants)	1 790	1 570	Heywood, 1989
Hawaii (plants)	956	861	Wagner *et al.*, 1990
Hawaii (all species)	17 591	4 465	Miller and Eldridge, 1996
Tristan de Cunha (plants)	70	97	Moore, 1983
Campbell Island (plants)	128	81	Moore, 1983
South Georgia (plants)	26	54	Moore, 1983
Southern Africa (freshwater fish)	176	52	De Moor and Bruton, 1988; Bruton and Van As, 1986
California (freshwater fish)	83	52	Moyle, 1976, 1998

In the cases where the direct cause of extinction is identifiable, introduced species head the list. For example, introduced mammals are responsible for all but one of the nine known extinctions of endemic vertebrate species or sub-species from the islands of northwest Mexico. Globally, almost 20 percent of the vertebrates thought to be in danger of extinction are threatened in some way by invasive species (Table 2.2). The single biggest tragedy is the probable loss of at least 200 of the 300 endemic cichlid species in Lake Victoria as a result of the introduction of the Nile perch, *Lates niloticus,* to the lake (Lowe-McConnell, 1993; Ogutu-Ohwayo 1998); this was exacerbated by eutrophication of the lake and the introduction of new fishing gear. The global effects of certain invasive species such as the European pig *Sus scrofa* (Oliver, 1994), rats *Rattus* spp. (Atkinson, 1985; Stuart and Collar, 1988; Brockie *et al.,* 1989) and the aquatic plants *Salvinia molesta* and *Eichhornia crassipes* (Ashton and Mitchell, 1989) also attest to the destructive power of invasives.

The general global picture is, then, one of tremendous mixing of species with unpredictable long-term results. While many introduced species have special cultivation requirements which restrict their spread, many other species are finding appropriate conditions in their new homes while many more may invade their new habitats and constantly extend their distribution,

Table 2.2 The percentage of threatened terrestrial vertebrate species affected by introductions in the continental landmasses of the different biogeographic realms and on the world's islands. The total number of threatened species in the realm is given in brackets. Source: Macdonald *et al.*, 1989.

Taxonomic group	Mainland areas		Insular areas	
	%	(n)	%	(n)
Mammals	19.4	(283)	11.5	(61)
Birds	5.2	(250)	38.2	(144)
Reptiles	15.5	(84)	32.9	(76)
Amphibians	3.3	(30)	30.8	(13)
Total for all groups considered	12.7	(647)	31.0	(294)

thereby representing a potential threat to local species. All of this calls into question the concept of "naturalness". The fauna and flora of any area at a specific point in time has some special characteristics that make it different from other times in history. Because chance factors, human influence (including species introductions) and small climatic variation can cause very substantial changes in vegetation and the associated fauna, the biodiversity for any given landscape will vary substantially over any significant time period - and no one variant is necessarily more "natural" than the others (Sprugel, 1991). The future is certain to bring considerable additional ecological shuffling as people influence ecosystems in various ways, not least through both purposeful and accidental introduction of species. This shuffling will have both winners and losers although the overall effect will likely be a global loss of biodiversity at species and genetic levels.

Global trade and species introductions: intentions and accidents

The trade-based global economy stimulates the spread of economically important species, often with funding from development agencies to establish plantations of rubber, oil palm, pineapples, and coffee, and fields of soybeans, cassava, maize, sugarcane, wheat, and other species in countries far from their place of origin. But it also stimulates the accidental spread of species through a variety of pathways. While it is difficult with present information to determine precisely how much of the invasives problem globally is due to conscious intent and how much to inadvertence, some hints are available:

- OTA (1993), in a comprehensive review, concludes that about 4 500 exotic species occur in a free-ranging condition in the United States, and that about 20% of them have caused serious economic or ecological harm. OTA concluded that 81% of the harmful new exotics detected from 1980 through 1993 where the pathway could be identified were from unintentional imports. Examples include the zebra mussel introduced into the Great Lakes via ship ballast water and the Asian tiger mosquito introduced via used tire imports.
- WWF reports that more than 60 introduced species have been found in the Baltic Sea, with many species apparently having arrived via ballast water. For example, the bristleworm *Marenzellaria viridis* has become a dominant species in the Vistula lagoon, constituting 97% of the biomass of bottom-living macrofauna.
- The rapid decline in frog populations in Queensland is attributed to a virus which is exotic to Australia and that may have been introduced through the thriving international trade in ornamental fish.
- OTA (1993) found that the importation of raw logs from Siberia to the west coast of the US carried with them pests with significant potential negative economic impacts. These included the Siberian gypsy moth, which is considered more damaging to coniferous forests than the European gypsy moth which has already caused significant damage. (As a result, imports of raw logs from Siberia have now been banned).
- With an estimated 3 000 species, on any one day, of freshwater, brackish water (estuarine) and marine protists, animals, and plants in motion around the world in the ballast of ocean-going ships, numerous opportunities are available for the invasion of aquatic environments by exotic organisms. Examples from the last decade include: the Japanese sea star *Astrias amurensis* has appeared in Australia, where it has broad potential impacts on the shell fish industry; the Japanese shore crab *Hemigrapsus sanguineus* has colonized Atlantic North America (where it is now becoming relatively common from Cape Cod to Chesapeake Bay); the American comb jelly fish *Mnemiopsis leidyi* has invaded the Black and Azov Seas and has been linked to the near-demise of regional anchovy fisheries; the Chinese estuarine clam *Potamocorbula amurensis* has become one of the most abundant benthic organisms in San Francisco Bay, where the disappearance of spring phytoplankton blooms in parts of the Bay and extensive decreases in zooplankton have been attributed to high densities of this clam; and the Indo-Pacific mussel *Perna perna* has colonized Caribbean mangrove ecosystems

> and Gulf of Mexico jetties, where it forms extensive monoculture-like beds. In the Great Lakes of Canada and the US, three European fish, two species of zebra mussels, and a carnivorous water flea, all unknown from North America in 1980, are now six of the most common species regionally or in large parts of those waters (OTA, 1993).

It appears, then, that "the problem of invasive species" has two very distinct elements: species that are introduced consciously, and for which management procedures such as environmental impact assessments are available; and inadvertent invasives, which may be far more pervasive and far less amenable to management intervention. I will return to this point later, but I would first like to digress slightly into another externality: climate change.

Globalization, climate change, and exotic species

The Intergovernmental Panel on Climate Change (IPCC, 1996) has concluded, on the basis of long and detailed studies, that human activities are having a discernable impact on the climate, primarily through the burning of fossil fuels which is increasing the amount of carbon dioxide in the air and thereby contributing to the so-called "greenhouse effect." Much of the global economy is based on these fossil fuels: global trade in fossil fuels was US$ 335 255 000 000 in 1990 (WRI, 1994). Without the cheap petroleum-based transport which subsidizes global trade, commodities would be far more expensive and trade would be greatly reduced, with a commensurate reduction in the threats from introduced species.

The expected climate change could well promote the interests of invasive species, especially since invasives are especially likely to become established in habitats disturbed by human or other factors (Mooney and Hofgaard, 1998). When CO_2 increases, seedlings in forest gaps may grow more quickly, so turnover rates in tropical forests may increase. It may well be that increased fire or other disturbances that break up the forest canopy would increase the likelihood of invasion by alien species.

Thus climate change could open up new opportunities for introduced species that could devastate native flora and fauna. For example, if the species which are dominant in the native vegetation are no longer adapted to the environmental conditions of their habitat, what species will replace them? It may well be that exotic species will find these "new" habitats especially attractive, and the increasing presence of new species and the decline of old ones will drastically change successional patterns, ecosystem function, and the distribution of resources. Thus concepts of global change need to include consideration of the behaviour and distribution of invasive species. It seems highly likely that invasive species are going to have far more opportunities in the future than they have at present.

Many variables may limit the distribution of a species in different parts of its range, and detailed studies are required to define the distribution limit of various invasive species as climates change. Sutherst *et al.* (1996), for example, used a computer programme for comparing the relative climatic potential for population and persistence of the invasive Cane toad (*Bufo marinus*) in relation to season and locality. This type of study is likely to be increasingly relevant and important.

Global solutions to the global problem of alien species

While this paper has supported the argument that global trade promotes invasive species, it is possible that agreements under the World Trade Organization could offer some help in dealing with exotic species, though bans and restrictions should be founded on science-based risks so that they will be less likely to be challenged before the WTO. But such risk assessment is often extremely expensive; for example, the risk assessment for the proposed importation of raw Siberian larch cost the US Government about US$ 500 000 (Jenkins, 1996).

As Yu (1996) points out, the GATT Treaty contains three important provisions to protect the environment and human health and these might be expanded to deal with exotic species. These provisions include the Agreement on Sanitary and Phyto-sanitary Measures, the Agreement on Technical Barriers to Trade, and Article 20: General Exceptions, which protects the right of members to take any measures "necessary to protect human, animal, or plant life or health".

Even so, the impact of trade on biodiversity remains poorly addressed. Free traders maintain that liberalized markets will solve environmental problems by promoting more efficient use of natural resources, while others maintain that global markets actually undermine efforts to protect the environment. The former argue that increased revenues will lead to decreased environmental damage, while the latter contend that increasing revenues are precisely the problem leading to over-consumption of biological resources. UNCED was relatively ineffective in addressing trade issues, much less trade's promotion of invasives. It endorsed the establishment of strong environmental rules on trade without exploring the basic principles of trade reform that would enable a balance to be struck between trade and environment. The language adopted by Agenda 21 in Rio generally adopted the line being promoted by the GATT Uruguay round of negotiations but ignored the possibility that GATT could itself undermine the environmental measures initiated by the Earth Summit (Prudencio, 1993).

Several global mechanisms already exist to address the problem of invasives carried in ballast water. For example, the United Nations Convention on the Law of the Sea requires states to take all measures necessary to prevent, reduce and control intentional or accidental introduction of alien or new species in the marine environment, and Agenda 21 directs States to consider adopting appropriate rules on ballast water discharge to prevent the spread of invasives (see also Carlton, 1998).

Jenkins (1996) calls for an international advisory panel under the Convention on Biological Diversity (CBD) to advise the various convention secretariats, trade regulation bodies, and national governments regarding risky trade routes, potentially threatening species, the nations and ecosystems most vulnerable to exotic threats to biodiversity, and improvements in prevention and control efforts. He calls for independent assessments to be conducted for the most vulnerable areas identified, based on at least the following elements:

- natural histories of exotic species impacts on native biodiversity;
- current and projected future pathways of exotic species introductions;
- laws, policies, and programmes for preventing and managing exotics affecting biodiversity;
- institutional and technological capabilities for preventing and managing exotics;
- public education and awareness strategies, as called for under Article 13 of the CBD.

As suggested above, other international conventions might also be brought to bear on the invasive species problem (see also Glowka and de Klemm, 1998). Particularly interesting in this regard is the Climate Change Convention, which has very broad government support, especially in the industrialized countries. It also has an associated scientific body, the Intergovernmental Panel on Climate Change; the latter has not yet had the issue of invasive species brought to its attention, but is increasingly recognizing the importance of climate change on biodiversity more generally. It too could be mobilized in support of a global effort to address the invasive species problem.

Another global solution might be through the World Health Organization, especially for invasives which are relevant to human health – primarily viruses and bacteria. While most papers in this volume are looking at insects, vertebrates, and plants, the bacteria and viruses may be much more interesting to governments and the general public, as indicated by the popularity of books and films dealing with the Ebola virus. Mechanisms designed to address these disease-causing invasives may also be relevant to other parts of the invasive species problem (cf. Bryan, 1998).

Costs and benefits of alien species

It is probably fair to say that most people who seek to introduce an exotic or alien species into a new habitat are doing so for an economic reason. They may wish to increase their profits from agriculture, they may believe that the public will like a new flower from a distant part of the globe, or they may think that exotic species will be able to carry out functions that native species cannot carry out as effectively (examples of these will be given below).

But it may also be fair to say that most of those introducing exotic species have not carried out a thorough cost-benefit analysis before initiating the introduction, at least partly because they may not have been aware of the advantages of such analyses. On the other hand, it is also possible that at least some people would prefer to ignore the negative impacts that may follow from species introductions, because they might be expected to compensate those who are negatively affected.

Similarly, those who have been responsible for inadvertently introducing species into new habitats may not have been willing to make the investment to prevent such accidents from occurring. They may not have realized the dangers, and in any case the dangers would be unlikely to have much economic impact on their own welfare. Rather, the costs of such accidents are borne by people other than those who are permitting the accidents to happen. Thus the costs of introducing alien species into new habitats are "externalized" in considerations of the costs of global trade. The line of responsibility is insufficiently clear to bring about the necessary changes in behaviour, so the general public – or future generations – ends up paying most of the costs.

This paper will introduce, in a preliminary way, some of the economic factors affecting the issue of alien species. It will quickly become clear that this field is still relatively immature, but that considerable benefits for biodiversity will come from a more inclusive consideration of economic factors, and the application of economic tools to deal with them.

Good intentions: somebody is going to make money

Many exotic species were introduced for economic purposes. Introduced fish can produce excellent sport fishing, introduced plants can provide food, fodder, and energy, and introduced insects can provide biological controls. A few examples (from among hundreds that could be quoted):

- Brush-tailed possums from Australia were introduced to New Zealand between 1858 and 1900 to establish a fur trade, but in New Zealand they have fewer competitors, fewer predators, and fewer parasites than in their native Australia, so they have successfully spread and have sometimes reached densities ten times greater than in their native Australia. They have been a bonanza for the fur industry.
- A number of woody plants from various parts of the world, such as acacias from Australia, were introduced into South Africa in the middle of the 19th century for purposes of dune stabilization, tannin extraction, and firewood. This appears to have been an economically successful invasion, with the greater Capetown region alone supporting a 30 million Rand charcoal and firewood industry.
- In the Thar Desert of India, the African tree *Prasopsis juliflora* was introduced 70 years ago and has become the dominant flora around human habitats. With its dense green vegetation, this tree is very useful in checking soil erosion, reducing the dryness of the desert air, giving shelter to several species of wild animals, and providing legumes which are relished by wild as well as domesticated animals. It meets 85 percent of firewood demands of rural people.
- The Triclad flatworm *Platydemus manokwari*, first described from New Guinea in 1963, is a successful predator of the giant African snail *Achatina fulica*, so it was transported as a biological control agent to areas where the African snail had become established in the Pacific.
- Water hyacinth *Eichhornia crassipes* was introduced into China from South America in the 1930s and was spread through mass campaigns in the 1950s to the 1970s as an ornamental plant, to provide livestock food, and to control pollution through absorbing heavy metals.

But something went wrong: somebody had to pay the costs

But we all know that there is no free lunch. Introduced species can carry a heavy pricetag, in terms of reduced crop and livestock production, loss of native biodiversity, increased production costs, and so forth. As just one example, OTA (1993) estimates that the total direct costs to the US of invasive species of weeds is US$ 3.6–5.4 billion per year. This figure does not include environmental, human health, regulatory, and other indirect costs of using herbicides on these weeds, though these costs are estimated to be at least US$ 1 billion per year; if herbicides were not available, the crop losses would jump to nearly US$ 20 billion per year.

All of the introductions listed above carried with them some hidden – or, in retrospect, obvious – costs:

- The Australian brush-tailed possums introduced into New Zealand have caused considerable damage to native forests, changing forest composition and structure through the defoliation and progressive elimination of favoured food plants. Note that none of these costs are particularly relevant to those interested primarily in the benefits from furs. In an effort to control these possums, New Zealand is working on bio-control agents, including the possibility of a genetically-engineered immunocontraceptive virus. This innovative approach could have profound implications elsewhere in the world, showing that some problems may lead to solutions which have considerable global value.
- As a result of the introduced species, South Africa's highly-endemic Cape Flora is under serious threat and the watersheds are becoming less productive, potentially causing a considerable increase in the price of water (Wilgen *et al.*, 1996; Huntley, 1998). (See below for more details).
- While *Prasopsis juliflora* has been a boon to people in the Thar Desert who need firewood and fodder, it overwhelms other flora in the area, thereby reducing the range of products available to local people and reducing biodiversity.
- The triclad flatworm now poses a serious threat to the native gastropod fauna of the Pacific region. This is especially troubling because the Pacific has seen a remarkable radiation of the snail family Partulidae, and some 24 of these are on the 1994 IUCN Red List of Threatened Animals. The triclad flatworm has become established on Guam, Saipan, Tinian, Rotar, and Palau.
- In China, the water hyacinth has become the worst weed in many aquatic habitats, leading the loss of species of both plants and animals. In Dianchi Lake, just outside of Kunming, Yunnan, the total number of fish species has declined from 68 to about 30 and Chinese scientists attribute this to water hyacinth. Reduction of the lake area as a result of the water hyacinth infestation has also caused notable climatic changes in Kunming (Kequinga, 1993).

Few of these examples have explicit costs attached to them, but qualitative costs are often available. For example, in the early 1900s, the most economically important hardwood species in eastern American forests was the chestnut (*Castanea dentata*), but the chestnut blight brought in on diseased horticultural stock from China killed nearly a billion trees and all but eliminated this species, leading to profound ecosystem changes in the eastern hardwood forests (USDA, 1991).

More quantitative cost estimates have also been made. For example, OTA (1993) estimated the cumulative losses to the USA from 79 alien species from 1905 to 1991 at more than US$ 96 billion, a conservative estimate since data were available for only a few years (Table 2.3).

OTA (1993) also projected potential future losses under a "worst-case scenario," estimating that just 15 species of plants, insects, plant pathogens, and aquatic invertebrates could eventually cause economic losses of over US$ 134 billion. Since the USA has over 4 500 alien species, the potential for considerable damages is very great.

Table 2.3 Estimated cumulative losses to the United States from selected harmful non-indigenous species, 1906-1991. Source: OTA (1993).

Category	Species analyzed (number)	Cumulative loss estimates (millions of US$, 1991)	Species not analyzed (number)
Plants (except agricultural weeds)	15	603	-
Terrestrial vertebrates	6	225	>39
Insects	43	92 658	>330
Fish	3	467	>30
Aquatic invertebrates	3	1 207	>35
Plant pathogens	5	867	>44
Other	4	917	-
Total	79	96 944	>478

Despite such figures, the issue of costs and benefits is not always clear, at least partly because different people have different perceptions of what these are. Luken and Thieret (1996), for example, report that within less than a century after its deliberate introduction into North America to improve habitat for birds, serve ornamental functions in landscape plantings, and stabilize and reclaim soil, the Amur honeysuckle had become established in at least 24 states in the eastern USA. While many resource managers perceive the plant as an undesirable element, gardeners and horticulturalists consider it an extremely useful plant. Thus the "noxious invasive" of one group is the "desirable addition" of other groups. How can costs and benefits be determined in such a case?

Diamond *et al.* (1991) have estimated the costs and benefits of controlling the invasive tree *Melaleuca quinquenervia* in Florida. Total annual benefits, based largely on tourism, of preventing infestation from the tree would be US$ 168.6 million, while the costs to honey producers to whom the tree provides nectar would amount to just US$ 15 million. Again, the costs and benefits in this case are differentially distributed: those who suffer losses are unlikely to be compensated, while those who benefit pay few of the costs.

Considerable work has now been done in the USA, at least, on the cost-benefit ratios for various forms of managing invasive species (though the distribution of these continue to be ignored). OTA's (1993) summary of these demonstrates a very wide range of cost-benefit ratios, though in nearly all cases, the benefits of control far outweigh the costs involved. This strongly suggests that significantly increased investments in managing invasive species is justified in economic terms, though again those paying the costs – usually the taxpaying public – may not always be the primary beneficiaries; and those who earned the benefits from the invasives in the first place are paying virtually none of the costs.

As an indication of one interesting approach to measuring costs and benefits, Wilgen *et al.* (1996) presented a case study showing how invasion by alien plants has affected water resources in the mountain catchment areas of the Western Cape Province, South Africa. They found that the sustained supply of high-quality water depends on maintaining the cover of fynbos (shrubland) vegetation. The fynbos binds the soil, preventing erosion, while its relatively low biomass ensures conservative water use and low-intensity fires, which in turn ensure high water yields and low impacts on the soil from periodic fires. Fynbos-clad mountain catchments fulfil approximately two-thirds of the Western Cape's water requirements, an ecosystem service that plays a crucial role in the region's economy and contributed to a gross domestic product of US$ 15.3 billion in 1992. The fynbos flora is widely harvested for cut flowers, dried flowers, and thatching grass, producing a combined value in 1993 of US$ 18–19.5 million and providing a livelihood for 20–30 000 people.

However, catchment management is complicated by the invasion of the fynbos vegetation by non-indigenous woody trees and shrubs, which increase biomass and reduce runoff. Fynbos ecosystems are remarkably prone to invasion by alien woody species which displace the native fynbos and increase biomass by between 50% and 1 000%. These invasive plants were introduced to South Africa to provide a source of fast-growing timber in the relatively treeless landscape, as hedge plants, as agents for binding the shifting dunes along the coast, and as ornamental plants. The most important invasive species originated in Australia and the Mediterranean-climate areas of Europe and North America. On the slopes of Table Mountain, above Cape Town, invasion by alien species has increased fire intensities, leading to severe soil erosion.

Wilgen *et al.* (1996) developed a computerized model that indicated that alien plants would invade approximately 40% of the area within 50 years and 80% after 100 years, with a corresponding increase in biomass of 150% or more. This invasion would result in an average decrease of 347 cubic metres per ha per year of water at the end of 100 years, resulting in average losses of more than 30% of the water supply to the city of Cape Town. In some years, when large areas would be covered by mature trees, losses would be much greater, exceeding 50% of the runoff from similar uninvaded areas. They concluded that investments in managing alien plants at a level that would ensure that they are no longer part of the ecosystem would lead to a net unit cost of water of US$ 12 per cubic metre, as compared to US$ 14 without the management of alien plants.

While it is important to identify the costs and benefits, such determination does not automatically determine a decision because value judgements and distributional questions are nearly always involved. Further, the magnitude of the costs may sometimes be so high as to render an action politically unacceptable, even when the benefits are likely to be even greater. Part of the problem is that the benefits may be widely spread throughout the public over a period of many years, while the costs of control may need to be paid rather quickly by tax payers.

Conclusions and recommendations

One of the main intentions of liberalizing trade as advocated by the World Trade Organization is to stimulate an even greater volume in materials traded, thereby offering – in addition to more goods for consumers – greater opportunities for introduction of exotics and ultimately greater homogenization of ecosystems. Despite the evidence of significant impact, the response of governments is inadequate to prevent the increased trade that it is promoting from resulting in more introductions of harmful species (Jenkins, 1996; 1998). And if even governments with relatively well-developed control structures are unable to deal with the problem, what of the many tropical countries which have even less capacity for dealing with such problems? Here are some suggestions to the global community:

- Use the Convention on Biological Diversity more effectively. As this paper, and many others at this volume, has indicated, the issue of invasive species is clearly also an issue of biosafety. Yet the Convention's biosafety focus is on living modified organisms (Article 19.3), which more properly is just a subset of the invasive species problem covered by Article 8(h). While the discussion on biosafety has already gone on for several years, it is perhaps not too late to bring the relevance of the broader invasive species issue to the attention of the Conference of the Parties of the CBD for inclusion in the biosafety protocol being negotiated. This carries the recommendations of Jenkins (1996) one step further, and while it is unlikely that the biosafety protocol discussion will be broadened, at least the process of addressing Article 8(h) will begin.

- Bring the issue of invasive species to the attention of the World Trade Organization, perhaps through a statement adopted by the Conference of the Parties of the Convention on Biological Diversity. Such a statement will need to be relatively concise and explicit, and contain within it specific responses that would be expected from the WTO.
- Build on the experience of countries faced with significant invasive species problems (e.g., USA, Australia, New Zealand) to develop further a body of principles and practice that could be transferred to developing countries.

It is apparent from the material presented in this paper, which is only a small sample of a much larger literature, that economics has much to contribute to programmes to address the problems of alien species. Decision-makers often find arguments couched in economic terms to be more convincing than those cast in emotive or ethical terms, and this paper has suggested some ways that economics-based arguments can be used to support stronger programmes to deal with invasive species. Without trying to be comprehensive, I suggest at least the following actions to be considered:

- Build an economics component into any international programme to deal with invasive species. This should include IUCN's Species Survival Commission recruiting economists onto its Invasive Species Specialist Group, and ICSU–SCOPE ensuring that economists are involved in their invasive species research programmes. A partnership between ecologists working on invasive species and economists could be extremely productive. Economic analysis can provide a useful and rigorous structure to guide policy makers who might otherwise "externalize" some of the most relevant factors. Applying numbers to the problem can highlight the areas of debate and uncertainty, particularly when they look at the distribution of costs and benefits.
- In each country, or as part of each programme to deal with one or more invasive species, seek to quantify the costs and benefits involved. Ensure that such quantification is as complete as possible, so that some costs (or benefits) are not "externalized". The ability of economists to provide useful analyses will depend to a large extent on how well biologists are able to estimate the probabilities of future impacts of alien species in a consistent, convincing, and comparable way. Because

economic models provide little assistance where they rest on vague or equivocal predictions of biological events, an effective partnership between ecologists and economists is essential.

- Mobilize economic instruments, including such incentives as grants, taxes, and fines to ensure better compliance with programmes dealing with invasive species. Economic analysis can help design appropriate economic instruments, such as incentives and disincentives, helping to determine appropriate levels of fines and penalties for those introducing alien species.

The problem of human-induced invasive species is as old as our own species. But the severity of the problem has grown tremendously as the global economy has reached into virtually all corners of our planet. As a significant global problem with negative impacts on all countries, it is deserving of a significant global response.

References

Ashton, P.S. and Mitchell, D.S. (1989) Aquatic plants: patterns and modes of invasion, attributes of invading species and assessment of control programmes, in *Biological Invasions: A Global Perspective,* (eds J.A. Drake, H.A. Mooney, F. di Castri, R.H. Groves, F.J. Kruger, M. Rejmánek and M. Williamson), SCOPE 37, John Wiley and Sons, New York, pp. 111–147.

Atkinson, S.F. (1985) Habitat-based methods for biological assessment. *The Environmental Professional,* **7**, 265–282.

Bratton, S.P. (1982) The effects of exotic plant and animal species on nature preserves. *Natural Areas Journal,* **2**, 3–12.

Brockie, R.E., Loope, L.L., Usher, M.B. and Hamann, O. (1989) Biological invasions of island nature reserves. *Biological Conservation,* **44**, 9–36.

Bryan, R. (1998) Allien species and emerging infectious diseases: past lessons and future implications, in *Invasive Species and Biodiversity Management*, (eds O. T. Sandlund, P. J. Schei and Å. Viken), Kluwer Academic Publishers, Dordrecht, The Netherlands.

Carlton, J.T. (1998) The scale and ecological consequences of biological invasions in the World's oceans, in *Invasive Species and Biodiversity Management*, (eds O. T. Sandlund, P. J. Schei and Å. Viken), Kluwer Academic Publishers, Dordrecht, The Netherlands.

Carlton, J.T. and Geller, J. B. (1993) Ecological roulette: the global transport of non-indigenous marine organisms. *Science,* **261**, 78–82.

Crosby, A. W. (1972) *The Colombian Exchange: Biological and Cultural Consequences of 1492.* Greenwood Press, West Port, CT, USA.

Crosby, A. W. (1986) *Ecological Imperialism: The Biological Expansion of Europe, 900–1900.* Cambridge University Press, Cambridge, UK.

Cuddihy, L.W. and Stone, C.P. (1990) *Alteration of Native Hawaiian Vegetation: Effects of Humans, Their Activities and Introductions.* University of Hawaii Press, Honolulu.

Diamond, C., Davis, D. and Schmitz, D.C. (1991) Economic impact statement: The addition of *Melaleuca quinquenervia* to the Florida Prohibited Aquatic Plant List, in *Proceedings of the Symposium on Exotic Pest Plants,* U.S. Department of the Interior, Washington, DC.

di Castri, F. (1989) History of biological invasions with special emphasis on the Old World, in *Biological Invasions: A Global Perspective,* (eds J.A. Drake, H.A. Mooney, F. di Castri, R.H. Groves, F.J. Kruger, M. Rejmánek and M. Williamson), SCOPE 37, John Wiley and Sons, New York, pp. 1–26.

Drake, J.A., Mooney, H.A., di Castri, F., Groves, R.H., Kruger, F.J., Rejmanek, M. and Williamson, M., eds (1989) *Biological Invasions. A Global Perspective.* John Wiley and Sons, New York.

Glowka, L. and de Klemm, C. (1998) International instruments, processes, organizations and non-indigenous species introductions: is a protocol to the convention on biological diversity necessary?, in *Invasive Species and Biodiversity Management*, (eds O. T. Sandlund, P. J. Schei and Å. Viken), Kluwer Academic Publishers, Dordrecht, The Netherlands.

Heywood, V. (1989) Patterns, extents and modes of invasions by terrestrial plants, in *Biological Invasions: A Global Perspective,* (eds J.A. Drake, H.A. Mooney, F. di Castri, R.H. Groves, F.J. Kruger, M. Rejmánek and M. Williamson), SCOPE 37, John Wiley and Sons, New York, pp. 31–51.

Hoyt, E. (1992) *Conserving the Wild Relatives of Crops.* IBPGR, IUCN and WWF, Gland, Switzerland, Second Edition.

Huntley, B. (1998) South Africa's experience regarding alien species: impacts and controls, in *Invasive Species and Biodiversity Management*, (eds O. T. Sandlund, P. J. Schei and Å. Viken), Kluwer Academic Publishers, Dordrecht, The Netherlands.

IPCC (1996) *Climate Change 1995: The Science of Climate Change.* Intergovernmental Panel on Climate Change, Cambridge University Press, Cambridge, UK.

Jacobs, J. (1975) Diversity, stability and maturity in ecosystems influenced by human activities, in *Unifying Concepts in Ecology,* (eds W.H. van Dobben and R.H. Lowe-McConnell), Junk, The Hague, The Netherlands, pp. 187–207.

Jarvis, P.H. (1979) The ecology of plant and animal introductions. *Progress in Physical Geography,* **3**, 187–214.

Jenkins, P.T. (1996) Free trade and exotic species introductions. *Conservation Biology,* **10**, 300–302.

Jenkins, P.T. (1998) Trade and exotic species introductions, in *Invasive Species and Biodiversity Management*, (eds O. T. Sandlund, P. J. Schei and Å. Viken), Kluwer Academic Publishers, Dordrecht, The Netherlands.

Kowarik, I. (1990) Some responses of flora and vegetation to urbanization in Central Europe, in *Urban Ecology: Plants and Plant Communities in Urban Environments,* (eds H. Sukopp, S. Mejny and I. Kowarik), SPB Academic Publishing, The Hague, The Netherlands, pp. 45–74.

Levin, S.A. (1989) Analysis of risk for invasions and control programmes, in *Biological Invasions: A Global Perspective,* (eds J.A. Drake, H.A. Mooney, F. di Castri, R.H. Groves, F.J. Kruger, M. Rejmánek and M. Williamson), SCOPE 37, John Wiley and Sons, New York, pp. 425–432.

Lowe-McConnell, R.H. (1993) Fish faunas of the African Great Lakes: origins, diversity, and vulnerability. *Conservation Biology,* **7**, 634–643.

Luken, J. O. and Thieret , J. W. (1996) Amour honeysuckle, its fall from grace. *BioScience,* **46**, 18–24.

Macdonald, I.A.W., Loope, L. L., Usher, M. B. and Hamann, O. (1989) Wildlife conservation and the invasion of nature reserves by introduced species: a global perspective, in *Biological Invasions: A Global Perspective*, (eds J.A. Drake, H.A. Mooney, F. di Castri, R.H. Groves, F.J. Kruger, M. Rejmánek and M. Williamson), SCOPE 37, John Wiley and Sons, New York, pp. 215–255.

Martin, P.S. and Klein, R.G., eds (1984) *Quarternary Extinctions: A Prehistoric Revolution.* University of Arizona Press, Tucson, AZ, USA.

McNeely, J.A. 1995. *Cities, nature, and protected areas: a general introduction.* Paper presented to Symposium on Natural Areas in Conurbations and on City Outskirts, Barcelona, Spain, 25–27 October.

Miller, S.E. and Eldridge, L.G. (1996) Numbers of Hawaiian species: Supplement one. *Bishop Museum Occasional Papers,* **45**, 8–17.

Mooney, H. and Drake, J.A. (1987) The ecology of biological invasions. *Environmentalist,* **19**, 10–37.

Mooney, H.A. and Hofgaard, A. (1998) Biological invasions and global change, in *Invasive Species and Biodiversity Management*, (eds O. T. Sandlund, P. J. Schei and Å. Viken), Kluwer Academic Publishers, Dordrecht, The Netherlands.

Moyle, P.B. (1976) Fish introductions in California: history and impact on native fishes. *Biological Conservation*, **9**, 101–118.

Moyle, P.B. (1998) Effects of invading species on freshwater and estuarine ecosystems, in *Invasive Species and Biodiversity Management*, (eds O. T. Sandlund, P. J. Schei and Å. Viken), Kluwer Academic Publishers, Dordrecht, The Netherlands.

OTA (1987) *Technologies to Maintain Biological Diversity.* Office of Technology Assessment, US Government Printing Office, Washington D.C.

OTA (1993) *Harmful Non-Indigenous Species in the United States.* Office of Technology Assessment, US Government Printing Office, Washington D.C.

Ogutu-Ohwayo, R. (1998) Nile perch in Lake Victoria: the balance between benefits and negative impacts of aliens, in *Invasive Species and Biodiversity Management*, (eds O. T. Sandlund, P. J. Schei and Å. Viken), Kluwer Academic Publishers, Dordrecht, The Netherlands.

Oliver, W.L.R., ed. (1993) *Pigs, Peccaries and Hippos. Status Survey and Conservation Action Plan.* IUCN, Gland, Switzerland.

Prudencio, R. J. (1993) Why UNCED failed on trade and environment. *Journal of Environment and Development,* **2**, 103–109.

Smith, S.H. (1972) Factors of ecologic succession in oligotrophic fish communities of the Laurentian Great Lakes. *Journal of the Fisheries Research Board of Canada,* **29**, 717–730.

Sprugel, D. G. (1991) Disturbance, equilibrium and environmental variability: what is "natural" vegetation in a changing environment? *Biological Conservation,* **58**, 1–18.

Stuart, S.N. and Collar, N.J.. (1988) *Birds at Risk in Africa and Related Islands: The Causes of Their Rarity and Decline.* Proceedings of the Sixth Pan-African Ornithological Congress.

Sutherst, R.W., Floyd, R.B. and Maywald, G.F. (1996) The potential geographical distribution of the Kaintode, *Bufo marinus* in Australia. *Conservation Biology,* **10**, 294–299.

USDA (1991) Pest risk assessment of the importation of larch from Siberia and the Soviet Far East. *U.S. Department of Agriculture, Forest Service Miscellaneous Publications,* 1495.

Waage, J.K. (1991) Biodiversity as a resource for biological control, in *The Biodiversity of Micro-organisms and Invertebrates: Its Role in Sustainable Agriculture*, (ed. D.L. Hawksworth), CAB International, Oxford, UK, pp. 149–163.

Wagner, W.L., Herbst, D.R. and Sohmer, S.H. (1990) *Manual of the Flowering Plants of Hawaii.* Bishop Museum Press, University of Hawaii Press, Honolulu, Hawaii.

Wilgen, B. W. van, Cowling, R. M. and Burgers, C. J. (1996) Valuation of ecosystems services. *BioScience,* **46**, 184–189.

WRI (1994) *World Resources: 1994—95*. World Resources Institute, Oxford University Press, New York.

Yu, D. W. 1996. New factor in free trade: Reply to Jenkins. *Conservation Biology,* **10**, 303–304.

Zaret, T.M., and Paine, R.T. (1973) Species introduction in a tropical lake. *Science,* **182**, 449–455.

3 Species invasions and the displacement of biological and cultural diversity

VANDANA SHIVA
Research Foundation for Science and Technology, New Dehli, India

Abstract

This paper approaches the issue of species invasions as a bio-cultural phenomenon. Bio-cultural invasiveness is analyzed as a combination of cultural invasiveness and biological invasiveness. These combinations lead to three categories of species invasions: 1) Culturally invasive, biologically non-invasive, 2) Culturally non-invasive, biologically invasive, 3) Culturally invasive, biologically invasive.

Introduction of Green Revolution crops is an example of the first category of invasiveness. Unintended spontaneous spread of plants such as the *Parthenium* and *Lantana* in India and the paper back tree in Florida are examples of the second. Genetically engineered organisms (GEOs) with potential to develop a competitive advantage such as the genetically engineered *Klepsiella planticola*, or the transfer of herbicide resistance genes to wild and weedy relatives are examples of the third category. The three categories are analyzed in the context of their cultural, ethical and ecological impacts on biocultural diversity. The categorization will also be used to challenge the dominant assumption that equates GEOs with domesticated species on the ground that mankind has been manipulating genes and introducing alien species in agriculture for centuries. The deliberate introduction of GEOs in agriculture combines the cultural invasiveness of the Green Revolution and the uncertainties related to biological invasiveness of exotic species, and hence requires a double caution in order to protect cultural and biological diversity.

Introduction

The protection of biological and cultural diversity is the biggest ecological and ethical challenge of our times. While the threats to biocultural diversity are many, I will focus on the problem of invasiveness in this paper.

Biocultural invasiveness has two dimensions – the cultural and the biological. The two dimensions cannot be fully isolated. However, for analytic purposes it is useful to divide invasive species phenomena into three categories.

1. Culturally invasive, biologically non-invasive, e.g. large scale introduction of agricultural crops and crop varieties.

O. T. Sandlund et al. (eds.), Invasive Species and Biodiversity Management, 33–45.

2. Culturally non-invasive, biologically invasive, e.g. spontaneous spread of exotic species.
3. Culturally invasive and biologically invasive, e.g. release of genetically engineered organisms in agriculture.

Further, while deliberately introduced species in agriculture may themselves be noninvasive, the ecological changes they induce can lead to "secondary invasions" in the form of pests and diseases or weeds. Species invasions therefore need to be assessed at both the primary and secondary level.

The Green Revolution: An example of culturally invasive, biologically non-invasive displacement of biological and cultural diversity

The Report on the State of World's Plant Genetic Resources (FAO, 1996) has identified replacement of local varieties as the most frequent cause for genetic erosion (Figure 3.1). The Green Revolution has been recognised as the most significant reason for the displacement of agricultural biodiversity by the IV Technical Conference on Plant Genetic Resources in Leipzig in June 1996. Green Revolution monocultures are a form of cultural invasion which displace local cultural and biological diversity (Shiva, 1992). Food cultures based on diverse cereals, millets, oilseeds and pulses were displaced by rice and wheat varieties released from international agricultural centres. Thousands of local rice varieties were displaced by a few International Rice Research Institute (IRRI) varieties. A culture of growing crops to feed the earth to feed animals and to feed humans was replaced by a culture ignoring the needs of nature and other species, and meeting human needs of nutrition only through long distance market chains.

Politics and culture were built into the Green Revolution because the technologies created were directed at capital intensive inputs for one-dimensional production by the best endowed farmers in the best endowed areas, and directed away from resource prudent diversity based options of the small farmer in resource scarce regions. The science and technology of the Green Revolution excluded diversity and poor regions and poor people as well as sustainable options. American advisors gave the slogan of "building on the best". The science of the Green Revolution was thus essentially a political and cultural choice which chose a chemical culture in place of a biological culture, a monoculture instead of diversity, a culture of profits in place of a culture of needs, a market oriented culture instead of a biospheric one.

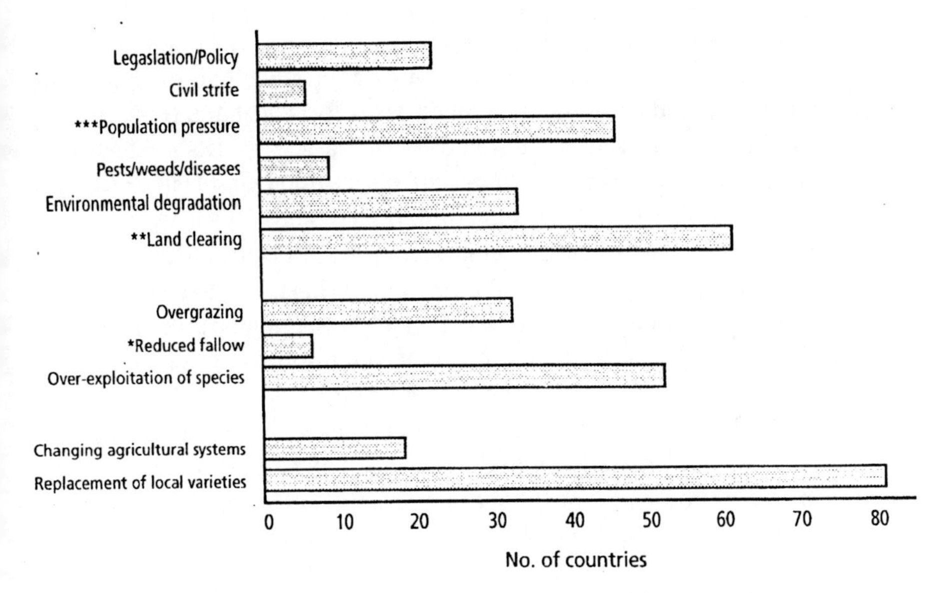

Figure 3.1 Causes of genetic erosion mentioned in country reports. * mentioned in the case of shifting cultivation systems. ** including deforestation and bush fires. *** including urbanization. From FAO (1996).

As Lappe and Collins (1982) have stated: "Historically, the Green Revolution represented a choice to breed seed varieties that produce high yields under optimum conditions. It was a choice not to start by developing seeds better able to withstand drought or pests. It was a choice not to concentrate first on improving traditional methods of increasing yields, such as mixed cropping. It was a choice not to develop technology that was productive, labour-intensive, and independent of foreign input supply. It was a choice not to concentrate on reinforcing the balanced, traditional diets of grain plus legumes."

The crop and varietal diversity of indigenous agriculture was replaced by a narrow genetic base and monocultures. The focus was on internationally traded grains, and a strategy of eliminating mixed and rotational cropping, and diverse varieties by varietal simplicity. While the new varieties reduced diversity, they increased resource use of water, and of chemical inputs such as pesticides and fertilisers.

The monocultures of the Green Revolution were a form of cultural invasion which resulted in the displacement of biological diversity. The new crop varieties were themselves not biologically invasive. Crops bred through conventional means have lost their competitive capacity and their natural ability to spread. They depend on humans to be sown and cultivated.

However, while the plants introduced as Green Revolution monocultures are not biologically invasive in themselves, they do set off a second order species invasions in the form of pests and diseases. The genetic uniformity of monocultures creates the ecological conditions for transforming harmless pests and diseases into devastating epidemics as has become evident with the creation and spread of new pests and diseases for wheat and rice.

In 1966, IRRI released a "miracle" rice variety – IR-8. This was particularly susceptible to a wide range of disease and pests: in 1968 and 1969 it was hit hard by bacterial blight and in 1970 and 1971 it was ravaged by another tropical disease called tungro. In 1975, Indonesian farmers lost half a million acres of Green Revolution rice varieties to leaf hoppers. In 1977, IR-36 was developed to be resistant to 8 major diseases and pests including bacterial blight and tungro. However this was attacked by two new viruses called "ragged stunt" and "wilted stunt".

The vulnerability of rice to new pests and disease due to monocropping and a narrow genetic base is very high, IR-8 is an advanced rice variety that came from a cross between an Indonesian variety called "Pea" and another from Taiwan called "Dee-Geo-Woo-Gen". IR-8, Taichung Native (TN1) and other varieties were brought to India and became the basis of the All India Coordinated Rice Improvement Project to evolve dwarf, photoinsensitive, short duration, high yielding varieties of rice suited to high fertility conditions. The large-scale spread of exotic strains of rice with a narrow genetic base was known to carry the risk of the large-scale spread of disease and pests. In India the introduction of high yielding rice varieties has brought about a marked change in the status of insect pests like gall midge, brown planthopper, leaf folder, whore maggot etc. Most of the high yielding varieties (HYVs) released are susceptible to major pests with a crop loss of 30 to 100%. Most of the HYVs are the derivatives of TN1 or IR-8 and therefore have the dwarfing gene of Dee-Geo-Woo-Gen. The narrow genetic base has created alarming uniformity. Most of the released varieties are not suitable for typical uplands or low-lands which together constitute about 75% of the total rice area of the country.

The "miracle" varieties displaced the diversity of traditionally grown crops, and through the erosion of diversity, the new seeds became a mechanism for introducing and fostering pests. Indigenous varieties or land races are resistant to locally occurring pests and diseases. Even if certain diseases occur, some of the strains may be susceptible, while others will have the resistance to survive. Crop rotations also help in pest control. Since many pests are specific to particular plants, planting crops in different seasons and different years causes large reductions in pest populations. On the other hand, planting the same crop over large areas year after year encourages pest build-ups. Cropping systems based on diversity have built-in protection.

The wheat seeds that spread world-wide from Centro Internacional de Mejoramiento de Maiz y Trigo (CIMMYT) through Norman Borlaug and his "wheat apostles" were the result of nine years of experimenting with Japanese "Norin" wheat. "Norin" released in Japan in 1935 was a cross between Japanese dwarf wheat called "Daruma" and American wheat called "Fultz" which the Japanese government had imported from the U.S. in 1887. The Norin wheat was brought to U.S. in 1946 by Dr. D.C. Salmon, an agriculturist acting as a U.S. military adviser in Japan, and further crossed with American seeds of the variety called "Bevor" by US Department of Agriculture (USDA) scientist Dr. Orville Vogel. Vogel in turn sent it to Mexico in the 1950s where it was used by Borlaug, who was on the Rockefeller Foundation staff, to develop his well-known Mexican varieties. Of the thousands of dwarf seeds created by Borlaug, only three went to create the "Green Revolution" wheat plants which were spread world-wide. On this narrow and alien genetic base, the food supplies of millions are precariously perched.

Large scale monocultures of exotic varieties have turned minor diseases such as Karnal Bunt into epidemic proportions. Green Revolution wheat varieties came to India as a project of cultural invasion. As if in an ecological backlash, the Karnal Bunt has invaded the wheat crop of the U.S. In combination with drought, the fungal disease has wiped out more than 50% of the wheat crop in the bread basket of the world.

These secondary invasions introduced through culturally invasive but biologically non-invasive monocultures show that even domesticated species can trigger species invasions by changing the local ecological conditions through displacement of biodiversity. At the level of secondary impacts large scale domesticated species introductions can have ecological effects similar to exotic species invasions.

Exotic Species Invasions: An example of culturally non-invasive, biologically invasive introductions

An Englishman brought the *Lantana camara* to India as a hedge plant from America via Sri Lanka (Bhandari, 1996). By 1941, *Lantana* had encroached on more than 40 000 hectares. It has started to invade agricultural and pasture lands, compelling villagers to abandon farming. It has also spread as a noxious weed through India's forests, preventing regeneration of native species. *Lantana* has thus become a serious problem in cropped and non-cropped areas in India. The seeds are carried by birds and the plant, once established, quickly covers open areas where it forms dense, thorny thickets. When these thickets are cleared, there will be regrowth of suckers and even if the roots are dug out numerous seedlings will appear. The *Lantana* is an example of a culturally non-invasive, biologically invasive introduction of an alien species. The intent was not to displace local diversity, but the consequence of the introduction was major displacement.

Parthenium hysterophorus (wild carrot weed) is another noxious weed which has spread to many parts of India covering approximately 5 million hectares (Jayachandra, 1970). A native of tropical America, it is reported that the seeds of this weed came to India with grain shipments from the U.S. In India the weed was first pointed out in Poona in 1951. Since then it has spread across the length and breadth of the country. It was first found in Karnataka in 1961, in Kashmir in 1963, in Madhya Pradesh in 1968, in the western Himalaya in 1970, and in Assam and Rahjastan in 1979.

Besides displacing local biodiversity, the weed also causes dermatitis and other forms of allergy. Because of its allelopathic effects it also affects agricultural crops such as maize, jowar and arhar. Biomass productivity is thus declining in both cultivated and uncultivated lands. A single plant produces more than 10 000 seeds which travel long distances and can propagate under all kinds of environmental conditions. It is stated that this weed has infested cropped and non-cropped areas like wild fire.

Exotic species which are not ecologically destabilising in their native habitat can become major threats to local diversity. The invasiveness of species is a probabilistic and uncertain outcome both because of the time lag involved between introduction and invasion (Table 3.1; Crooks and Soulé, 1998), as well as the inherently uncertain nature of a species becoming invasive in a particular habitat. It has been shown in many cases that there is a considerable time-lag between the introduction of a non-native species and its spontaneous dissemination into the new ecosystem.

Kowarik (1995) presents data for the time that passed between the introduction and the spontaneous spreading of non-native trees and shrubs. The average time lag was 147 years, the extremes ranging from 8 to 388 years. These results have a very high statistical value since they are based on a nearly complete survey of 184 introductions in the area of Berlin and Brandenburg.

Simberloff (1981) investigated world-wide 854 cases of naturalisation of predominantly animal species and discovered that in 71 cases (8.3%) this meant the extinction of a native species.

Table 3.1 Observed time lags from introduction to spreading in various shrubs and trees introduced to the area around Berlin and Brandenburg, Germany. From Kowarik (1995).

Trees	Time lag (years)	Shrubs	Time lag (years)
Robinia pseudacacia	152	*Mahonia acuifolium*	38
Acer negundo	183	*Syringa vulgans*	124
Prunus serotina	29	*Symphoricarpos albus*	65
Aesculus hippocastanum	124	*Philadelphus coronarius*	183
Quercus rubra	114	*Lycium barbarurn*	70
Ailanthus altissima	122	*Cornus stolonifera*	76
Populus canadensis	165	*Lonicers tatarica*	94
Prunus mahaleb	54	*Ribes aureum*	61
Laburnum anagyroides	198	*Colutea arborescens*	265
Salix intermedia	112	*Cornus alba*	84

Deliberate release of genetically engineered organisms: Examples of culturally invasive and biologically invasive introductions

Genetically engineered organisms (GEOs) deliberately released into the environment for applications in agriculture combine the cultural invasiveness of the Green Revolution with a new level of cultural invasiveness reflected in breaking species barriers without concern for ecological implications. Further, since genetic engineering goes hand in hand with patents on life, it also reflects a new cultural arrogance of man as the creator of life (Shiva, 1993). GEOs also continue second order biological invasiveness in the form of pests and diseases of Green Revolution introductions and the biological invasiveness of exotic species with additional risks created by transgenic organisms. They thus combine three kinds of risks of biological invasion.

It has often been argued that genetically engineered organisms are like domesticated species. However, as Phil Regal has shown the domesticated species model is inappropriate for assessing biosafety or the impact of GEOs on biodiversity (Regal, 1994).

Biosafety assessment needs to be based on the exotic species model with the additional recognition of the fluidity of the genome. The transfer of foreign genes in creating transgenic plants and animals, which are subsequently released on a large scale into the environment is a major perturbation both to the organisms to which the transgenes are introduced, and to the ecological community into which the organisms are released. The experience with transgenic organisms has shown that genetic engineering is not an exact and predictable technology. Genetically engineered organisms can exhibit behaviour and impact that was not anticipated.

Cotton engineered with genes coding for Bt toxin has exhibited high rates of shedding of bolls leading to millions of dollars of costs

Round-up Ready Canola seeds produced for planting on 600 000 acres in 1996 had to be withdrawn because they had the wrong genes. Estimates of the cost to farmers of the recall was US$ 12 million. US$ 24 million was the cost in lost sales (Rance, 1997; Anonymous, 1997).

Flavr Savr, the first genetically engineered food failed commercially. The Flavr Savr Tomato was developed to have a characteristic of delayed softening. When the Flavr Savr tomato was being taken off the plants it got seriously damaged. Flavr Savr also had disappointing yields and insufficient disease resistance. Consumer resistance also contributed to its failure. The company that had launched Flavr Savr suffered huge losses (Anonymous, 1996a, b; 1997).

A bacterium engineered to degrade a persistent herbicide 2,5-D formed a by-product 2,4-DCP which turned out to be toxic to soil fungi even in low concentrations (Doyle *et al.,* 1995).

The image of exactness, predictability and certainty associated with genetic engineering is falsified by the empirical data now available on the functioning of transgenics.

The old genetic paradigm has perpetrated an erroneous reductionistic view of organic wholeness and complexity. It regards DNA or the genes as the most important, constant and stable essences of organisms. It regards genes as the determiners, in a simple, linear way, of the characteristics of organisms. Molecular genetics since the 1970s has provided steadily growing evidence to the contrary, showing that genes are unstable, and may respond directly to the environment, when the environment is perturbed. These findings reveal hitherto undreamt of complexity and dynamism in cellular and genetic processes involved in gene expression, many of which serve to destabilise and alter genomes within the lifetime of all organisms (Pollard, 1984). These processes so impressed molecular geneticists that they coined the phrase "the fluid genome" more than ten years ago (Dover, 1982). The main lesson to be learned from these fluid genome processes is that the stability and repeatability of development – which we recognise as heredity – does not reside solely in the genes, but is distributed over the whole complex of inter-relationships between an organism and its ecological community.

On the basis of lab experiments which dealt with ecologically crippled organisms it was stated that GEOs are ecological cripples or non-invasive like domesticated crops.

However, as Lenski (1993) has stated, "while most experimental studies support the general hypothesis that genetically modified microorganisms are less fit than their unmodified progenitors. But there are a few .... exceptions in which genetic modifications unexpectedly enhance competitive fitness. Also, many genetic modifications that are disadvantageous under certain environmental conditions are favourable under others."

At the 1994 annual meeting of the Ecological Society of America researchers from Oregon State University reported oil tests to evaluate a genetically engineered bacterium designed to convert crop waste into ethanol (Holmes and Ingham, 1995). A typical root-zone inhabiting bacterium, *Klebsiella planticola,* was engineered with the root-zone novel ability to produce ethanol, and the engineered bacterium was added to enclosed soil chambers in which a wheat plant was growing. In one soil type, all the plants with the genetically engineered microorganism (GEM) treatments died, while those in the parent and no-addition treatments remained healthy. In all cases as well, mycorrhizal fungi in the root system was reduced by more than half, which ruined nutrient uptake and plant growth. This result was unpredicted. Reduction in this vital fungus is known to result in plants being less competitive with weeds, or being more susceptible to disease. In low organic matter sandy soil, the plant died from ethanol produced by the GEM in the root system, while in high organic matter sandy or clay soil, changes in nematode density and species composition resulted in significantly decreased plant growth. It was concluded that these results imply that there can be significant and serious effects resulting from the addition of a GEM to soil. The tests, using a new and comprehensive system, disproved earlier suggestions that no significant ecological effects have been seen when GEMs are added to test systems. Based on the results, the following was emphasized:

- Only 14 genetically engineered organisms have actually been tested for ecological effects, a minimal set from which to broadly apply any principle, or to state that other engineered organisms (with extremely different genetic modifications) will not havc impacts.
- The test systems to determine whether addition of these engineered organisms result in ecologically significant effects have often consisted of sterile soil, soils with no plants, or systems without the organisms present that could be affected or impacted.

- There was often inadequate food resources in such test systems and the engineered organisms often did not reproduce during the course of the test, and did not carry out their engineered function.

The Danish researchers Jørgensen and Andersen (1994) reported strong evidence that an oilseed rape plant genetically engineered to be herbicide tolerant transmitted its transgene to a weedy natural relative, *Brassica campestris* ssp. *campestris.* This transfer can take place in just two generations of the plant. In Denmark, *B. campestris* is a common weed in cultivated oilseed rape fields, where selective elimination by herbicides is now impossible. The wild relative of this weed is spread over large parts of the world. One way to assess the risk of releasing transgenic oilseed rape is to measure the rate of natural hybridisation with *B. campestris,* because certain transgenes could make its wild relative a more aggressive weed, and even harder to control. Although crosses with *B. campestris* have been used in the breeding of oilseed rape, natural interspecific crosses with oilseed rape was genetically thought to be rare. Artificial crosses by hand pollination carried out in a risk assessment project in the UK were reported to be unsuccessful. However, a few studies have reported spontaneous hybridisation between oilseed rape and the parental species *B. campestris* in field experiments. As early as 1962 hybridisation rates of 0.3–88% were measured for oilseed rape and wild *B. campestris.* The results of the Danish team showed that high levels of hybridisation can occur in the field. Their field tests revealed that between 9% and 93% of hybrid seeds were produced under different conditions.

Jørgensen and Andersen (1994) also warn that as the gene for herbicide resistance is likely to be transferred to the weed, this herbicide strategy will be useless after a few years. Like many other weeds, *B. campestris* is characterised by seed dormancy and longevity of the seeds. Therefore, *B. campestris* with transgenes from oilseed rape may be preserved for many years in spite of efforts to exterminate it. They conclude that weedy *B. campestris,* with this herbicide tolerant transgene may present economic risks to farmers and the biotechnology industry. Finally, natural ecosystems may also be affected.

Other concerned scientists add that the potential spread of the transgene will indeed be wide because oilseed rape is insect-pollinated, and bees are known to fly far distances. The existence of the wild relative of *B. campestris* in large parts of the world poses serious hazards once the transgenic oilseed rape is marketed commercially. In response to the Danish findings, the governments of Denmark and Norway have acted against the commercial planting of the engineered plant, but the UK government has approved its marketing.

Existing field tests are not designed to collect environmental data, and test conditions do not proximate production conditions that include commercial scale, varying environments and time scale. Given that invasiveness can occur with a time lag of more than a hundred years, field tests done over a few weeks or months are clearly not a reliable indicator for biosafety.

The argument that the safety of field trials predicts safety at the commercial scale is thus untrue. One cannot claim that since plants in small confined and ecologically irrelevant field plots (used to study commercial features) have not "caused problems" or have not "caused surprises" then it will be safe to commercially release any transgenic forms. It is often claimed that there have been no adverse consequences from over 500 filed releases in the U.S. However, the term "releases" is completely misleading (Regal, 1994). Those tests were largely not scientific tests of realistic ecological concerns, yet "this sort of non-data on non-releases has been cited in policy circles as though 500 true releases have now informed scientists that there are no legitimate scientific concerns."

Recently, for the first time, the data from the USDA field trials were evaluated to see whether they support the safety claims. The Union of Concerned Scientists (UCS) which conducted the evaluation found that the data collected by the USDA on small scale tests have little value for commercial risk assessment. Many reports fail to even mention – much less measure – environmental risks. Of those reports that allude to environmental risk, most have only visually scanned field plots looking for stray plants or isolated test crops from relatives (Rissler and Mellon, 1993). The UCS concluded that the observations that "nothing happened" in those hundreds of tests do not say much. In many cases, adverse impacts are subtle and would never be registered by scanning a field. In other cases, failure to observe evidence for the risk is due to the contained conditions of the tests. Many test crops are routinely isolated from wild relatives, a situation that guarantees no outcrossing. The UCS cautioned that ".....care should be taken in citing the field test record as strong evidence for the safety of genetically engineered crops."

The same concern is echoed by a commentary in Nature (Kareiva, 1993): ".....it is a pity that opportunities to obtain appropriate data have been missed in the hundreds of completed field trials, which have emphasised agronomic performance and have been managed in a way which discouraged multigeneration observations on transgenic populations. So although more than 300 field trials have been carried out and no evidence of "weediness" has yet emerged, that should not be interpreted as an especially comforting observation – we have been so thorough in containing or destroying all material in field trials that we could hardly expect to see any hint of problems from these studies, the real question is what will happen when transgenic seeds are widely broadcast year after year in many different habitats, as would happen if genetically engineered crops are planted commercially." Even where testing on ecosystem is adequate, it does not allow extrapolation of risk assessment to other dissimiliar ecosystems.

Conclusion

The erosion of local cultural and biological diversity has been a characteristic of biocultural invasions, whether the displacement was deliberate and intended in the case of the Green Revolution or it was unintended as in the case of exotic species. In the case of deliberate release of genetically engineered organisms, the cultural invasiveness of the Green Revolution combines with the biological invasiveness of exotic species. Further, the genetic modification across species barriers creates new levels of unpredictability and uncertainty with respect to the invasiveness of GEOs. Genetic ecology needs to be developed following the exotic species model to assess the impact of introduction of GEOs on other species and ecosystems. Until these tools of predictive ecology are evolved to ensure biosafety, there is need for a moratorium on large-scale commercial releases. A combination of cultural invasiveness and biological invasiveness can have serious ecological impacts. A combination of cultural arrogance and ecological ignorance can be devastating.

References

Anonymous (1995) Battling on two fronts, U.S. *National Biotechnology Impacts Assessment Programme Newsletter*, May, 1995.

Anonynous (1996a) Whether the Flavr Savr. U.S. *National Biotechnology Impacts Assessment Programme Newsletter*, March, 1996.

Anonymous (1996b) High tech tomato hits low tech problems. *The Splice of Life*, April, 1996.

Anonymous (1997) Monsanto recalls GM seeds in regulation score. *Farmers Weekly*, May 2, 1997.

Bhandari, N. (1996) Curbing a growing menace, *Hindustan Times*, May 2, 1996.

Crooks, J.A. and Soulé, M.E. (1998) Lag times in population explosions of invasive species: Causes and implications, in *Invasive Species and Biodiversity Management*, (eds O.T. Sandlund, P.J. Schei and Å. Viken), Kluwer Academic Publishers, Dordrecht, The Netherlands.

Dover, G. A. and Flavell R. B., eds (1982) *Genome Evolution*. Academic Press, London, UK.

Doyle, J.D., Stotzky, G., Clung, M.C. and Hendericks, C.W. (1995) Effects of genetically engineered microorganisms on microbial populations and processes in natural habitats. *Advances in Applied Microbiology*, **40**, 237.

FAO (1996) *Report on the State of the World's Plant Genetic Resources for Food and Agricul*ture. Food and Agriculture Organisation of the United Nations, Rome.

Holmes, M.T. and Ingham, E.R. (1995) *The effects of genetically engineered microorganisms on soil foodwebs.* 79th Annual Meeting of the Ecological Society of America, Knoxville, Tennessee, August 7 –11, 1994.

Jayachandra (1970) Parthenium weed in Mysore State and its control. *Current Science*, **40**, 568–569.

Jørgensen, R. B. and Andersen, B. (1994) Spontaneous hybridization between oilseed rape (*Brassica napus*) and weedy *B. campestris* (Brassicaceae): A risk of growing genetically modified oilseed rape. *American Journal of Botany*, **81** (12), 1620–1626.

Kareiva, P. (1993) Transgenic plants on trial. *Nature*, **363**, 580–581.

Kowarik, I. (1995) Time lags in biological invasions with regard to the success and failure of alien species, in *Plant Invasions*, (eds P Pysek, K. Prach, M. Rejmánek and P.M. Wade), SPB Academic Publishing, The Hague, pp. 15–38.

Lappe, F. M. and Collins, J. (1982) *Food First.* Abacus, London, UK.

Lenski, R. E. (1993) Evaluating the fate of genetically modified microorganisms in the environment: Are they inherently less fit? *Experientia*, **49**.

Pollard, J. W. (1984) Is Weismann's barrier absolute?, in *Beyond Neo-Darwinism*, (eds M-W. Ho and P.T. Saunders), Academic Press, London.

Rance, L. (1997) Mix-up prompts recall. *Manitoba Cooperator*, April 24, 1997.

Regal, P.J. (1994) Scientific principles for ecologically based risk assessment of transgenic organisms. *Molecular Ecology*, **3**, 5–13.

Rissler, I. and Mellon, M. (1993) *Perils amidst the Promise – Ecological Risks of Transgenic Crops in a Global Market.* Union of Concerned Scientists, USA.

Shiva, V. (1991) *The Violence of Green Revolution: Third World Agriculture, Ecology and Politics*. Third World Network, Penang, Malaysia.

Shiva, V. (1992) The seed and the spinning wheel: technology development and biodiversity conservation, in *Conservation of Biodiversity for Sustainable Development,* (eds. O.T. Sandlund, K. Hindar and A.H.D. Brown), Scandinavian University Press, Oslo, pp. 280–289.

Shiva, V. (1993) *Monocultures of the Mind: Biodiversity, Biotechnology and the Third World*. Third World Network, Penang, Malaysia.

Simberloff, D. S. (1981) Community effects of introduced species, in *Biotic Crises in Ecological and Evolutionary Time*, (ed. M.H. Nitecki), Academic Press, New York, pp. 53–81.

Williamson, M.H. and Brown, K.C. (1986) The analysis and modelling of British invasions. *Philosophical Transactions of the Royal Society of London, Series B*, **314**, 505–522.

4 Nile perch in Lake Victoria: balancing the costs and benefits of aliens

RICHARD OGUTU-OHWAYO
Fisheries Research Institute, Jinja, Uganda

Abstract

Lake Victoria had high diversity of endemic fishes which were important as food, in evolutionary studies and in sustaining a stable ecosystem. Most of these fishes were depleted following introduction of a large piscivore, Nile perch and this was concurrent with deterioration in water quality. The lake basin has a rapidly growing population of about 30 million people and this has put great pressure on the resources of the lake. The loss in fish species diversity attributed not only to predation by the Nile perch but is also due to human over-exploitation of the fisheries and loss of habitats critical for fish survival. Deterioration in water quality and loss of fish habitat has been attributed to increased nutrient inputs, eutrophication, depletion of oxygen, invasion of the lake by the water hyacinth and changes in food-web structure associated with Nile perch predation. The lake is located in one of the poorest regions of the world and conservation of biodiversity has to be evaluated with the needs of the people around it. Introduction of Nile perch resulted in rapid increases in fish production which has benefited the poor riparian communities through provision of food, employment and revenue. Some of the species which have been lost from the lake are found in satellite lakes and in refugia within the lake and can be conserved by protecting some of these areas. Biodiversity and the overall health of the ecosystem will also benefit from reducing increases in human population, regulation of fishing effort and nutrient inputs.

Introduction

Lake Victoria is among the African Great Lakes (Figure 4.1). It is the largest tropical lake and the second largest lake in the world. It has a surface area of 68 800 km^2, a shoreline 3 450 km long, a catchment area of 193 000 km^2, a mean depth of 40 m and a maximum depth of 84 m. The lake is estimated to have a water residence time of 140 years. It is shared between Kenya (6%), Uganda (43%) and Tanzania 51%. The lake is an important source of protein food, clean water, transportation, and is globally unique for evolutionary and ecological studies. It is also important as a modifier of local climate, especially rainfall. Lake Victoria and the other African Great

O. T. Sandlund et al. (eds.), Invasive Species and Biodiversity Management, 47–63.

Lakes (Malawi and Tanganyika), are under great pressure from growing human populations. The areas around these lakes are among the most heavily populated in the East African region with the Victoria lake basin having the highest population of about 30 million people. This population is growing at a very fast rate of about 3.0% per annum compared to many developed countries where human populations are increasing at about 0.5% annually. This high human population density and that of domestic animals have put pressure on the land and lake's resources. Human population growth has increased fishing effort leading to over-exploitation of the fisheries. It has also intensified land use for agriculture and enhanced deforestation for fuel wood. High population densities also increase sewage inputs that lead to nutrient enrichment of aquatic systems.

The first manifestation of increasing human pressure on the fisheries resources of Lake Victoria was a decrease in stocks of several native fish species. Lake Victoria had a diverse fish fauna represented by many species endemic to the Victoria and Kyoga lake basins. Two tilapiine cichlids, *Oreochromis esculentus* and *O. variabilis* were originally the most important commercial species in the two lake basins but other taxa such as *Protopterus aethiopicus*, *Bagrus docmac*, *Clarias gariepinus*, *Barbus* spp., Mormyrids, *Synodontis* spp. and *Schilbe intermedius* were also commercially important. Selective fishing had strongly reduced stocks of the native tilapiines by the 1960s (Jackson, 1971; Ogutu-Ohwayo, 1990a). *Labeo victorianus* which was the most important anadromous fish species of the Victoria and Kyoga lake basins was depleted through intensive gill netting and basket trapping of gravid individuals on breeding migrations (Cadwalladr, 1965).

Like the other African Great Lakes, Lake Victoria had one of the richest fish faunas on earth most of which were endemic. One group, the haplochromine cichlids comprised more than 300 species, of which 99% evolved in the lake within the recent 12 000 years (Johnson *et al.*, 1996). Haplochromines were not originally exploited by fisheries due to their small size. However, the number of species and the rate at which they had evolved, made the lake a renown world heritage that has illustrated how organisms undergo adaptive radiation to produce new species. Haplochromines remained the most abundant species and formed up to 83% of the demersal fish community in the lake until the 1980s (Kudhongania and Cordone, 1974; Okaronon *et al.*, 1985).

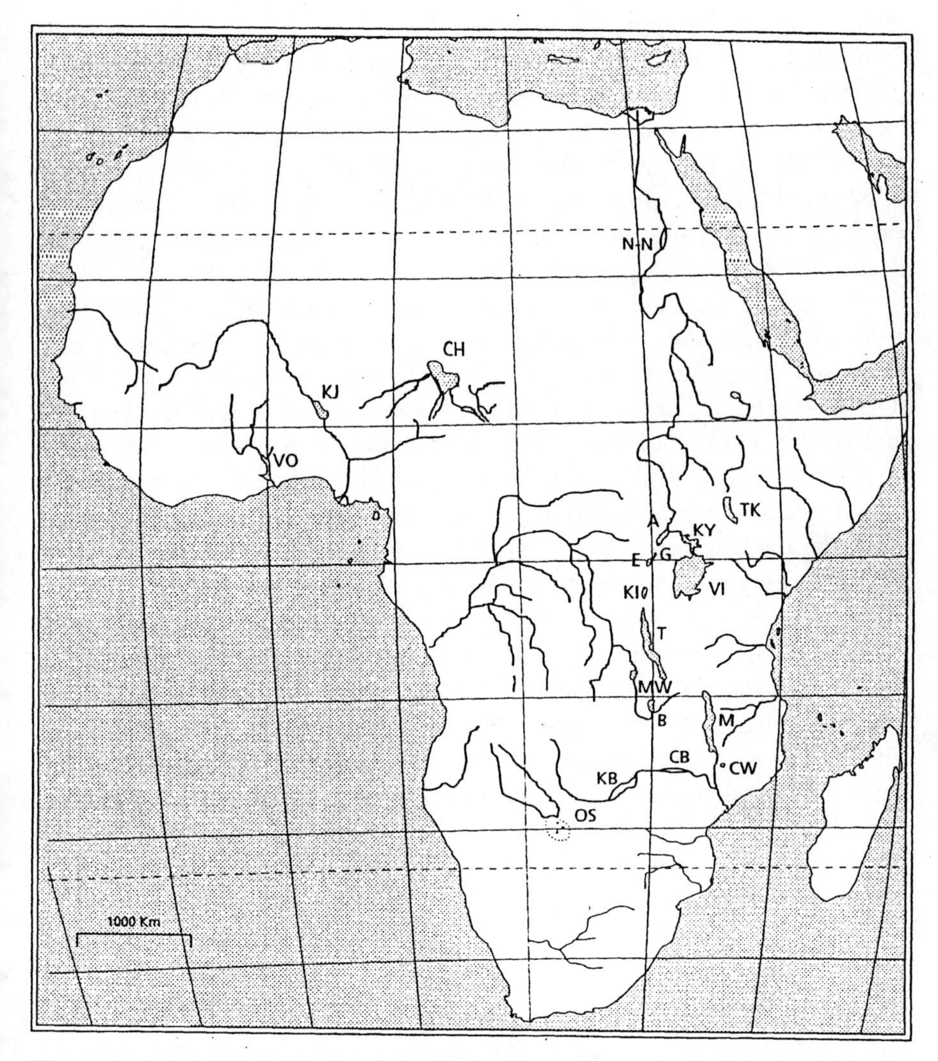

Figure 4.1 Map of Africa showing the main lakes and rivers. NN = Nasser-Nubia, CH = Chad, KJ = Kainji, VO = Volta, TK = Turkana, KY = Kyoga, A = Albert, G = George, E = Edward, VI = Victoria, KI = Kivu, T = Tanganyika, M = Malawi, MW = Mweru, B = Bangweulu, CB = Cahora Bassa, KB = Kariba and OS = Okavango swamp

Introduction of alien fish species

The need to sustain and improve fish production to satisfy the demands of the increasing human population prompted in introduction of exotic fish species into lakes Victoria, Kyoga, Nabugabo, and several other East African lakes (Graham, 1929; Gee, 1969). Nile perch, *Lates niloticus* and three herbivorous tilapiine cichlids, *Oreochromis niloticus* (Nile tilapia), *O.*

leucostictus and *Tilapia zilli* were introduced into Lake Victoria during the 1950s and early 1960s (Gee, 1969). Nile perch is a predatory fish which grows to a length of 2 m and a weight of 200 kg. It was expected to feed on haplochromines which were abundant at the time but not exploited and convert them into a larger fish of greater commercial and recreational value (Graham, 1929; Worthington, 1929; Anderson, 1961). The tilapiines were expected to improve stocks of the native tilapiines whose stocks had declined due to over-fishing.

Dramatic changes were observed in Lake Victoria during the 1980s and early 1990s following increases in Nile perch stocks. Fish species diversity decreased. There were frequent algal blooms and periodic mass fish kills (Ochumba and Kibaara, 1989). Algal blooms indicated that the lake was undergoing eutrophication. This could have been caused directly by increased nutrient loading or indirectly through a cascade of events arising out of reduced phytoplanktivory following depletion of haplochromines stocks by the Nile perch. Mass fish kills could have been due to pollution or the lack of oxygen that was symptomatic of eutrophication.

In Uganda, the Canadian International Development Research Centre (IDRC) provided funding to investigate these changes with the Uganda Fisheries Research Organisation (UFRO), now the Fisheries Research Institute (FIRI), in partnership with the Canadian Freshwater Institute. These efforts also involved collaborations with scientists from the University of Michigan, New England Aquarium and other North American institutions. A similar initiative took place in the Kenyan and Tanzanian regions of the lake. These efforts provided information on the changes in the structure and functioning of the Lake Victoria ecosystem.

Impact of predation by Nile perch

Nile perch started to increase rapidly in the late 1970s and early 1980s in Lake Victoria. As the stocks of Nile perch increased, fish species diversity, especially of haplochromines, decreased rapidly (Ogari and Dadzie, 1988; Ligtvoet and Mkumbo, 1990; Ogutu-Ohwayo, 1990a, b). The contribution of haplochromines to fish biomass in the lake decreased rapidly with increase in Nile perch (Figure 4.2). More than 60% of haplochromine species are believed to have become extinct and others reduced to negligible densities during this period (Witte *et al.*, 1992a, b). The contribution of haplochromines to experimental trawl catches in the northern waters of Lake Victoria declined dramatically between 1981 and 1995 (Figure 4.2). Haplochromines comprised up to 83% of the demersal fish stocks in Lake Victoria in 1970 (Kudhongania and Cordone, 1974) and remained the most abundant taxa in the northern part of the lake near Jinja up to 1982 (Okaroron *et al.,* 1985). After 1982, the proportion by weight of haplochromines declined from 76.2% in 1983 to 6.9% in 1985 while that of Nile perch increased from 16.8% in 1983 to 90.4%. By 1985, haplochromines were virtually absent from trawl catches. It is a widely held notion that Nile perch was responsible for the rapid depletion of haplochromine stocks at the time that its populations increased rapidly (Ogutu-Ohwayo, 1990a, b; Witte *et al.,* 1992a, b). Haplochromines formed the main food of Nile perch soon after its introduction (Figure 4.3) and their population started to decrease rapidly only after increase in Nile perch stocks. After depletion of haplochromines, Nile perch survived by switching to other prey notably the native zooplanktivorous cyprinid, *Rastrineobola argentea*, young Nile perch and the prawn, *Caridina nilotica*.

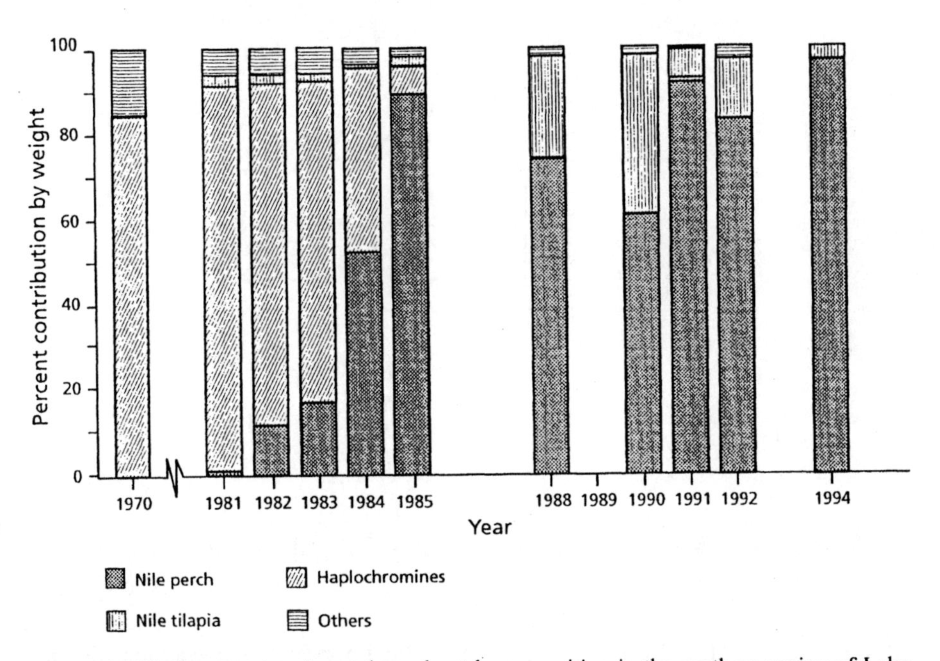

Figure 4.2 Changes in experimental trawl catch composition in the northern region of Lake Victoria between 1970 and 1994.

After the depletion of haplochromines, Lake Victoria became dominated by three species; Nile perch, Nile tilapia and *R. argentea*. Experimental fishing in lakes Victoria, Kyoga and Nabugabo, to which Nile perch were introduced, shows that these three fish species were among the very few taxa that were able to survive in the presence of Nile perch (Ogutu-Ohwayo, 1994). One zooplanktivorous haplochromine *Yssichromis laparogramma*

(Greenwood and Gee, 1969) has recently been observed to be abundant in offshore waters of Lake Victoria (Tumwebaze, 1997). The relative abundance of this species increases as that of *R. argentea* decreases from inshore to offshore. Open waters of Lake Victoria therefore became dominated by only four species; one piscivore (Nile perch), one detritivore/phytoplanktivore (*O. niloticus*) and two spatially segregated zooplanktivores (*R. argentea* and *Y. laparogramma*). A number of rock dwelling haplochromines and *Brycinus* spp. have also been observed to survive among rock and macrophyte refugia especially along the margins of the lake (Wanink, 1991; Namulemo, 1998).

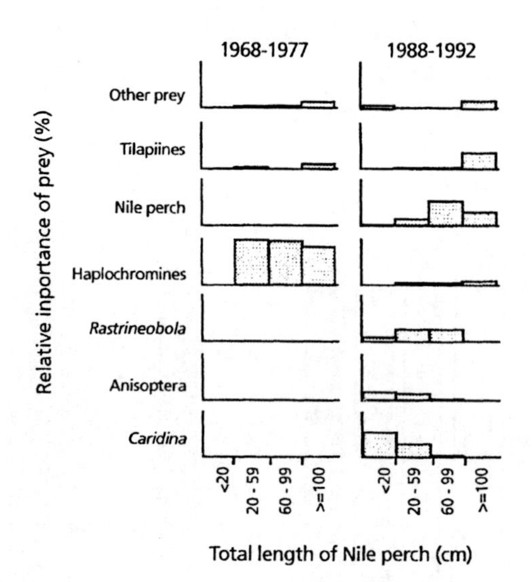

Figure 4.3 The prey eaten by Nile perch in Lake Victoria in 1968–1977 and 1988–1992.

The total absence of some of the originally abundant native species, especially *O. esculentus* in parts of lakes Victoria, Kyoga and Nabugabo to which Nile perch and non-native tilapiines were introduced, may not be purely due to predation by Nile perch. The only tilapiine species that has become abundant in the lakes to which Nile perch was introduced was *O. niloticus*. This species has some similarities in distribution, feeding, and breeding habits with the native tilapiines which disappeared from lakes Victoria, Kyoga and Nabugabo. It has a wider food spectrum, it is more fecund, grows to a larger size, has a faster growth rate, and a longer life span than the native tilapiines of lakes Victoria and Kyoga (Fryer and Iles, 1972). Competition with *O. niloticus* is likely to have contributed to the elimination of native tilapiine species. These tilapiine species have been observed to hybridize. *O. esculentus* and *O. variabilis* could also have lost ground to *O. niloticus* as a consequence of genetic dominance.

Changes in lake productivity mechanisms

Work done on Lake Victoria during the first half of the 20th century showed that the phytoplankton community was dominated by large diatoms especially *Aulacoseira (Melosira)* and *Stephanodiscus*. The lake was well-mixed and had adequate oxygen throughout the water column for most of the year (Worthington, 1931; Fish, 1957; Talling, 1966). The zooplankton composition was dominated by larger calanoid copepods and cladocerans (Worthington, 1931; Rzoska, 1957).

Changes in lake productivity occurred concurrently with the increase in Nile perch stocks in Lake Victoria. Concentration of phosphorus, a key nutrient in plant production, doubled from the 1960s to the 1990s (Talling, 1966; Hecky, 1993; Mugidde, 1993). Eighty percent of the water input into Lake Victoria comes from direct precipitation. It was therefore thought that the phosphorus loading into the lake comes from the watershed and the air (Hecky, 1993; Lehman and Branstrator, 1994; Gophen *et al.*, 1995). This was supported by the observation that the concentration of phosphorus in the rain increased during this period probably due to atmospheric inputs from wind erosion. Although there was excess phosphorus, nitrogen remained deficient. The concentration of silicon decreased by a factor of 10 probably as a result of increased phosphorus loading (Hecky, 1993).

The changes in nutrient chemistry initiated changes in lake productivity. Between the 1960s and the 1990s phytoplankton production doubled and algal biomass increased four to five times. These changes were accompanied by a four-fold decrease in water transparency (Talling, 1966; Hecky, 1993; Mugidde, 1993). Phytoplankton composition also changed from dominance of diatoms (*Aulacoseira*) to nitrogen fixing cyanobacteria of the genus *Cylindrospermopsis* and *Planktolyngbya*. Diatom composition changed from dominance of *Aulacoseira* to *Nitzschia*. These changes in algal community are likely to have affected stocks of some fish species such as *O. esculentus* that depended on *Aulacoseira* for food (Graham, 1929) and may also have affected relative abundance among filter feeding zooplankton. The zooplankton community in the lake changed from dominance of larger calanoid copepods and cladocerans to smaller cyclopoid copepods probably as a consequence of size selective predation (Worthington, 1931; Rzoska, 1957; Mwebaza-Ndawula, 1994). This zooplankton community was not effective in reducing phytoplankton biomass (Lehman and Branstrator, 1993). The benthic invertebrate community became dominated by chironomid and chaoborid midges and the prawn *C. nilotica* (Mbahinzireki pers. comm.). The population of molluscs increased due to decreased abundance of molluscivorous haplochromines.

Oxygen concentration in the hypolimnion during the period of stratification decreased and anoxia started occurring for longer periods and extended to shallower depths (Hecky, 1993; Hecky *at al.*, 1994). This reduced habitable space for many fish species and is thought to have accelerated depletion of the deep water haplochromines by confining them to shallower waters where they were more vulnerable to Nile perch (Hecky, 1993). The reduction of haplochromines may also have reduced competition for invertebrate prey thus allowing juvenile Nile perch and *R. argentea* to proliferate.

Paleolimnological information (Hecky, 1993) has revealed that changes in physico-chemical conditions and lake productivity mechanisms in Lake Victoria started to occur before upsurge of the Nile perch as human activities in the catchment area intensified. The changes seem to have started at the turn of the century with increased opening up of land for agriculture as human populations increased. At this time, species of the diatom *Cyclostephanos* begun to increase while some species of green algae, e.g., *Botryococcus,* begun to decline. The changes accelerated after 1960 when *Aulacoseira* began to disappear and was replaced by *Nitzschia.* The rapid changes in the ecosystem following establishment of the Nile perch seem to have made it difficult for the ecosystem to absorb the changes and accelerated their manifestation.

Interaction between changes in food web structure and eutrophication

Haplochromines formed the main food of Nile perch soon after its establishment (Figure 4.3) but then declined rapidly (Figure 4.2). The rapid decline of haplochromine stocks (Ogari and Dadzie, 1988; Ogutu-Ohwayo, 1990a, b; Ligtvoet and Mkumbo, 1990) seems to have favoured proliferation of zooplanktivores (*R. argentea* and *Y. laparogramma*). The increase in zooplanktivores could have increased predation pressure on the larger calanoid and cladoceran zooplankton and reduced their abundance. Haplochromines had played an important role in the flow of organic matter in the lake (Goldschmidt and Witte, 1992). Each species had its own unique combination of food and habitat preference (van Oijen, 1982; Witte *et al.*, 1992a, b). There were up to eleven trophic groups which included phytoplanktivores, detritivores, epilithic and epiphytic algal grazers, plant eaters, molluscivores, zooplanktivores, insectivores, piscivores, parasite eaters, paedophages and scale eaters (Witte and van Oijen, 1990). Virtually all these trophic groups were depleted, except those which lived among refugia such as rocks and macrophyte cover to which Nile perch had no access (Witte *et al.*, 1992a, b). The detrivorous-phytoplanktivorous haplochromines and the pelagic phytoplanktivores fed on and controlled the algal biomass produced in the lake. These two trophic groups constituted about 50% of the total haplochromine biomass in the lake (Goldschmidt *et al.,* 1993). The depletion of the trophically complex haplochromine community and the changes in zooplankton community reduced grazing pressure which allowed organic matter to accumulate in that lake (Ogutu-Ohwayo and Hecky, 1991). The food web of the lake thus became greatly simplified. The reduction of the detritivorous haplochromines by the Nile perch and the larger herbivorous zooplankton by the *R. argentea* and *Y. laparogramma* reduced the grazing pressure which left much of the primary production in Lake Victoria unconsumed. Decay of the excess organic matter depletes the water column of oxygen. This led to oxygen deficiencies in parts of the lake > 40 m deep.

The development of a deep water anoxic layer favoured certain organisms especially *C. nilotica*, Chironomidae and Chaoboridae which became abundant. When the whole water column was well oxygenated, there was no refugium for *C. nilotica*, chironomid and chaoborid larvae and their populations may have been kept low due to predation by haplochromines and other native predators like *B. docmac.* The development of deep water hypoxia provided refugium for those benthic invertebrates which are tolerant of low oxygen tensions. Live *C. nilotica* have been observed in waters with < 1.0 mg/l of oxygen (Gophen *et al.*, 1995). Nile perch is not tolerant to such low oxygen tensions (Fish, 1956). It therefore only feeds on *C. nilotica* when they are out of the low oxygen refugia. This has allowed *C. nilotica*, chironomid and chaoborid larvae to proliferate. *C. nilotica* is the main food of juvenile Nile perch. The abundance of *C. nilotica* has favoured proliferation of Nile perch stocks. Since *C. nilotica* feeds on detritus, much of the primary production in the Lake Victoria is channelled to fish production through the detrital food chain.

Other factors affecting Lake Victoria ecosystem

One of the factors that may be contributing to increased water column stability and enhanced anoxia is climate change. The climate of the African Great Lake's region in general has become warmer due to global and regional climatic changes (Hastenrath and Kruss, 1992). Higher temperatures were recorded in Lake Victoria in 1989 to 1990 compared to 30 years ago (Worthington, 1931; Fish, 1957; Talling, 1965, 1966; Hecky, 1993) An increase in temperature would make the lake more stable and less able to mix effectively (Hecky, 1993). This would promote persistence of anoxia observed in Lake Victoria. Fuelwood was the most important source of energy and accounts for 94% of total energy consumption in Uganda. The demand for fuelwood in the heavily populated Victoria lake basin would have had an effect on the lake's environment through deforestation which exposes the soil to erosion and also increases in atmospheric CO_2 inputs through combustion processes.

Another problem that has developed in Lake Victoria since alien fish species have established in the lake is the invasion of the lake by an obnoxious alien water weed, the water hyacinth, *Eichhornia crassipes*. This is a beautiful plant which has been moved to different parts of the world due to its ornamental beauty. Water hyacinth thrives in shallow, sheltered bays that are breeding, nursery and feeding grounds for fish. It can, therefore, affect breeding success and feeding by juvenile fishes. The zone below extensive water hyacinth mats is low in oxygen, which reduces habitable space for most fish and other aquatic organisms upon which fish feed. Water hyacinth forms dense populations in areas where nutrients are high and the high phosphorous level in Lake Victoria has enhanced its growth. For instance on Lake Victoria, in Uganda, the biggest hyacinth expanses are in the areas where the city of Kampala released sewage into the lake.

Economic benefits of Nile perch introduction

The countries bordering Lake Victoria are low income societies and some of them are among the poorest in the world with a *per capita* income of about US$ 250. The economies of these countries are intimately dependent on agriculture and fisheries. An increase in fisheries yield, therefore has an important bearing on improving the livelihood of the people in these countries.

Considerable economic benefits have been derived from increased catches arising out of the establishment of Nile perch. Prior to 1960's, the catch from Lake Victoria was 30 000 to 50 000 metric tonnes of native species. Fishery yield increased rapidly with the establishment of Nile perch. Current estimates show that 400 000 to 500 000 metric tonnes of fish valued at US$ 300 to 400 million are landed annually from the entire lake. More than 90% of the catch is Nile perch. In the Ugandan portion of the lake, total fish catches decreased to about 10 000 tonnes by 1980 (Figure

4.4). As the Nile perch became established, fish catches increased from 17 000 tonnes in 1981 to 132 000 tonnes by 1989. The export market for fish, especially Nile perch has expanded rapidly since establishment of the predator into the lake. In Uganda, up to 20 factories with the capacity to process 20 tonnes of fresh fish per day have been licensed and half of these are already operational. In 1995 fish was the second most important export commodity in Uganda and fetched up to US$ 40 million. The increase in industrial fish processing has brought positive change in the balance of payment of riparian countries. The fishing effort has more than doubled. Total fishery yield has started to decline and there is need to control the current fishing effort if the fishery is to sustain production.

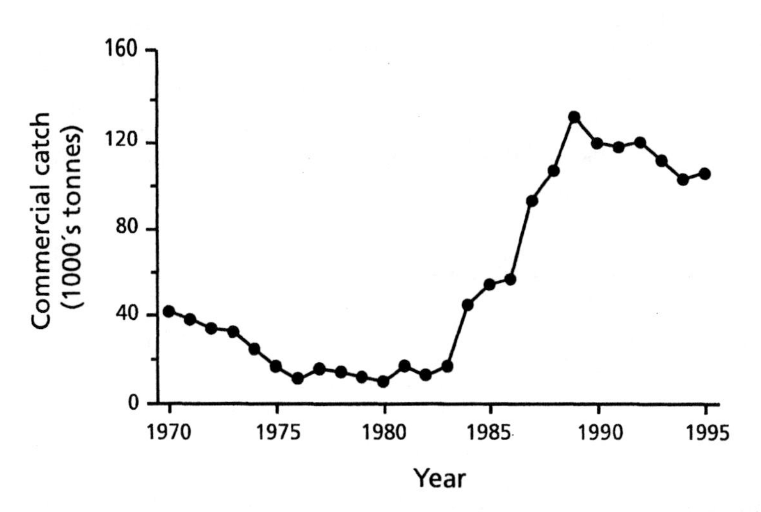

Figure 4.4 Commercial fish catches in the Ugandan portion of Lake Victoria just before and after establishment of Nile perch in the lake.

The expansion of the Nile perch fishery, especially the rapid development of fish processing plants has caused socio-economic problems. Fish has historically been and is still the cheapest source of animal protein for the average Ugandan. The demand by the fish processing plants has driven the price of fish to levels which cannot be afforded by the majority of the local people. In Uganda, the price of Nile perch more than doubled from about US$ 0.5 to US$ 1.2 per kg between 1990 and 1996 even when there was no significant inflation in the country.

Managing the changes

The introduction of the Nile perch brought both positive (increased availability of table fish, income and employment) and negative (reduced fish species diversity, environmental stress, and increased fish prices on local markets) effects. Efforts to manage the Lake Victoria ecosystem with respect to Nile perch requires proper management of its positive and negative effects. There are also other factors especially eutrophication which are impacting the ecosystem that need to be managed.

The governments of Uganda, Kenya and Tanzania have initiated a multi-sectoral Lake Victoria Environmental Management Program under sponsorship of GEF/World Bank, EU, IDRC and other donor agencies to strengthen, harmonize and integrate rational utilization, management and conservation of the resources of Lake Victoria and its catchment area. This program will implement projects aimed at controlling the loading of nutrients and contaminants into the lake, management of land use in the catchment area including wetlands, managing introduced species, conserving biodiversity and controlling the spread of the water hyacinth. The EU project is specifically intended to prepare a fisheries management plan for the lake. These efforts should assist in management of the lake.

Conservation of biodiversity

One of the biggest challenges facing the Lake Victoria and the other lakes (Kyoga and Nabugabo) to which Nile perch was introduced, is to restore and improve stocks of those fish species which have been depleted by the Nile perch. It was initially suggested that the diversity of species in lakes Victoria, Kyoga and Nabugabo could be preserved by collecting those species that were still surviving and keeping them in captivity until the populations of the Nile perch had been reduced to levels where these species could be re-introduced into the lakes (Ribbink, 1987). It was also suggested that comprehensive museum collections of existing species should be made to form permanent collections of the diversity that was existing at that time. Since then some fish, especially haplochromines and *O. esculentus*, have been collected from Lake Victoria and associated lakes and kept in captivity mainly in temperate Europe and North America. At that time, there was hope that the species maintained in captivity could be re-introduced into the lakes from which they had been collected after the Nile perch populations had been depleted from these lakes. At present, there is no hope of completely removing Nile perch from lakes Victoria, Kyoga and Nabugabo. At the same time, it will be economically undesirable and politically unacceptable to do so because of the large increases in fish catches and income realised due to establishment of Nile perch.

As long as there are still high populations of Nile perch in lakes Victoria, Kyoga and Nabugabo, it will not be possible to increase stocks of native species which are not resilient to Nile perch. The surviving species can only be conserved in areas where they are protected from Nile perch. The habitats that have been found to serve as refugia for the endangered species include; satellite lakes within the Victoria lake basin, rocky and macrophyte refugia within the lakes, swampy areas and associated lagoons bordering the lakes, and macrophyte zones along rivers (Witte *et al.,* 1992a, b; Ogutu-Ohwayo, 1990b, 1993; Chapman *et al.,* 1995).

There are many satellite lakes in the Victoria and Kyoga lake basins which originally had fish faunas similar to those of lakes Victoria and Kyoga. Nile perch was not introduced to many of these lakes and many of them are completely separated from the lakes to which Nile perch was introduced. A fish faunal survey of some of them; lakes Kayanja, Kayugi, Manywa and Kanyaboli in the Victoria lake basin (Ogutu-Ohwayo, 1993; Chapman *et al.,* 1995) and the Kyoga minor lakes of the Kyoga lake basin (Wandera pers. comm.) has shown that these lakes contain some of the species such as *O. esculentus* and many trophic groups of haplochromines that have disappeared from lakes Victoria, Kyoga and Nabugabo. Surveys of these lakes are being carried out to identify those that should be protected for conservation purposes.

In Lake Tanganyika, which has four *Lates* spp. and a species flock of haplochromine species, most of the haplochromines live among rocky areas where they are probably able to evade predation by *Lates* species (Fryer and Iles, 1972). Examination of rocky areas within Lake Victoria has revealed that the rock-dwelling species have been least affected by Nile perch predation (Ogutu-Ohwayo, 1990b; Witte *et al.,* 1992a, b). This suggests that rocky areas can serve as refugia for some haplochromines. Surveys carried out in the lakes to which Nile perch was introduced (Ogutu-Ohwayo, 1993, 1994; Chapman *et al.,* 1995) have shown that macrophyte and rocky refugia are important in conservation of endangered species. Stocks of haplochromines have increased in Lake Kyoga following expansion of cover due to the spread of the water hyacinth and reduction in predation pressure due to over-fishing of Nile perch. Papyrus swamps and other fringing macrophytes also act as barriers to the spread of the Nile perch since the species cannot survive under low oxygen tension. Some native species which can survive under these conditions can therefore take refuge in these areas. This is, however, not an effective barrier to some of the introduced species such as *O. leucostictus* which can tolerate low oxygen tensions and can migrate across swamps into water bodies that are separated from the main lake by the swamps (Welcomme, 1964). In order to protect some of the fish species, it will be necessary to prevent clearing of the papyrus swamps and vegetation along the affected lakes to stop the spread of the introduced species to the protected areas. For instance, the Yala swamp which separates Lake Kanyaboli from Lake Victoria should be protected to prevent the spread of Nile perch into Lake Kanyaboli. Similarly, the swamps separating lakes Manywa and Kayugi from Lake Nabugabo, and Lake Kayanja from Lake Victoria and those separating the Kyoga satellite lakes from the main lake should be protected. These refugia, if protected, can serve as future sources for the endangered species. Stocks of these species could further be improved and sustained by stocking them in dams and through aquaculture. The best strategy is to protect environmental degradation of these lakes, and to maintain their isolation, as much as possible.

Conservation of biodiversity will also benefit from control of increase in human population and control of eutrophication through regulation of nutrient loading, especially of phosphorus into the lake. The increased loading of phosphorus in the lake is thought to come from poor land use practices and sewage discharge. Reduction of phosphorus loading can be achieved through improved management of land use, adequate disposal and treatment of sewage and proper management of wetlands which reduce nutrient inputs to lakes.

What has happened in Lake Victoria should act as a lesson and a warning to what can happen to other African Great lakes. Like Lake Victoria, the major commercial fisheries of most of the African Great lakes had declined due to over-fishing. There have been proposals to introduce non-native species in some of the other lakes. The water hyacinth has spread to many of the lakes. Actions should be taken to prevent what has happened in Lake Victoria from occurring in the other Great Lakes of Africa. The changes were manifested first in Lake Victoria because its catchment is the most densely populated of the East African Great Lakes, it is the shallowest and has long water residence time. Many of the processes which have contributed to the changes in Lake Victoria are at work in all the African Great Lakes. It is apparent that the effects of the alien species such as the Nile perch may interact with other factors such as over-fishing and increase in eutrophication to cause changes that could have either taken longer to manifest themselves or could have been absorbed by the system.

Acknowledgements

The work on which this paper is based was supported by a grant from the International Development Research Centre (IDRC), Ottawa, Canada. The author is grateful to IDRC for financial assistance which made the work possible.

References

Anderson, A.M. (1961) Further observations concerning the proposed introduction of the Nile perch into Lake Victoria. *East African Agricultural Journal,* **26**, 195–201

Cadwalladr, D.A. (1965) Notes on breeding biology and ecology of *Labeo victorianus* Boulenger (Pisces: Cyprinidae) of Lake Victoria. *Revue de Zoologie et de Botanique Africaines,* **72**, 109–134.

Chapman, L.J., Chapman, C.A., Ogutu-Ohwayo, R., Chandler, M., Kaufman, L., and Keiter, A.E. (1995) Refugia for endangered fishes from an introduced predator in Lake Nabugabo, Uganda. *Conservation Biology,* **10**, 554–561.

Fish, G.R. (1956) Some aspects of the respiration of six species of fish from Uganda. *Journal of Experimental Biology,* **33**, 186–195.

Fish, G.R. (1957) A seiche movement and its effects on the hydrology of Lake Victoria. *Fisheries Publications, London,* **10,** 68 pp.

Fryer, G., and Iles, T.D. (1972) *The Cichlid Fishes of the Great Lakes of Africa: Their Biology and Evolution*. Oliver & Boyd, London.

Gee, J.M. (1969) A comparison of certain aspects of the biology of *Lates niloticus* (Linnaeus) in some East African lakes. *Revue de Zoologie et de Botanique Africaines,* **80**, 244–262.

Goldschmidt, T. and Witte, F. (1992) Explosive speciation and adaptive radiation of the haplochromine cichlids from Lake Victoria: an illustration of the scientific value of a lost species flock. *Mitteilungen der Internationale Vereinigung der Limnologie,* **23**, 101–107.

Goldschmidt, T., Witte, F. and Wanink, J. (1993) Cascading effects of the introduced Nile perch on the detritivorous/phytoplanktivorous species in the sublittoral areas of Lake Victoria. *Conservation Biology,* **7**, 686–700.

Gophen, M., Ochumba, P.B.O. and Kaufman, L.S. (1995) Some aspects of perturbation in the structure and biodiversity of the ecosystem of Lake Victoria. *Aquatic Living Resources,* **8**, 27–41.

Graham, M. (1929) *The Victoria Nyanza and its fisheries. A report on the fish survey of Lake Victoria 1972–1928 and Appendices*. Crown Agents for the Colonies, London.

Hastenrath, S., and Kruss, P.D. (1992) Greenhouse indicators in Kenya. *Nature,* **355**, 503.

Hecky, R.E. (1993) The eutrophication of Lake Victoria. *Verhandlungen der Internationale Vereinigung der Limnologie*, **25**, 39–48.

Hecky, R.E., Bugenyi, F.W.B., Ochumba, P.B.O., Talling, J.F., Mugidde, R., Gophen, M. and Kaufman, L.S. (1994) Deoxygenation of the deep waters of Lake Victoria, East Africa. *Limnology and Oceanography,* **39**, 1476–1481.

Jackson, P.B.N. (1971) The African Great Lakes Fisheries: past, present and future. *African Journal of Tropical Hydrobiology and Fisheries,* **1**, 35–49.

Johnson, T.C., Scholz, C.A., Tabolt, M.R., K. Kelts, Ricketts, R.D., Ngobi, G., Beuning, K., Ssemanda I. and McGill, J.W. (1996) Late pleistocene desiccation of Lake Victoria and rapid evolution of cichlids. *Science,* **273**, 1091–1093.

Kudhongania, A.W. and Cordone, A.J. (1974) Batho-spatial distribution patterns and biomass estimate of the major demersal fishes in Lake Victoria. *African Journal of Tropical Hydrobiology and Fisheries*, **3**, 15–31.

Lehman, J.T. and Branstrator, D.K. 1993 Effects of nutrients and grazing on phytoplankton of Lake Victoria. *Verhandlungen der Internationale Vereinigung der Limnologie*, **25**, 850–855.

Lehman, J.T. and Branstrator, D.K. (1994) Nutrient dynamics and turnover rates of phosphates and sulphate in Lake Victoria, East Africa. *Limnology and Oceanography*, **39**, 227–233.

Ligtvoet, W., and Mkumbo, O.C. (1990) Stock assessment of Nile perch in Lake Victoria. *FAO Fisheries Report*, **430**, 35–74.

Mugidde, R. (1993) The increase in phytoplankton production and biomass in Lake Victoria (Uganda). *Verhandlungen der Internationale Vereinigung der Limnologie*, **25**, 846–849.

Mwebaza-Ndawula (1994) Changes in relative abundance of zooplankton in Lake Victoria, East Africa. *Hydrobiologia*, **272**, 259–264.

Namulemo, G. (1998) *Distribution, relative abundance, population structure and food of surviving haplochromine cichlids in the littoral areas of Napoleon Gulf (Lake Victoria).* MSc thesis, Makerere University, Uganda.

Ochumba, P.B.O. and Kibaara, D.I. (1989) Observations on blue-green algal blooms in the open waters of Lake Victoria, Kenya. *African Journal of Ecology*, **27**, 23–34.

Ogari, J. and Dadzie, S. (1988) The food of the Nile perch, *Lates niloticus* (L.) after disappearance of the haplochromine cichlids in the Nyanza Gulf of Lake Victoria (Kenya). *Journal of Fish Biology*, **32**, 571–577.

Ogutu-Ohwayo, R. (1990a) The decline of the native fishes of lakes Victoria and Kyoga (East Africa) and the impact of introduced species, especially the Nile perch, *Lates niloticus*, and the Nile tilapia, *Oreochromis niloticus*. *Environmental Biology of Fishes*, **27**, 81–96.

Ogutu-Ohwayo, R. (1990b) Changes in the prey ingested and the variations in Nile perch and other fish stocks in Lake Kyoga and the northern waters of Lake Victoria (Uganda). *Journal of Fish Biology*, **37**, 55–63.

Ogutu-Ohwayo, R. (1993) The impact of predation by Nile perch, *Lates niloticus* on the fishes of Lake Nabugabo: with suggestions to conserve endangered native species. *Conservation Biology*, **7**, 701–711.

Ogutu-Ohwayo, R. (1994) *Adjustments in fish stocks and in life history characteristics of the Nile perch,* Lates niloticus *L. in lakes Victoria, Kyoga and Nabugabo*. PhD thesis, University of Manitoba, Canada.

Ogutu-Ohwayo, R. and Hecky, R.E. (1991) Fish introductions in Africa and some of their implications. *Canadian Journal of Fisheries and Aquatic Sciences*, **48** (Suppl. 1), 8–12.

Oijen van, M.J.P. (1982) Ecological differentiation among the haplochromine piscivorous species of Lake Victoria. *Netherlands Journal of Zoology*, **32**, 336–363.

Okaronon, J., Acere, T. and Ocenodongo, D. (1985) The current state of the fisheries in the northern portion of Lake Victoria (Uganda). *FAO Fisheries Report*, **335**, 89–98.

Ribbink, A.J. (1987) African lakes and their fishes: conservation scenarios and suggestions. *Environmental Biology of Fishes*, **19**, 3–26.

Rzoska, J. (1957) Notes on the crustacean plankton of Lake Victoria. *Proceedings of the Linnean Society of London*, **168**, 116–125.

Talling, J. (1965) The photosynthetic activity of phytoplankton in East African Lakes. *Internationale Revue der gesamten Hydrobiologie*, **50**, 1–32.

Talling, J.F. (1966) The annual cycle of stratification and phytoplankton growth in Lake Victoria (East Africa). *Internationale Revue des gesamten Hydrobiologie*, **51**, 545–621.

Tumwebaze, R. (1997) *Application of hydroacoustics in fish stock assessment of Lake Victoria.* MPhil Thesis, University of Bergen, Norway.

Wanink, J.H. (1991) Survival in a perturbed environment: the effects of Nile perch introduction on the zooplanktivorous fish community of Lake Victoria, in *Terrestrial and Aquatic Ecosystems: Perturbation and Recovery,* (ed. O. Ravera), Ellis Horwood Ltd., Chichester, pp. 269–275.

Welcomme, R.L., 1964 Notes on the present distribution and habitat of the non-endemic species of tilapia which have been introduced into Lake Victoria. *Annual Report of the East African Freshwater Fisheries Research Organisation,* **(1962–63)**, 36–39.

Witte, F. and van Oijen, M.J.P. (1990) Taxonomy, ecology and fishery of Lake Victoria haplochromine trophic groups. *Zoologische Verhandlungen, Leiden,* **262**, 1–47.

Witte, F., Goldschmidt, T., Wanink, J., van Oijen, M., Goudswaard, K., Witte-Maas, E. and Bouton, N. (1992a) The destruction of an endemic species flock: quantitative data on the decline of the haplochromine cichlids of Lake Victoria. *Environmental Biology of Fishes,* **34**, 1–28.

Witte, F., Goldschmidt, T., Goudswaard, P.C., Ligtvoet, W., van Oijen, M.J.P. and Wanink, J.H. (1992b) Species extinction and concomitant ecological changes in Lake Victoria. *Netherlands Journal of Zoology,* **42**, 214–232.

Worthington, E.B. (1929) *A report on the fishing survey of Lake Albert and Kyoga.* Crown Agents, London.

Worthington, E.B. (1931) Vertical movements of freshwater macroplankton. *Internationale Revue der gesamten Hydrobiologie und Hydrographie,* **25**, 394–436.

5 An alliance of biodiversity, agriculture, health, and business interests for improved alien species management in Hawaii

ALAN HOLT
The Nature Conservancy of Hawaii, Honolulu, Hawaii, USA

Abstract

Hawaii is in the midst of an invasive species crisis affecting the archipelago's highly-endemic biota, overall environmental and human health, and the viability of its tourism- and agriculture-based economy. Each year, an average of 20 alien invertebrates become newly established in the islands, compared to an estimated natural colonization rate of one new invertebrate every 25–100 000 years. Half of these alien invertebrates are known pests. More than one third of the threatened and endangered plants and birds in the United States live only in Hawaii. The primary threat to these taxa is from invasive species. The islands remain free of venomous snakes, most biting insects, and many diseases because of a long-established quarantine program, but this status is threatened by potential invasions of the brown tree snake, biting midges, mosquitoes and other pests via the large and expanding international traffic utilizing Hawaiian ports.

A special alliance of biodiversity, agriculture, health, and business interests is emerging which has the potential to address this pest crisis. The Hawaii alliance has focused on the early formation of partnerships among parties regarded as key to any successful pest management program and on assessing the full cost of the impact of alien pests on the Hawaiian economy. The group is conducting a major public awareness campaign to build political support for new tools needed to stem the flow of new invasives and more effectively control those that enter the islands. The most serious need is for tools which help target problem species, especially in the form of pest risk assessment to identify potential pests, sampling systems to identify and monitor "leaks" in port-of-entry inspections, and surveillance to detect newly-established pests while eradication or containment is still possible. The Hawaii program may serve as a useful test of these or other elements of any proposed global strategy for invasive species management.

Introduction

The Hawaiian Islands are in the midst of an invasive species crisis affecting the archipelago's highly-endemic biota, overall environmental and human health, and the viability of its tourism- and agriculture-based economy. This

O. T. Sandlund et al. (eds.), Invasive Species and Biodiversity Management, 65–75.

crisis is occurring in spite of the fact that Hawaii has one of the world's longest-standing and most comprehensive quarantine systems. This paper briefly describes the nature and extent of the alien species threat in Hawaii, the strategy currently underway to address it, and some of the main improvements needed in the Hawaiian pest prevention and control systems.

The impact of alien species in Hawaii

Two major factors have combined to bring about Hawaii's alien species crisis.

First, the archipelago offers an extraordinarily wide range of environments to potential invaders, as well as relatively mild competition for these habitats from native organisms. Before the arrival of humans some 1 500 years ago, Hawaii's isolation in the middle of the Pacific Ocean severely limited the rate of colonization by plants and animals, as well as the kind of colonists that could cross 2 000 miles (3 200 km) of salt water. As a result, Hawaii's native biota is famous both for its unequaled levels of endemism, and for its complete lack of terrestrial reptiles, amphibians, many major invertebrate groups including social Hymenoptera (e.g., ants and wasps), and virtual absence of terrestrial mammals (certainly one and possibly two species of bats are the only native land mammals). These native taxa once occupied and, in about one quarter of the archipelago's land area, still occupy habitats ranging in elevation from sea level to nearly 14 000 feet (4 270 m), in rainfall from 10 inches (25 cm) per year to over 500 inches (1 270 cm), and in substrates from newly erupted lava and cinders to highly-weathered wet clays (HDLNR, USFWS and TNCH, 1992). Any colonizing species that survived the ocean crossing to become established in Hawaii found a range of climates, fertile soils, relatively few competitors, and fewer diseases or predators than in most continental settings. Today, invading alien species benefit from the same favorable conditions.

The second major factor in Hawaii's alien species crisis is the breakdown of the extreme isolation once provided by the Pacific Ocean. Hawaii is the primary shipping link between North American, Asian, and other Pacific Rim ports, handling nearly 19 million tons of shipped cargo each year (HDOT, 1994). Honolulu International is the 17th busiest airport in the world, averaging one arriving flight every 1.3 minutes, and carrying 7 million tourists to the islands each year. Hawaii itself is reliant on these links; over 80% of the goods consumed in Hawaii are imported. Inevitably, however, cargo shipments, passenger flights, military transports, mail, and other traffic entering Hawaii bring with them living plants, animals, and microbes that would have been unable to reach the islands on their own. Figure 5.1 summarizes data for alien invertebrates intercepted at Hawaii ports of entry in 1994. The negative impact of alien pests has increased continually since first European contact with the islands in 1778, and is very serious today.

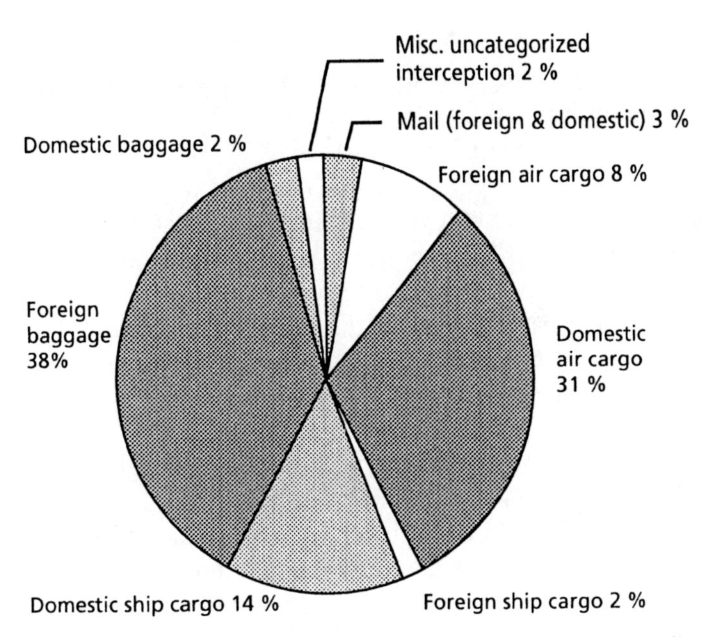

Figure 5.1 Alien invertebrate interceptions at Hawaii's borders, according to means of transport, as reported by U.S. Department of Agriculture (USDA) and Hawaii Department of Agriculture (HDOA) in 1994. Invertebrates are reportable by USDA if: a) it is a plant pest that will impact agriculture and/or the economy, and it does not occur on the US mainland, or b) it is a plant pest that occurs on the U.S. mainland and is currently a target of an eradication program.

Alien species are the chief threat to Hawaii's native biota, including an estimated 10 000 endemic life forms. Native habitats are threatened by alien ungulates such as pigs, goats, and deer that destroy vegetation, accelerate soil erosion, and facilitate the spread of alien weeds and insects. Our native birds suffer from introduced predators, loss of habitat to feral ungulates, and alien diseases spread by alien mosquitoes. Hawaii is now home to 38% of the United States' threatened and endangered plants and 41% of its endangered birds, in spite of the fact that these islands make up only 0.2%

of the nation's land area (HDLNR, USFWS and TNCH, 1992). For more than 95% of these 282 imperiled Hawaiian species, alien competitors, diseases, or predators are a primary threat.

Hawaii's agricultural sector, the third largest revenue producer behind tourism and military spending, estimates it is losing US$ 300 million per year in revenue from potential markets that now refuse Hawaii exports because of alien fruit flies that infest many island crops. Sugarcane and pineapple, the long-standing forces of Hawaiian plantation agriculture, are rapidly scaling down, creating an opportunity and a need for crop diversification. Many of Hawaii's most promising crops, however, are struggling under a siege of alien pests. In recent years these have included the papaya ringspot virus, banana bunchytop disease, bacterial blight of anthuriums and others. Each year, an average of 20 new alien invertebrates become established in the islands (Beardsley, 1979). This is a rate of one successful colonization every 18 days, compared to the estimated natural rate of once every 25-100 000 years (Zimmerman, 1970). Moreover, in the average year, half of the newly established invertebrates are taxa with known pest potential.

A single alien pest that entered the islands in the early 1900's – the Formosan subterranean termite – now causes nearly US$ 150 million in treatment and damage repair costs annually, most of which is paid by private homeowners (Tamashiro *et al.*, 1990). By comparison, the combined budget of all government pest prevention programs in Hawaii is only US$ 25 million (TNCH and NRDC, 1992).

Hawaii's US$ 18.9 billion visitor industry and island residents are increasingly concerned about new pests which threaten to invade. Hawaii has no snakes (except the harmless blind snake, introduced from the orient), no malaria, rabies, or dengue fever, and few biting insects; these facts are a large part of what makes Hawaii such a pleasant place to live or visit. The interception of brown tree snakes (*Boiga irregularis*) in Hawaii on seven occasions since 1971, however, has made Hawaii's people extremely concerned (see Table 5.1). Brown tree snakes intercepted in Hawaii have been on aircraft or in cargo from Guam. Experts do not believe this snake is

Table 5.1 Impacts of the brown tree snake (*Boiga irregularis*) on Guam (from Fritts *et al.*, 1995).

- Over 200 snakebite victims, 84% bitten while sleeping
- Power outages average once every four days
- Virtually all birdlife destroyed
- 9 endemic birds extinct in the wild

established in Hawaii at the present time. However, the risk of snake invasion, as well as recent interceptions of imported piranhas (*Serrasalmus* spp.), red fire ants (*Solenopsis invicta*), and emperor scorpions (*Heterometrus* sp.) in mail parcels have created increasing concerns regarding the threat of alien species to human health and safety and the overall quality of life in Hawaii.

Current pest prevention systems

Hawaii has been actively involved in alien pest prevention since 1888, when King David Kalakaua declared a quarantine on imported coffee to prevent the introduction of coffee rust and other diseases. Today, more than 20 state, federal, and private organizations and a number of volunteer groups dedicate a major part of their resources to designing, implementing, and improving alien pest prevention and control programs (TNCH and NRDC, 1992). The primary prevention agencies are the state and federal departments of agriculture. In general, federal agencies in Hawaii are concerned with preventing the introduction of noxious pests into the U.S. from foreign sources and preventing pests established in Hawaii from reaching the U.S. mainland. The primary task of the U.S. Department of Agriculture's inspection branch in Hawaii is to prevent the spread of Mediterranean and other fruit flies to major U.S. agricultural areas by inspecting passengers and flights leaving Hawaii for mainland destinations. The State Department of Agriculture, meanwhile, is mandated to protect Hawaii against pests from both domestic U.S. and foreign sources. Although state and federal agencies support each other to some extent in these inspections, the lists of restricted pests for which each agency has the authority to inspect differ dramatically, placing major limits on the sharing of inspection duties. Hawaii's list of prohibited or restricted taxa is longer than the federal list and includes vertebrates for which USDA has no inspection authority. State resources for inspection, however, do not reflect this broader inspection mandate. In 1992, federal agricultural inspection staff was double the size of the state's counterpart agency (TNCH and NRDC, 1992).

The control of established or newly escaped pests in Hawaii is primarily the responsibility of the state government, although federal agencies carry out pest control operations on federal lands, enforce endangered species laws, and carry out research to improve control methods. Private and non-governmental organizations are also actively involved in pest research and control. Hawaii has been a center for biological control research focused mostly on agriculture, and is actively engaged in the management of invasions for the protection of biodiversity. Over 75% of the management costs at Hawaii's national, state, and private nature reserves are for alien species control.

Strategy for improving Hawaii's protection against harmful alien species

The current effort to strengthen Hawaii's quarantine systems has developed in three stages. During 1991 and 1992, two non-governmental organizations (The Nature Conservancy of Hawaii and the Natural Resources Defense Council) prepared a report entitled "The Alien Pest Species Invasion in Hawaii: Background Study and Recommendations for Interagency Planning". This report described the roles, legal mandates, and resources of each agency or organization involved in preventing pests from becoming established in Hawaii or in controlling established pests. It identified at a general level the major problems in the current system, and recommended a process for developing plans to resolve these problems. The report highlighted two major needs above all others. First, it characterized the current system as "a set of programs that are generally effective within their own jurisdictions but which, together, leave many gaps and leaks for pest entry and establishment." The report called for a comprehensive pest management strategy linking the various players in a coordinated system. Second, it named strong public support and high-level political leadership as essential ingredients for success that, in 1992, did not exist. In preparing this report, the authors took special steps to work closely with the staff of the agencies whose work they were describing, in order to foster a constructive working relationship for future collaboration. For the public release of the document, key constituencies (e.g. the Hawaii Visitors Bureau, legislative leaders, agency heads) were briefed in advance and asked to prepare supporting statements for the media. The report was well received by the media and the community in general as a practical approach to an issue of real concern.

The 1992 background report set the stage for multi-agency development of an Alien Species Action Plan in 1993–94. This effort involved over 80 individuals from more than 40 government, non-profit, and private agencies, organizations, and businesses, who worked in professionally facilitated topic groups to prepare the plan. These topic groups submitted 34 more or less specific proposals for improvements to an oversight committee made up of leaders of key agencies and organizations. This committee then prepared the final plan, described as its commitment to "a first set of actions to improve pest prevention and control for Hawaii." The Oversight Committee's first action was to re-form itself as a permanent Coordinating Group on Alien Pest Species (CGAPS). CGAPS' most important feature is the broad set of interests it represents beyond the expected state and federal quarantine agencies. These include the state transportation and health departments, the Hawaii Visitors Bureau, the Hawaii Farm Bureau Federation which also represents horticultural interests, the U.S. Postal Service, the military, and state, federal, and non-profit biodiversity conservation agencies. The group is "held together by the voluntary efforts and enlightened self-interest of its members rather than by any formal authority," although formal agreements may be made for certain joint programs. Its purpose is "to expedite communications, problem-solving, and decision-making for more effective implementation of pest prevention and control work." The group is administered by the Hawaii Department of Agriculture, with additional staff support from The Nature Conservancy, and has held half-day, quarterly meetings since January 1995.

During its first 18 months, CGAPS faced two significant challenges in becoming an effective, multi-agency team. First, the launch of CGAPS coincided with the sharpest cutbacks in government budgets since statehood. This heightened member interest in collaboration and combining resources, but, more often, left key members with insufficient funding and personnel to pursue the desired alien species management actions. Second, many of the individuals sitting on CGAPS as agency representatives are unable to make major commitments for their agency. CGAPS can develop excellent strategies and resolve problems that require little new funding and no major legislative work. Major improvements, however, require political leadership of the highest level, and this depends upon widespread public support.

With this in mind, CGAPS launched a major public awareness campaign in late 1996. The campaign's centerpiece is a report entitled "The Silent Invasion" co-authored by all 14 CGAPS member agencies. The report is intended for elected officials and other community leaders, the media, and schoolteachers, and takes a bold approach to show how much Hawaii stands to lose from further pest introductions. It leads with the potential impacts on tourism, by far the state's leading industry, and describes the impact of pest species on people's lives. It includes culture as a potential victim of alien species invasion, and addresses the reader as an ally against this threat. The report provides the reader with the facts about why Hawaii is so vulnerable to invasion, and describes the main shortcomings in the current prevention system. It lays out a generalized 10-point plan that will serve as a framework for the many specific tasks needed to address the invasion problem. A goal in the campaign is to give the public a sense of the magnitude of the problem without leaving them feeling hopeless in the face of its complexity. For this reason, the report concludes with a list of the 10 Most Unwanted Pests, and a list of actions that every individual can take to reduce the chances that they or their friends and family will introduce a damaging pest species.

The campaign also includes lesson plans on alien pests for use by primary school teachers, an advertising campaign directed primarily at travelers, and polling to measure the effectiveness of the campaign in altering public knowledge and behavior. CGAPS will continue the highly successful Operation Miconia, a statewide media campaign to engage the public in locating and controlling *Miconia calvescens* (Melastomacae), a neotropical weed that has already overwhelmed major portions of Tahiti's native forests and which is now established on four Hawaiian islands. Other projects modeled after Operation Miconia will expand the opportunities for direct public involvement.

CGAPS is using the increased public awareness from this campaign to support specific alien species management legislative measures.

Priorities for improvement

The beginning of a major public awareness campaign brings all of CGAPS' members face to face with the obvious question: "What – specifically – do we want the public and our elected officials to do once they become aware of the magnitude of the alien species problem?" Like any highly complex problem, some parts of the solution are apparent and relatively simple while others are not yet clear even to the experts. More precisely, for the more difficult parts of the solution, the desired end result is clear but we cannot yet describe a practical approach for achieving it. This is also reflected in the SCOPE project "A Global Strategy for Alien Invasive Species" (Mooney, 1998) with its heavy focus on problem assessment in Phase I of the proposed planning project. In Hawaii's case, however, we have chosen to organize CGAPS and undertake some specific improvements before the problem assessment phase is fully completed. We recognized in 1992 that the relevant agencies had neither the necessary analytical capacity for full problem assessment nor the political support to develop it at that time. We also recognized a widespread sense of hopelessness in most agencies about being able to do anything to reduce significantly the alien species problem, and felt we had to get started with simple tasks, register some victories and public enthusiasm, and build our combined strength and commitment for the tougher challenges. Operation Miconia, for example, was carried out in 1996 to test CGAPS' ability to enlist the community in a pest containment effort. The overwhelming positive response from all sectors of the community not only accelerated *Miconia* control statewide but gave CGAPS members a strong boost to undertake additional projects.

CGAPS regards the following as the areas most in need of improvement:

Self-sustaining public education program – We are convinced that our greatest opportunity for improved pest prevention lies in educating the public. CGAPS' goal is to establish a dedicated funding source for continuous, high-quality public education messages delivered through a wide range of vehicles (e.g., tourist information, in-flight print and video materials, baggage claim area signage, school curricula, etc.). We are investigating the use of commercial advertising associated with alien species prevention messages in airports and other public facilities; the commercial ads are intended to pay for the public education program. State policy currently prohibits commercial advertising in most areas of the airport, and there are other legal complications to overcome.

Developing the ability to inspect all pest pathways – A large proportion of the total passenger, cargo, and other traffic entering Hawaii is currently uninspected, including materials known to be significant sources of new alien species. Domestic U.S. arrivals are very lightly inspected, and the state relies on voluntary declaration in order to foster a friendly, welcoming atmosphere for visitors. There are significant logistical and financial constraints on instituting mandatory domestic inspection, which would probably require pre-clearance of Hawaii-bound traffic at ports of origin to avoid redesign of Hawaii airports. State inspections are further hampered by the lack of x-ray equipment, and by questions about the state's legal authority to use x-ray to inspect baggage without probable cause. Moreover, some known pest pathways are legally protected against inspection. The U.S. First Class and air mails are common vehicles for transport of illegal animals and plants (TNCH and NRDC, 1992). Both California and Hawaii are working now with the U.S. Postal Service to find a way to stop these pests without violating the Fourth Amendment of the U.S. Constitution which protects these classes of mail. Until a remedy is found, an inspector must either have the permission of the sender or recipient to open the package, or a warrant from a federal magistrate for each package, and must complete the inspection without delaying the mail. The best hope for near-term improvement is probably through education reinforced by strict prosecution of violators.

Systems to monitor total pest traffic – Neither the federal nor state inspection agencies maintain consistent protocols to monitor the total pest traffic through a particular pathway as a gauge on the effectiveness of quarantine programs. Those data which are collected on pest interceptions are not fully utilized to improve inspection efficacy due to the lack of personnel dedicated to data analysis. In some cases (e.g., state monitoring of domestic U.S. pathways), the ability of agencies to conduct monitoring is constrained by the same laws which constrain inspection (above). Quarantine agencies do not currently have the resources to investigate newly detected pests to determine how they entered the state in order to detect leaks in the prevention system. Until greater monitoring capacity is developed, our appeal to the public and elected officials for quarantine improvements will be negated by our inability to tell them how well we are doing with present resources (i.e., what percentage of the estimated total alien species traffic are we intercepting). Or worse, our only gauge on the effectiveness of quarantine systems will be the number of newly established pest species, most of which are detected only after they have caused significant damage. The U.S. Department of Agriculture Animal and Plant Health Inspection Service initiated a sampling and data analysis program in 1997 to begin to fill this monitoring gap.

Technical support and timely processing of import permit review decisions – Although the Hawaii Department of Agriculture has the most comprehensive regulations in the U.S. for review of animal, plant, and microorganism imports (OTA, 1993), the expert committees that recommend permit decisions to the Board of Agriculture lack ready access to information relevant to assessing the subject taxon's disruptive potential. Decision-making is an inconsistent and time-consuming process because of this, and is made worse by state legal requirements for multiple reviews and public hearings that bring the standard processing time for many permits to over 12 months.

Early detection and eradication of new pest infestations – This is the most neglected phase of the invasion process, in that virtually all pest management effort is directed at port-of-entry inspections and the control of widespread pests (TNCH and NRDC, 1992). The U.S. Geological Survey's Biological Resource Division and others in Hawaii are now working on a prototype database to organize information from diverse sources on established pests. One application of these data will be to identify infestations that may be vulnerable to containment or eradication on a statewide, whole-island, or island region scale. These will be identified first for plants, and presented to weed control agencies in an attempt to organize range-wide containment or eradication projects. Remote sensing and other survey methods will have to be improved in order to support these projects and strengthen our ability to detect new pests.

In addition to range and biology data, eradication of incipient invasions requires better training for managers in pest control strategies to maximize the chances for success. Too often, the initial treatment of an infestation is intense but short-lived, and without precautions to prevent reinfestation or spread to other sites through contaminated equipment. Most projects also lack thc long-term follow-up to ensure complete eradication. A commitment to better training and planning is expected to improve the rate at which these projects succeed. Some Hawaii managers have suggested establishing a statewide team of pest control experts as trainers and information sources for natural area managers, much as the Cooperative Extension Service does for farmers. The Cooperative Extension Service and associated field agents from other agricultural agencies are the only team in Hawaii currently dedicated to early detection of pests, compilation of pest information, and dissemination of the best available control methods to field practitioners. Their approach needs to be applied to natural areas.

References

Beardsley, J.W., Jr. (1979) New immigrant insects in Hawaii: 1962 through 1976. *Proceedings of the Hawaiian Entomological Society*, **13** (1).

Fritts, T.H., Rodda, G.H. and Kosaka, E.F. (1995) *Unpublished update to Brown Tree Snake Cooperators.*

HDLNR, USFWS and TNCH (1992) *Hawaii's Extinction Crisis: A Call to Action*. Joint agency report, Hawaii State Department of Land and Natural Resources, United States Fish and Wildlife Service, and The Nature Conservancy of Hawaii.

HDOT (1994). *Annual report*. Hawaii State Department of Transportation.

Mooney, H.A. (1998) A Global Strategy for dealing with alien invasive species, in *Invasive Species and Biodiversity Management*, (eds O.T. Sandlund, PJ. Schei and Å. Viken), Kluwer Academic Publishers, Dordrecht, The Netherlands..

OTA (1993) *Harmful Non-Indigenous Species in the United States*. OTA-F-565, Office of Technology Assessment, U.S. Government Printing Office, Washington, DC.

Tamashiro, M., Yates, J.R. and Ebesu, R.H. (1987) The Formosan termite in Hawaii: problems and Control, in *Proceedings of the International Symposium on the Formosan Subterranean Termite*, University of Hawaii, College of Tropical Agriculture and Human Resources, Research Extension Series **083**.

TNCH and NRDC (1992) *The Alien Pest Species Invasion in Hawaii: Background Study and Recommendations for Interagency Planning*. Joint agency report, The Nature Conservancy of Hawaii and the Natural Resources Defense Council.

Zimmerman, E.C. (1970) Adaptive radiation in Hawaii with special reference to insects, in *A Natural History of the Hawaiian Islands*, University Press of Hawaii, Honolulu.

Part 2

Ecology of introductions

6 Invasive plant species and invasible ecosystems

MARCEL REJMÁNEK
University of California, Davis, California, USA

Abstract

The velocity with which an invading species spreads over space depends on the attributes of the individuals and populations making up that species. The ability of an organism to maintain relatively constant fitness over a range of environments, and population genetic polymorphism, contribute to species invasiveness. Small genome size seems to be a result of selection for short minimum generation time and, as it is also associated with small seed size and high leaf area ratio, may be an ultimate determinant of plant invasiveness in disturbed environments. Invasiveness of woody species in disturbed landscapes is associated with small seed mass, short minimum generation time, and short mean interval between large seed crops. Vertebrate dispersal is responsible for the success of many woody invaders in disturbed as well as 'undisturbed' habitats. Native latitudinal range of herbaceous species is a promising predictor of their invasiveness. Vegetative reproduction is essential for establishment and spread of many species in terrestrial environments and even more for dispersal in aquatic habitats. Disturbance of successionally advanced communities and their slow recovery rate promote invasions of many introduced species. A predictive theory of seed plant invasiveness which takes into account all these points seems to be emerging.

Introduction

Biological invasions are an important component of human-caused global environmental change (Bright 1996; Vitousek *et al.*, 1997a; Mooney and Hofgaard, 1998). We therefore need predictive theories that can help us set priorities for the control of introduced invasive species and allow us to predict the risk of future invasions. Attempts to predict which species will become invaders and which ecosystems will be invaded represent a fascinating area of basic research with far-reaching applications.

Until recently, problems with exotic plants have been treated individually, case by case, using a more or less trial and error approach. Substantial progress has been made, however, in the last few years. A concise conceptual framework of invasion biology was formulated (Williamson, 1996), the first screening systems for the prediction of

O. T. Sandlund et al. (eds.), Invasive Species and Biodiversity Management, 79–102.

invasive species have been developed (Panetta, 1993; Rejmánek, 1995; Tucker and Richardson, 1995; Rejmánek and Richardson, 1996; Williams, 1996; Reichard and Hamilton, 1997), and a predictive theory of seed plant invasiveness seems to be emerging (Rejmánek, 1996a). Still, many recognized relationships between species attributes and invasiveness remain unquantified and several assumed relationships have to be tested. Nontrivial generalizations about invasibility of natural and seminatural ecosystems remain in the first sketch stage.

In this chapter, I define invasive species as species spreading into areas where they are not native. I focus on invasiveness of plants in terms of their establishment and spread without explicitly considering their impacts on other biota and abiotic environments. Obviously, there is a continuum between invasive and "non-invasive" species. Species which do not exhibit any invasive behavior 100 years after introduction may turn out to be somewhat invasive after 300 years. The predictions I am trying to develop are on time scales of years and decades. This scale limitation, therefore, allows the possibility that a "delayed" invader (one not considered on this time scale) could have, one day, greater impacts than many "ready-to-spread" species.

What makes the difference between invasive and non-invasive species?

To find out what characters or combinations of characters are responsible for species invasiveness we need to know who is who first. Usually, it is easy to put together a list of "invasive" species upon which most biologists and managers in a given region would agree. To put together a list of "non-invasive" species is, as a rule, more difficult. The reason is that our knowledge of successful species invasions is much better than our knowledge of failed invasions. In short, we see what is invasive around us now but know only very little about the history of all the plants introduced to the area. There have been several attempts to overcome this difficulty by comparing alien invaders with native species (Mazer, 1989; Andersen, 1995; Pysek *et al.*, 1995; Baruch and Gomez, 1996; Williamson and Fitter, 1996). This approach can bring some interesting insights but is not necessarily the most powerful: among "native" species there are usually some which are invasive somewhere else. Fortunately, there are other feasible options.

Among the most important invasive trees in the Southern Hemisphere are several species of pines (Richardson, 1998). Pines are economically important and many species have been introduced to almost all countries with climates reasonably similar to those in their native lands. Thanks to that, reliable records about individual introductions and their subsequent failures, survival, growth, regeneration, and spread are extensive. Life history characters of many species have been studied in detail (Richardson *et al.*, 1998). Finally, pine reproductive biology is relatively simple, so underlying trends which can be masked in angiosperms by the intricacies of pollination and seed dispersal can be more easily detected. The genus *Pinus* therefore represents a unique opportunity for determining species characteristics responsible for invasiveness.

Our analyses (Rejmánek and Richardson, 1996) were based on data available for 24 well known and frequently cultivated pine species. We classified 12 of them as invasive a priori (reported as spontaneously spreading on at least two continents) and 12 as non-invasive (planted on at least three continents but never reported as spreading). A simple discriminant analysis (Selvin, 1995) was performed using ten potentially relevant life-history characters as predictors of membership in the two groups. Only three characters contributed significantly to the discriminant function and consistently maximized the difference between the two groups: mean seed mass, minimum juvenile period, and mean interval between large seed crops (Figure 6.1).

Two variables incorporated into the discriminant function are rather easy to interpret: short juvenile period and short interval between large seed crops mean early and consistent reproduction. Small mean seed mass seems to be associated with several potentially important phenomena: larger number of seeds produced, better dispersal, high initial germinability, and shorter chilling period needed to overcome dormancy (references in Mazer, 1989 and Rejmánek and Richardson, 1996). Short juvenile period may be related to fast growth in general and fast growth may be related to high leaf area ratio (LAR – leaf area/total biomass) which at least some of the invasive pine species exhibit (Strauss and Ledig, 1985; Grotkopp and Rejmánek, unpublished). Similarly, a short juvenile period (earlier bolting) was recognized as a critical factor determining invasiveness of a herbaceous species, *Conyza sumatrensis* (Thebaud *et al.*, 1996).

In short, early and fairly constant reproduction, as well as high rates of population growth and dispersal should allow alien species to escape the low numbers at which extinction is likely. The three variables significantly contributing to the discriminant function point to an r-K selection continuum (early–late successional species) along which invasive and non-invasive pines are situated (e.g., in the absence of disturbance *Pinus concorta* can be replaced by *P. ponderosa* and the last one by *P. lambertiana*; Fowells and Schubert, 1956; Pfister and Daubenmire, 1975; Yeaton, 1983). The fact that "r-strategists" are the best invaders is not surprising because the overwhelming majority of biological invasions take place in human- and/or naturally-disturbed habitats (Rejmánek, 1989, 1996b; Whitmore, 1991; Hobbs and Wuenneke, 1992; Schiffman, 1997). Our modern landscapes are mainly disturbed landscapes (Hannah *et al.*, 1994).

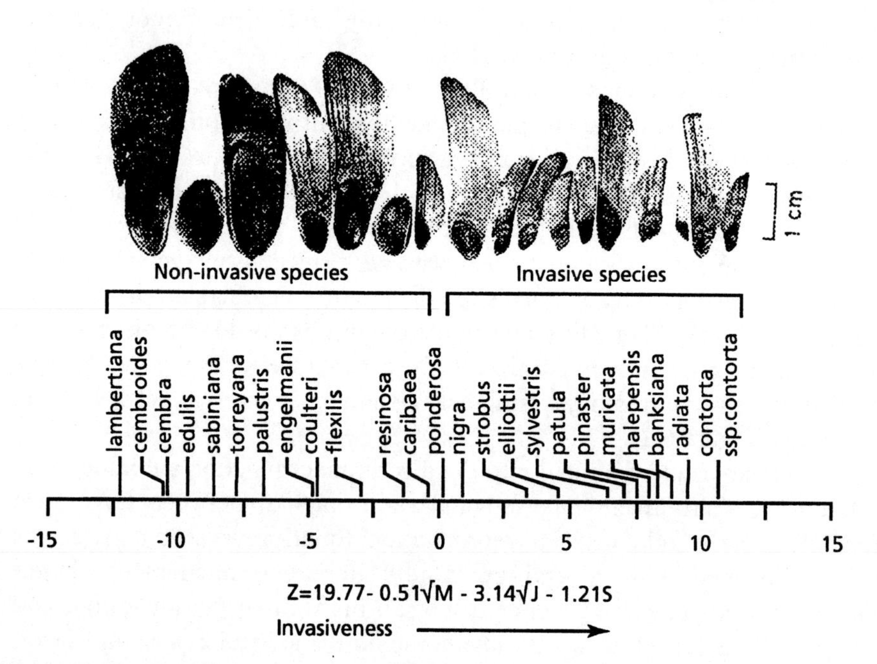

Figure 6.1 Simple discriminant analysis of invasiveness in frequently cultivated pine (*Pinus*) species based on mean seed mass in mg (M), mean interval between large seed crops in years (S), and minimum juvenile period in years (J). $F = 23.38$, $p < 0.001$, dividing point = 0.00. Relative contributions of $\sqrt{M}$, $\sqrt{J}$, and S are 42.7, 40.2, and 17.1%, respectively. Above are seeds of selected species as they appear when shed naturally from their cones.

We next applied the discriminant function (Z, Figure 6.1) to 40 of the most invasive woody angiosperm species from 40 different genera and obtained correct classification of 38 species (Rejmánek and Richardson, 1996). Only *Melia azedarach* and *Maesopsis eminii* were incorrectly classified as non-invasive. Efficient bird, bat, and primate dispersal of these species is responsible for this discrepancy. The discriminant function also correctly predicts that even some officially recommended exotic woody species for revegetation in the U.S. (USDA, 1992) are invasive (*Acer ginala, Lonicera maackii*).

Using the "pine discriminant function" (Figure 6.1) in analyses of other groups of species, many frequently cultivated but usually non-invasive angiosperm species are correctly classified as non-invasive (*Acer saccharum, Araucaria* spp., *Bertholletia excelsa, Camellia japonica, Corylus* spp., *Fagus* spp., *Juglans* spp., *Magnolia* spp., many species of *Quercus, Thevetia peruviana, Swietenia macrophylla, Aesculus hippocastanum, Aleurites molucana*). (The last two species are examples of plants which become naturalized in many places due to extensive human help but which can not be called invasive.) Using the same function, however, some non-invasive species of *Populus* (*P. tremula, P. tremuloides*) are classified as invasive. The short seed viability and high seedling mortality brought about by the slow growth of seedling primary roots in these *Populus* species probably prevent them from becoming invasive. In general, it appears that invasiveness of woody species with dry fruits and mean seed mass < 2.0 mg (*Populus* spp., *Salix* spp., *Betula* spp., *Alnus* spp., *Eucalyptus* spp., *Melaleuca quinquenervia, Tamarix* spp.) is often limited to wet environments and mineral substrates. On the other hand, vertebrate dispersal is responsible for the success of many woody invaders in disturbed as well as "undisturbed" habitats (Bingelli, 1996; Rejmánek, 1996a, b; Williams and Kral, 1996; Ehrenfeld, 1997). This fact is taken into account in Table 6.1, which summarizes general rules for detection of invasive woody seed plants, assuming compatibility of recipient habitats.

For many species, however, not all data used in Table 6.1 are readily available. Using some easily accessible characters, Reichard and Hamilton (1997) developed a decision tree for admission or rejection of proposed woody exotic introductions to North America. Their first question "Does the species invade elsewhere?" is a logical implementation of the U.S. Department of Agriculture general strategy on assessment of potentially important foreign weeds (Reed, 1977; Gun and Ritchie, 1988). Pragmatically, this is the most important question and with development of relevant data-bases (e.g., Wells *et al.*, 1986; Swarbrick and Skarratt, 1994; Randall, 1995; Poschlod *et al.*, 1996) should lead to immediate rejection of many species. Some of the branches of Reichard and Hamilton's decision tree, however, may lead to our Table 6.1 again (Figure 6.2). The resulting decision process is rather conservative but it is the best one for woody taxa admission at the moment. It can be easily modified for other continents.

Besides woody species, grasses represent the second most important group of plant invaders (Labouriau, 1966; Parsons, 1970; Baker, 1978; Huenneke and Mooney, 1989; D'Antonio and Vitousek, 1992). Low construction costs associated with high leaf area ratio (LAR – leaf area/total biomass) and its component, high specific leaf area (SLA – leaf area/leaf biomass) are certainly factors contributing to invasiveness of some grasses, namely in disturbed areas (Baruch, 1996; Baruch and Gomez, 1996; Ryser and Notz, 1996). This is compatible with the conclusion based on evaluation of experimental trials with introduced pasture species in Australia: "... a useful exotic pasture species is almost certain to become a weed in some circumstances" (Lonsdale, 1994). Shorter intervals between reproductive phenophases may be another important factor (Baruch *et al.*, 1989).

Table 6.1 General rules for detection of invasive woody seed plants based on values of the discriminant function Z (Figure 6.1), seed mass values, and presence or absence of opportunities for vertebrate dispersal (Rejmánek and Richardson, 1996). For examples of species in individual categories see Rejmánek (1996a).

		Opportunities for vertebrate dispersal	
	Fruit characters	**Absent**	**Present**
Z > 0	Dry fruits and seed mass > 2 mg	Likely invasive	Very likely invasive
	Dry fruits and seed mass < 2 mg	Likely invasive in wet habitats	
	Fleshy fruits	Unlikely invasive	Very likely invasive
Z < 0		Non-invasive unless dispersed by water	Possibly invasive

Nuclear DNA content

The amount of nuclear DNA varies over a thousand-fold among angiosperms, from about 0.2 picograms in *Arabidopsis thaliana* to > 200 picograms in some Liliaceae (Bennett, 1987). Polyploidy accounts for some of this variation but most of it is due to differences in genome size (the amount of DNA in one complete set of chromosomes). Differences between species within individual genera are smaller but may still be responsible for important life-history contrasts. Wakamiya *et al.* (1993) reported significant positive correlations between nuclear DNA content (genome size) and both length of juvenile period and seed mass for 18 North American *Pinus* species. Consequently, values of our discriminant function (Z in Figure 6.1) for these species are significantly negatively correlated with their nuclear DNA content (Rejmánek, 1996a). This result is not surprising. The majority of the DNA in the nuclei of most vascular plant species consists of repetitive

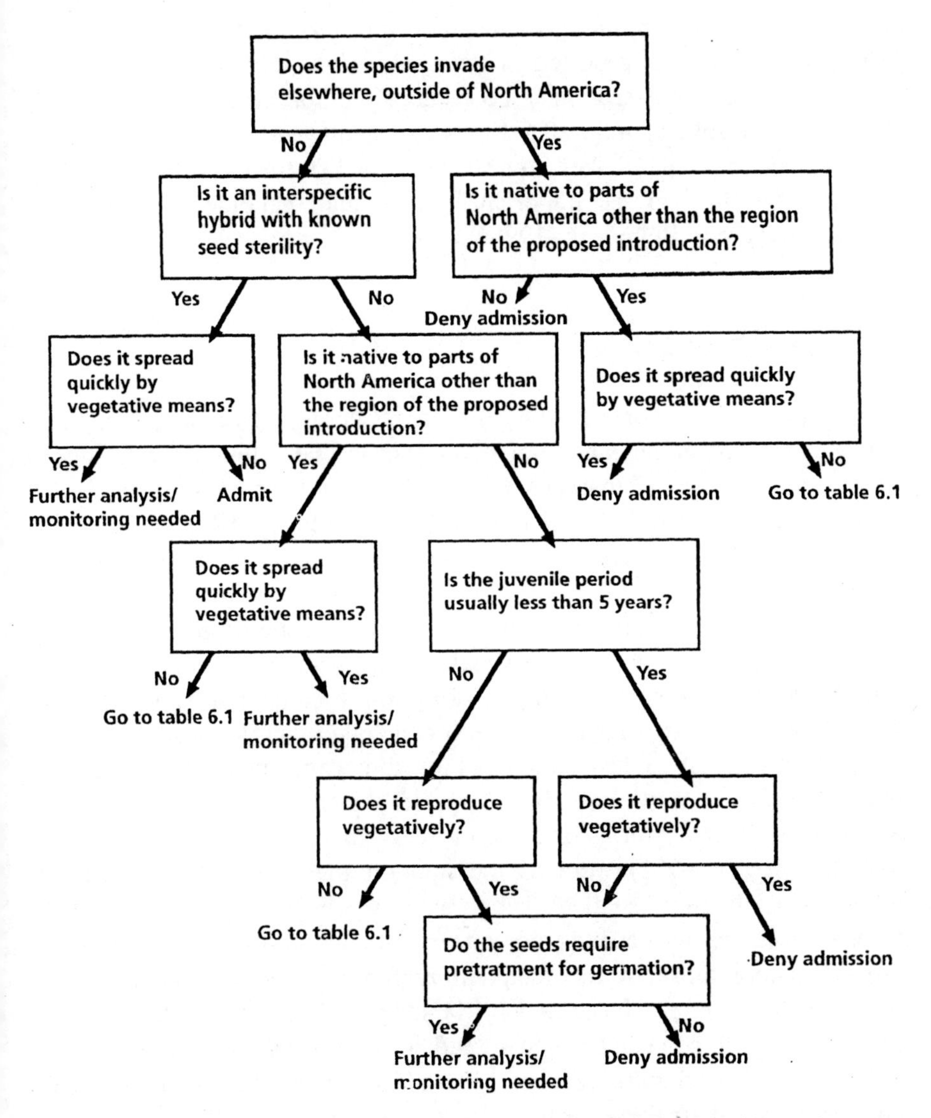

Figure 6.2 Decision tree for admission of exotic woody species to North America. After Reichard and Hamilton (1997), modified.

base-sequences that are not transcribed to proteins (Cavalier-Smith, 1985). Copying these sequences increases the duration of cell cycles and may increase the length of minimum generation time (juvenile period). A small genome size seems to be the result of selection for a short minimum generation time in time-limited environments (Bachmann *et al.*, 1985; Bennett, 1987).

A significant positive correlation between nuclear DNA content and seed size have been observed in individual genera, families, and surprisingly even in a random sample of 131 herbaceous angiosperms from 28 families (Thompson, 1990). In addition, as a significant positive correlation between nuclear DNA content and volume of both leaf palisade and spongy parenchyma cells was reported in the genus *Helianthus* (Sims and Price, 1985), small genome size may be responsible for high SLA and LAR. Therefore, for several interrelated reasons, small genome size may be an ultimate determinant of plant invasiveness in disturbed landscapes. Other examples pointing to the importance of a small genome size for plant invasiveness are in Rejmánek (1996a).

Geographic ranges

Significant positive correlations between primary (native) and secondary (adventive) geographic ranges of plant species were found for European herbaceous legumes (Fabaceae), grasses (Poaceae), and sunflower family (Asteraceae) members naturalized in North America (Rejmánek, 1995, 1996a). Another example is the significant positive dependence of secondary latitudinal ranges of *Ranunculus* species introduced to North America on their primary latitudinal ranges in Europe (Figure 6.3). There are two explanations why there should be a positive correlation between primary and secondary geographic ranges. Forcella and Wood (1984), Forcella *et al.* (1986), and Jäger (1988) concluded that the positive relation between area of native distribution and invading capacity arose from the fact that the propagules of widespread species have a higher probability of transport to other countries or continents. On the other hand, Noble (1989) and Roy *et al.* (1991) were of the opinion that, with the considerable increase in intercontinental exchange since the beginning of this century, invasion by a species depends more on the interaction between its biological properties and those of the recipient region than on the probability of reaching that region. Roy *et al.* (1991) suggested that the same biological traits that enable some species to spread across their native continents (and across different climatic zones) also make them able to invade new continents. What are these traits?

Seed production seems to be onc of them (Peart and Fitter, 1994; Rejmánek, 1996a). However, the whole story is undoubtedly more complicated. The ability of an organism or population to maintain relatively constant fitness over a range of environments can be called "fitness homoeostasis" (Hoffman and Parsons, 1991). The phenotypic (physiological and morphological) plasticity responsible for "individual fitness homoeostasis" represents one particularly important factor in this context. Backer's "general purpose genotype" (Baker, 1974) seems to be equivalent with this concept. Similarly, we may distinguish "population fitness homoeostasis" which is a result of both individual fitness homoeostasis and population genetic polymorphism. Population fitness homoeostasis can then control several other factors which influence species invasiveness and its geographic range (Figure 6.4).

Analyses of primary and secondary geographic ranges can generate stimulating hypotheses, but these should be used with caution for identifying potential invaders. In Europe, for example, one of the most successful invaders, *Impatiens parviflora*, was introduced from a rather limited area in Central Asia (Meusel *et al.*, 1978). Returning to pines, *Pinus radiata* and *P. patula* are among the most invasive trees of the Southern Hemisphere (Richardson *et al.*, 1994) although their native geographic ranges are very narrow (Critchfield and Little, 1966).

Phylogenetic distance from resident species

It is reasonable to assume that species possessing traits different from those of resident species should be more successful invaders (Moulton and Pimm, 1987; Brown, 1995). That is the essence of Darwin's mostly forgotten theory of plant invasions (Darwin, 1859). From Alphonse de Candole he learned that "floras gain by naturalization, proportionally with the number of native genera and species, far more in new genera than in new species" (p. 114). Darwin illustrated this statement by an example: "... in the last edition of Asa Gray's "Manual of the Flora of the Northern United States," 260 naturalized species are enumerated, and these belong to 162 genera. ... out of the 162 genera, no less than 100 genera are not there indigenous, and thus a large proportional addition is made to the genera of these States". He then inferred that differences in structure and functioning (i.e., "place in the economy of nature") are greater between genera than between species within a genus and hence "... the struggle will generally be more severe between species of the same genus, than between species of distinct genera" (p. 76).

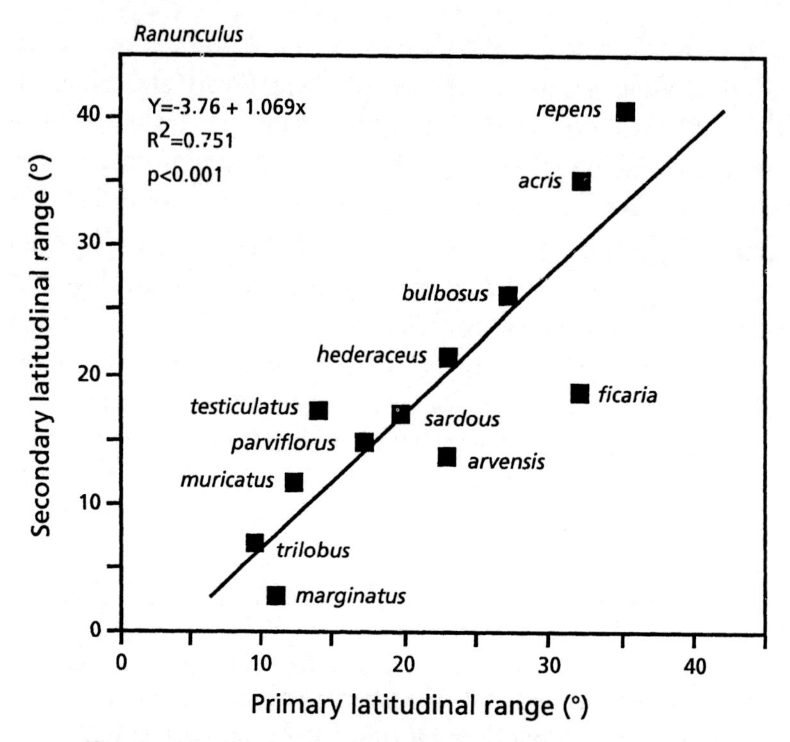

Figure 6.3 Relationship between primary (European) and secondary (North American) latitudinal ranges of buttercups (*Ranunculus*) naturalized in North America. Based on data in Jalas and Suominen (1989) and Whittemore (1997).

However, there is a problem with the null hypothesis in Darwin's theory because there are many genera available in other countries that are not native in the northern United States. To investigate this problem further, we may take as an example the 112 European grass species (Poaceae) naturalized in California (Hickman *et al.*, 1993; Rejmánek and Randall, 1994). From this number, 43 species belong to 16 genera native to California and 69 species belong to 39 genera new to California. This seems to be in agreement with the Darwin's theory. However, while there are 31 genera native to both California and Europe, there are 108 European genera originally not present in California (Tutin *et al.*, 1980; Hickman *et al.*, 1993), the result seems less impressive. Nevertheless, when we take into account numbers of European species available in the two categories (509 in shared genera and 326 in European genera), the result is again in agreement with Darwin: species belonging to genera not native to California are significantly over-represented (χ^2= 27.7, p < 0.001). This seems to be a rather robust result because the trend is significant even after exclusion of the largest shared genera (*Festuca* with 170 and *Poa* with 44 European species) from thc data (χ^2= 11.23, $p < 0.001$; see Table 6.2). Analyses of Asteraceae and Brassicaceae in California (Table 6.2) and preliminary analysis of naturalized Poaceae in Australia (not shown) yield similar results.

Table 6.2 Contingency table analysis of European species in families Poaceae, Asteraceae, and Brassicaceae naturalized in California.

Plant Family		**Number of European species not naturalized in California**	**Number of European species naturalized in California**	
Poaceae	in shared genera	466	43	χ^2=27.7, p<0.0001 (After exclusion of *Festuca* (170 Europ. species) and *Poa* (44 Europ. species) χ^2=11.2, p<0.001)
	in European genera	257	69	
Asteraceae	in shared genera	681	22	χ^2=16.45, p<0.0001 (After exclusion of *Hieracium* (260 Europ. species) and *Centaurea* (221 Europ. species) χ^2=7.12, p<0.01)
	in European genera	762	66	
Brassicaceae	in shared genera	230	12	χ^2=4.31, p<0.05 (After exclusion of *Alyssum* (64 Europ. species) χ^2=5.33, p<0.03)
	in European genera	342	36	

One potentially important aspect of this theory is that the success of species belonging to non-native genera and families may be partly due to the limited number of resident herbivores and pathogens able to switch to species phylogenetically distant from their native hosts. In this context, an important hypothesis was recently formulated by Blossey and Nötzold (1995; Blossey, 1997): Invasiveness of nonindigenous plants is seen as a result of shifts in biomass allocation patterns. In the absence of herbivores, selection will favor genotypes with improved competitive abilities and reduced resource allocation to herbivore defense (Figure 6.4). This may also contribute to the explanation of why successful alien plants are, on average, taller than natives, at least in some habitats (Pysek *et al.*, 1995; Crawley *et al.*, 1996). Reduced herbivore resistance in introduced *Spartina alterniflora* after a century of herbivore-free growth was recently reported by Daehler and Strong (1997). They found, however, that plants with faster growth rates had also higher resistance to herbivory.

Additional anecdotal support for the Darwin's theory is found in the fact that the majority of the most successful and most influential invasive plant species in North America seem to belong to non-native genera (e.g., *Aegilops, Ailanthus, Alhagi, Apera, Arundo, Avena, Brachypodium, Cardaria, Carduus, Carpobrotus, Casuarina, Chondrilla, Conium, Cortaderia, Cynara, Cytisus, Eichhornia, Genista, Halogeton, Hydrilla, Melaleuca, Melilotus, Mesembryanthemum, Pueraria, Salsola, Sapium, Schinus, Sorghum, Striga, Tamarix, Taeniatherum, Tribulus, Ulex; Ammophila* and *Cakile* on the west coast; *Elaeagnus* in southern states; *Centaurea* in California and the Pacific Northwest). As we have already seen, several species of pines (*Pinus* – a North Hemisphere genus) are among the most important invasive trees in temperate areas of the Southern Hemisphere where native pines are absent.

Other factors and an attempt to put it all together

There are certainly more pieces of the whole story and many important factors are species- or habitat-specific. For example, ants may assist in invasions of plants that produce seeds with elaiosomes (Smith, 1989; Bossard, 1991). Self-pollination, pollination niche separation, breakdown of pollinator specificity, or apomixis can be important, especially in initial stages of some invasions (Baker, 1967; Costas Lippmann, 1976; Parrish and Bazzaz, 1978; Ware and Compton, 1992). Special germination requirements are not beneficial in many situations and are therefore often associated with non-invasive species (Reichard and Hamilton, 1997). Long fruiting periods seem to be associated with invasiveness of both woody and herbaceous plants (Parsons and Cuthbertson, 1992; Reichard, 1994). Several structural and physiological characters like bark thickness (Richardson *et al.*, 1990), symbiotic nitrogen fixation (Vitousek and Walker, 1989), rapid root growth (Hulbert, 1955), shade tolerance (Jones and McLeod, 1989), resistance to browsing (Ledgard, 1988), stem photosynthesis (Bossard and Rejmánek, 1992), production of allelochemicals (Lawrence *et al.*, 1991; Gentle and Duggin, 1997), or association with mycorrhizal fungi (Goodwin, 1992) may contribute to the success of particular invaders in particular environments. In California, for example, rapid root growth of *Centaurea solstitialis* during the first half of the season allows this species a long flowering and fruiting period during the second half of the season which is usually without any rain.

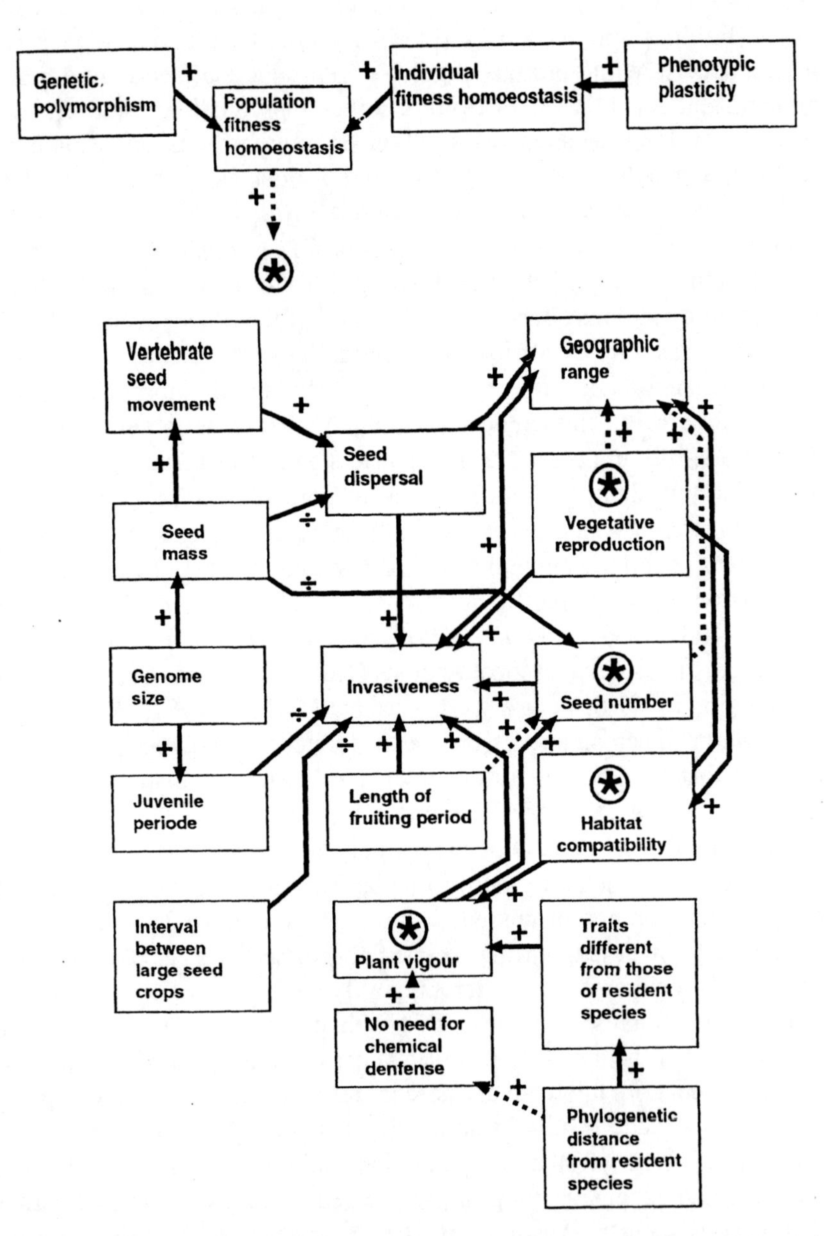

Figure 6.4 Recognized (full lines) and assumed (dashed lines) causal and correlative relationships involved in the determination of seed plant (herbaceous and woody) invasiveness. Arrows with "+" and "÷" indicate positive and negative effects, respectively.

Vegetative reproduction can be essential for establishment and short-distance dispersal of many species in terrestrial environments (Hu 1979; Auld, 1983; Chipping, 1993; Andersen, 1995; Thompson *et al.*, 1995; Pysek, 1997). Incidentally, the *Ranunculus* species with the broadest primary (Europe) and secondary (North America) latitudinal ranges is *R. repens*, a species with vigorous vegetative reproduction (see Figure 6.3 and Sarukhan and Gadgil, 1974). Vegetative reproduction is even more important for long-distance dispersal in aquatic habitats and floodplains (Ashton and Mitchell, 1989; Spencer and Rejmánek, 1989; Pieterse and Murphy, 1990; Bramley *et al.*, 1995; Brock *et al.*, 1995). The well known invasiveness of water hyacinth (*Eichhornia crassipes*) seems to be essentially the result of its free-floating life form and prolific asexual reproduction by stolons (Barrett, 1992). None of the other seven *Eichhornia* species possess these attributes. Asexually reproducing populations, however, often contain rather low levels of genetic polymorphism (Barrett, 1992). Theoretically, this could limit their invasiveness to some extent. In general, trade-offs in resource allocation to sexual and asexual reproduction (Sutherland and Vickery, 1988) can be responsible for important habitat-dependent differences in invasiveness of particular species and ecotypes.

Recipient habitat compatibility is usually treated as a necessary condition for all invasions (Sanders, 1976; Chicoine *et al.*, 1985; Panetta and Dodd, 1987; Beerling *et al.*, 1995). The match of primary and secondary environments is usually reasonably close (Hulten and Fries 1986; Hickman, 1993, Hügin, 1995) but not always perfect (Michael, 1981; Wilson *et al.*, 1992). Major discrepancies have been found for aquatic plants where secondary distributions are often much less restricted than their primary distributions (Cook, 1985; Thomas and Room, 1986). Vegetative reproduction of many aquatic plants is likely an important factor.

The sketch of a general theory of plant invasiveness, which takes into account major points discussed so far, is summarized in Figure 6.4. Unfortunately, data supporting particular connections in the inevitably complicated web are of very different quality and some connections are only assumed. Nevertheless, I believe this web provides at least a temporary framework for a theory of plant invasiveness. Some parts of this web may form a coherent core upon which more mature theories can be developed.

Predictions made on the basis of the presented theory (Figure 6.4) can be, to some extent, tested in quarantine field trials. This can be time-consuming and expensive but promising when dealing with limited numbers of herbaceous species (Austin *et al.*, 1985; Crawley *et al.*, 1993; Lonsdale, 1994). Mostly unexplained time lags in invasions of some woody species (Kowarik, 1995), however, make experimental testing less appealing as a universal tool. For example, eastern white pine (*Pinus strobus*) was recognized as an important invader in Central Europe only in the last ten years, more than 250 years after its introduction into forest plantations (Hadincová *et al*., 1997). Lag times can be considerable even for herbaceous species (40 years for *Impatiens glandulifera* in central Europe; Pysek and Prach, 1995). Lag times of invasive species remind us of the asymptomatic period of typical HIV infection (Nowak and McMichael, 1995). How much more could we learn from immunologists?

Finally, we should not forget that the spread of many alien species is heavily dependent on human activities which may change over time. For example, *Agrostema githago* was formerly one of the most common weeds in many countries. This species, however, is dependent upon repeated reintroduction to field in contaminated cereal seed. Since grain today is very effectively cleaned, the species has been driven to extinction in many areas where it was a serious weed some 50 years ago. The European species, *Centaurea solstitialis*, spreads slowly on it own but for a long time has been widely dispersed as a contaminant in alfalfa seed in Chile and California (John Gerlach, personal communication). Thanks to that and its preadaptation to summer dry climates, which was mentioned earlier, this species has become the number one rangeland weed in California, more than 130 years after its introduction. Currently, larger and larger volumes of soil are moved around by people (incl. mud on cars). Species with numerous, relatively small, persistent seeds are pre-adapted for this kind of transport (UCPE, 1996).

Invasibility

All analyses of ecosystem invasibility based on just one-point-in-time observations (a posteriori) are unsatisfactory (Rejmánek, 1989). In most of the cases we do not know anything about the quality, quantity, and regime of introduction of imported propagules. Usually, it is impossible to separate the resistance of biotic communities from resistance determined by abiotic environments. Nevertheless, available evidence indicates that only a very few alien species invade successionally advanced plant communities (Rejmánek, 1989, 1996b). Plant communities in mesic environments seem to be more invasible than communities in extreme environments. Apparently xeric environments are not favorable for germination and seedling survival of many introduced species and undisturbed wet terrestrial habitats do not provide open space for invaders because of the high biomass of resident species. Open water, however, is notoriously open to all kinds of exotic aquatic plants (Ashton and Mitchell, 1989).

In general, frequent disturbance, slow recovery rate, and fragmentation of successionally advanced communities promote plant invasions (Rejmánek, 1989; Hobbs and Huenneke, 1992; Huston, 1994; Robertson *et al.*, 1994; Clark and Ji, 1995; Bastl *et al.*, 1997; Schiffman, 1997). Disturbance-mediated susceptibility to invasions seems particularly pronounced in fertile/eutrophic environments (Burke and Grime, 1996; Fensham and Cowie, 1998). This is apparently one of the reasons why many riparian habitats are so heavily infested by exotic plants (Planty-Tabacchi *et al.*, 1996). Increasing world-wide eutrophication and, in particular, atmospheric nitrogen fertilization (Vitousek *et al.*, 1997b) provides a frightening prospect since many aggresive and highiy competitive invaders are preadapted to nutrient enriched environments (Odum, 1998).

How important the role of species richness is in biotic resistance to invasions still remains to be systematically tested (Mcintyre *et al.*, 1988; Robinson *et al.*, 1995; Allison, 1996; Planty-Tabacchi *et al.*, 1996; Rejmánek, 1996b; Kwiatkowska *et al.*, 1997; Tilman, 1997). Although some studies conclude that there is a negative relationship between species richness and invasibility, there is a question whether this is the result of species richness *per se*. Is it not possible that with increasing species richness the likelihood of presence of one or more important "protector species" (e.g., large plants or C4 grasses) increases? Huston's (1997) discussion of "hidden treatments" is particularly relevant here. It is also possible that experiments in different spacial scales will provide different answers (Stohlgren *et al.*, 1997). Formulation of both general and habitat-specific rules for assessment of invasiveness of species and ecosystem invasibility remain the two interconnected goals of research in the immediate future.

Summary

A predictive theory of seed plant invasiveness seems to be emerging:

(1) Phenotypic plasticity which is responsible for individual fitness homoeostasis (ability of an organism or population to maintain relatively constant fitness over a range of environments) represents one particularly important factor.

(2) Population fitness homoeostasis, a result of both individual fitness homoeostasis and population genetic polymorphism can control species invasiveness and its geographic range.

(3) Small genome size seems to be a result of selection for short minimum generation time and, as it is also associated with small seed

size and high leaf area ratio, may be an ultimate determinant of plant invasiveness in disturbed landscapes.

(4) Invasiveness of woody species in disturbed landscapes is associated with small seed mass, short juvenile period, long seed dispersal period, and short mean interval between large seed crops.

(5) Vertebrate dispersal is responsible for the success of many woody invaders in disturbed as well as "undisturbed" habitats.

(6) Native latitudinal range of herbaceous species is a promising predictor of their invasiveness. Both population fitness homoeostasis and dispersal abilities seem to be behind this generalization.

(7) Vegetative reproduction is essential for establishment and spread of many species in terrestrial environments and even more important for long-distance dispersal in aquatic habitats.

(8) Analysis of Poaceae, Asteraceae, and Brassicaceae introduced from Europe to California supports Darwin's suggestion that alien species belonging to exotic genera (and, therefore, possessing traits different from those of resident species) are more likely to be invasive than alien species with native congeners.

(9) Recipient habitat compatibility is important. However, the match of primary and secondary environments is often imperfect.

(10) In general, disturbance, slow recovery rate, eutrophication, and fragmentation of successionally advanced communities promote plant invasions.

Acknowledgments

I thank John Randall, Eva Grotkopp, and David Richardson for their suggestions, discussions, and/or critical comments on earlier drafts of this chapter.

References

Allison, G.W. (1996) Does high diversity facilitate or supress species invasions? *Bulletin of the Ecological Society of America,* **77/3**, Suppl. 9.

Andersen, U.V. (1995) Comparison of dispersal strategies of alien and native species in the Danish flora, in *Plant Invasions*, (eds P. Pysek, K. Prach, M. Rejmánek and P.M. Wade), SPB Academic Publishing, The Hague, pp. 61–70.

Ashton, P.J. and Mitchell, D.S. (1989) Aquatic plants: patterns and modes of invasion, attributes of invading species and assessment of control programmes, in *Biological Invasions. A Global Perspective*, (eds J.A. Drake, H.A. Mooney, F. di Castri, R.H. Groves, F.J. Kruger, M. Rejmanek and M. Williamson), John Wiley, Chichester, UK, pp. 111–154.

Auld, B.A., Hosking, J. and McFadyen, R.E. (1983) Analysis of the spread of tiger pear and parthenium weed in Australia. *Australian Weeds,* **2**, 56–60.

Austin, M.P., Groves, R.H., Fresco, L.M.F. and Kaye, P.E. (1985) Relative growth of six thistle species along a nutrient gradient with multispecies competition. *Journal of Ecology*, **73**, 667–684.

Bachmann, K., Chambers, K.L. and Price, H.J. (1985) Genome size and natural selection: observations and experiments in plants, in *The Evolution of Genome Size*, (ed. T. Cavalier-Smith), John Wiley & Sons, Chichester, pp. 267–276.

Baker, H.G. (1967) The evolution of weedy taxa in the *Eupatorium microstemon* species aggregate. *Taxon*, **16**, 293–300.

Baker, H.G. (1974) The evolution of weeds. *Annual Review in Ecology and Systematics*, **5**, 1–24.

Baker, H.G. (1978) Invasion and replacement in Californian and Neotropical grasslands, in *Plant Relations in Pastures*, (ed. J.R. Wilson), CSIRO, Melbourne, pp. 368–384.

Barrett, S.C.H. (1992) Genetics of weed invasions, in *Applied Population Biology*, (eds S.K. Jain and L.W. Botsford), Kluwer Academic Publishers, Dordrecht, pp. 91–119.

Baruch, Z. (1996) Ecophysiological aspects of the invasion by African grasses and their impact on biodiversity and function of Neotropical savannas, in *Biodiversity and Savanna Ecosystem Processes*, (eds O.T. Solbrig, E. Medina and J.F. Silva), Springer-Verlag, Berlin, pp. 79–93.

Baruch, Z. and Gomez, J.A. (1996) Dynamics of energy and nutrient concentration and construction cost in a native and two alien C4 grasses from two neotropical savannas. *Plant and Soil*, **181**, 175–184.

Baruch, Z., Hernandez, A.B. and Montila, M.G. (1989) Dinámica del crecimiento, fenología y reparticion de biomasa en gramineas nativas e introducidas de una sabana Neotropical. *Ecotropicos*, **2**, 1–13.

Bastl, M., Kocar, P., Prach, K. and Pysek, P. (1997) The effect of successional age and disturbance on the establishment of alien plants in man-made sites: an experimental approach, in *Plant Invasions*, (eds J.H. Brock, M. Wade, P. Pysek and D. Green), Backhuys Publishers, Leiden, pp. 191–201.

Beerling, D.J., Huntley, B. and Bailey, J.P. (1995) Climate and the distribution of *Fallopia japonica*: use of an introduced species to test the predictive capacity of response surfaces. *Journal of Vegetation Science*, **6**, 269–282 .

Bennett, M.D. (1987) Variation in genomic form in plants and its ecological implications. *New Phytologist*, **106** (Suppl.), 177–200.

Bingelli, P. (1996) A taxonomic, biogeographical and ecological overview of invasive woody plants. *Journal of Vegetation Science*, **7**, 121–124.

Blossey, B. (1997) The search for patterns or what determines the increased competitive ability of invasive non-indigenous plants?, in *Proceedings of the California Exotic Pest Plant Council Symposium. Vol. 2:1996*, (eds J.E. Lovich, J. Randall and M.D. Kelly), California Exotic Pest Plant Council, pp. 39–45.

Blossey, B. and Nötzold, R. (1995) Evolution of increased competitive ability in invasive nonindigenous plants: a hypothesis. *Journal of Ecology*, **83**, 887–889.

Bossard, C.C. (1991) The role of habitat disturbance, seed predation and ant dispersal on establishment of the exotic shrub *Cytisus scoparius* in California. *American Midland Naturalist*, **126**, 1–13.

Bossard, C.C. and Rejmánek, M. (1992) Why have green stems? *Functional Ecology*, **6**, 197–205.

Bramley, J.L., Reeve, J.T. and Dussart, G.B.J. (1995) The distribution of *Lemna minuta* within the British isles: identification, dispersal and niche constraints, in *Plant Invasions*, (eds P. Pysek, K. Prach, M. Rejmánek and P.M. Wade), SPB Academic Publishing, The Hague, pp. 181–185.

Bright, C. (1996) Understanding the threat of bioinvasions, in *State of the World*, (eds L.R. Brown and J. Abramovitz), W.W. Norton & Co., New York, pp. 95–113.

Brock, J.H., Child, L.E., de Waal, L.C. and Wade, M. (1995) The invasive nature of *Fallopia japonica* is enhanced by vegetative regeneration from stem tissues, in *Plant Invasions*, (eds P. Pysek, K. Prach, M. Rejmánek and P.M. Wade), SPB Academic Publishing, The Hague, pp. 131–139.

Brown, J.H. (1995) *Macroecology*. The University of Chicago Press, Chicago.

Burke, M.J.W. and Grime, J.P. (1996) An experimental study of plant community invasibility. *Ecology*, **77**, 776–790.

Cavalier-Smith, T. (1985) Cell volume and the evolution of eucaryotic genome size, in *The Evolution of Genome Size*, (ed. T. Cavalier-Smith), John Wiley & Sons, Chichester, pp. 105–184.

Chicoine, T.K., Fay, P.K. and Nielsen, G.A. (1985) Predicting weed migration from soil and climate maps. *Weed Science*, **34**, 57–61.

Chipping, D. (1993) German ivy infestation in San Luis Obispo. *CalEPPC News*, **1**, 15–16.

Clark, J.S. and Ji, Y. (1995) Fecundity and dispersal in plant populations: implications for structure and diversity. *American Naturalist*, **146**, 72–111.

Cook, C.D.K. (1985) Range extension of aquatic vascular plant species. *Journal of Aquatic Plant Management*, **23**, 1–6.

Costas Lippmann, M.A. (1976) *Ecology and reproductive biology of the genus* Cortaderia *in California*. Ph.D. Thesis, University of California, Berkeley, CA, USA.

Crawley, M.J., Hails, R.S., Rees, M., Kohn D and Buxton, J. (1993) Ecology of transgenic oilseed rape in natural habitats. *Nature*, **363**, 620–623.

Crawley, M.J., Harvey, P.H. and Purvis, A. (1996) Comparative ecology of the native and alien floras of the British Isles. *Philosophical Transactions of the Royal Society of London, Ser. B*, **351**, 1251–1259.

Critchfield, W.B. and Little, E.L. (1966) *Geographic distribution of the pines of the world*. Misc. Publ. 991. U.S. Department of Agriculture, Forest Service, Washington, DC.

D'Antonio, C.M. and Vitousek, P.M. (1992) Biological invasions by exotic grasses, the grass/fire cycle, and global change. *Annual Review in Ecology and Systematics*, **23**, 63–87.

Daehler, C.C. and Strong, D.R. (1997) Reduced herbivore resistance in introduced smooth cordgrass (*Spartina alterniflora*) after a century of herbivore-free growth. *Oecologia*, **110**, 99–108.

Darwin, C. (1859) *The Origin of Species by Means of Natural Selection*. Murray, London.

Ehrenfeld, J.G. (1997) Invasion of deciduous forest reserves in the New York metropolitan region by Japanese barberry (*Berberis thunbergii* DC.). *Journal of the Torrey Botanical Society*, **124**, 210–215.

Fensham, R.J. and Cowie, I.D. (1998) Alien plant invasions on the Tiwi Islands. Extent, implications and priorities for control. *Biological Conservation*, **83**, 55–68.

Forcella, F. and Wood, J.T. (1984) Colonization potentials of alien weeds are related to their "native" distributions: Implications for plant quarantine. *Journal of the Australian Institute of Agricultural Science*, **50**, 36–40.

Forcella, F., Wood, J.T. and Dillon, S.P. (1986) Characteristics distinguishing invasive weeds within *Echium*. *Weed Research*, **26**, 351–364.

Fowells, H.A. and Schubert, G.H. (1956) Seed crops of forest trees in the pine region of California. *U.S.D.A. Technical Bulletin,* **1150**, 1–48.

Gentle, C.B. and Duggin J.A. (1997) Allelopathy as a competitive strategy in persistent thickets of *Lantana camara* L. in three Australian forest communities. *Plant Ecology,* **132**, 85–95.

Goodwin, J. (1992) The role of mycorrhizal fungi in competitive interactions among native bunchgrasses and alien weeds: a review and synthesis. *Northwest Science,* **66**, 251–260.

Gunn, C.R. and Ritchie, C.A (1988) Identification of Disseminules Listed in the Federal Noxious Weed Act. *Technical Bulletin* No. **1719**. U.S. Department of Agriculture, Washington, DC.

Hadincová, V., Dobry, J. and Hanzélyová, D. (1997) Invazní druh *Pinus strobus* v Labskych pískovcích [Invasion of *Pinus strobus* in the Labské pískovce sandstone area]. *Zpravy Ces. Bot. Spolec. Praha, Mater.*, **14**, 63–79.'

Hannah, L., Lohse, D., Hutchinson, C., Carr, J.L. and Lankerani, A. (1994). A preliminary inventory of human disturbance of world ecosystems. *Ambio,* **23**, 246–250.

Hickman, J.C., ed. (1993) *The Jepson Manual. Higher Plants of California.* University of California Press, Berkeley.

Hobbs, R.K. and Huenneke, L.F. (1992) Disturbance, diversity, and invasion: implications for conservation. *Conservation Biology,* **6**: 324–337.

Hoffmann, A.A. and Parsons, P.A. (1991) *Evolutionary Genetics and Environmental Stress.* Oxford University Press, Oxford.

Huenneke, L.F. and Mooney, H.A., eds. (1989) *Grassland Structure and Function. California Annual Grassland.* Kluwer Academic Publishers, Dordrecht.

Hügin, G. (1995) Höhengrenzen von Rureral- und Segetalpflanzen in der Alpen. *Flora,* **190**, 169–188.

Hulbert, L.C. (1955) Ecological studies of *Bromus tectorum* and other annual bromegrasses. *Ecological Monographs,* **25**, 181–213.

Hulten, E. and Fries, M. (1986*) Atlas of Northern European Vascular Plants North of the Tropic of Cancer. Vols. 1,2,3.* Koeltz Scientific Books, Königstein.

Huston, M.A. (1994) *Biological Diversity.* Cambridge University Press, Cambridge.

Huston, M.A. (1997) Hidden treatments in ecological experiments: re-evaluating the ecosystem function of biodiversity. *Oecologia,* **110**, 449–460.

Jäger, E.J. (1988) Möglichkeiten der Prognose synanthroper Pflanzenausbreitungen. *Flora,* **180**, 101–131.

Jalas, J. and Suominen, J. (1989) *Atlas Florae Europaeae*, Vol. 8. Qkateeminen Kirjakaupa, Helsinki.

Jones, R.H. and McLeod, K.W. (1990) Growth and photosynthetic responses to a range of light environments in Chinese tallowtree and Carolina ash seedlings. *Forest Science,* **36**, 851–862.

Kowarik, I. (1995) Time lags in biological invasions with regard to the success and failure of alien species, in *Plant Invasions,* (eds P Pysek, K. Prach, M. Rejmánek and P.M. Wade), SPB Academic Publishing, The Hague, pp. 15–38.

Kwiatkowska, A.J., Spalik, K., Michalak, E., Palinska A. and Panufnik D. (1997) Influence of the size and density of *Carpinus betulus* on the spatial distribution and rate of deletion of forest-floor species in thermophilous oak forest. *Plant Ecology,* **129**, 1–10.

Labouriau, L.G. (1966) Revisao da situacao da ecologia vegetal nos Cerrados. *Anais da Academia Brasileira de Ciencias,* **38** (Suplemento), 5–58.

Lawrence, J.G., Colwell, A. and Sexton, O.J. (1991) The ecological impact of allelopathy in *Ailanthus altissima* (Simaroubaceae). *American Journal of Botany,* **78**, 948–958.

Ledgard, N.J. (1988) The spread of introduced trees in New Zealand's rangelands – South Island high country experience. *Tussock Grasslands and Mountain Lands Institute Review*, **44**, 1–8.

Lonsdale, W.M. (1994) Inviting trouble: introduced pasture species in northern Australia. *Australian Journal of Ecology*, **19**, 345–354.

Mazer, S.J. (1989) Ecological, taxonomic, and life histrory correlates of seed mass among Indiana Dune angiosperms. *Ecological Monographs*, **59**, 153–175.

Mcintyre, S., Ladiges, P.Y. and Adams, G. (1988) Plant species-richness and invasion by exotics in relation to disturbance of wetland communities on the River Plain, NSW. *Australian Journal of Ecology*, **13**, 361–373.

Meusel, H., Jäger, E., Rauschert, S. and Weinert, E. (1978) *Vergleichende Chorologie der Zentraleuropäischen Flora*. Band II. VEB Gustav Fischer Verlag, Jena.

Michael, P.W. (1981) Alien plants, in *Australian Vegetation*, (ed. R.H. Groves), Cambridge University Press, Cambridge, pp. 44–64.

Mooney, H.A. and Hofgaard, A. (1998) Biological invasions and global change, in *Invasive Species and Biodiversity Management*, (eds O.T. Sandlund, P.J. Schei and Å. Viken), Kluwer Academic Publishers, Dordrecht, The Netherlands.

Moulton, M.P. and Pimm, S.L. (1987) Morphological assortment in introduced Hawaiian passerines. *Evolutionary Ecology*, **1**, 113–24.

Noble, I.R. (1989) Attributes of invaders and the invading process: terrestrial vascular plants, in *Biological Invasions. A Global Perspective*, (eds J.A. Drake, H.A. Mooney, F. di Castri, R.H. Groves, F.J. Kruger, M. Rejmanek and M. Williamson), John Wiley, Chichester, UK, pp. 301–313.

Nowak, M.A. and McMichael, A.J. (1995) How HIV defeats the immune system. *Scientific American*, **271**, 58–65.

Odum, E.P. (1998) Productivity and biodiversity: a two-way relationship. *Bulletin of the Ecological Society of America*, **79/1**, 125.

Panetta, F.D. (1993) A system of assessing proposed plant introductions for weed potential. *Plant Protection Quarterly*, **8**, 10–14.

Panetta, F.D. and Dodd, J. (1987) Bioclimatic prediction of the potential distribution of skeleton weed (*Chondrilla juncea* L.) in Western Australia. *Journal of the Australian Institute of Agricultural Science*, **53**, 11–16.

Parrish, J.A.D. and Bazzaz, F.A. (1978) Pollination niche separation in a winter annual community. *Oecologia*, **35**, 133–140.

Parsons, J.J. (1970) The "Africanization" of the New World tropical grasslands. *Tübinger Geographische Studien*, **34**, 141–153.

Parsons, W.T. and Cuthbertson, E.G. (1992) *Noxious Weeds of Australia*. Inkata Press, Melbourne.

Peart, H.J. and Fitter, A.H. (1994) Comparative analyses of ecological characteristics of British angiosperms. *Biological Review*, **69**, 95–115.

Pfister, R.D. and Daubenmire, R. (1975) Ecology of lodgepole pine *Pinus concorta* Dougl, in *Management of Lodgepole Pine Ecosystems*, (ed. D.M. Baumgartner), Washington State University Cooperative Extension Service, Pullman, pp. 27–46.

Pieterse, A.H. and Murphy, K.J. eds (1990). *Aquatic Weeds*. Oxford University Press, Oxford.

Poschlod, P.J Matthies, D., Jordan, S. and Mengel, C. (1996) The biological flora of Central Europe – an ecological bibliography. *Bulletin of the Geobotanical Institute ETH*, **62**, 89–108.

Planty-Tabacchi, A., Tabacchi, E., Naiman, R.J., Defarrari, C. and Decamps, H. (1996) Invasibility of species-rich communities in riparian zones. *Conservation Biology*, **10**, 598–607.

Pysek, P. (1997) Clonality and plant invasions: can a trait make a difference?, in *The Ecology and Evolution of Clonal Plants*, (eds H. de Kroon and J. van Groenendael), Backhuys Publishers, Leiden., pp. 405–427.

Pysek, P. and Prach, K. (1995) Invasion dynamics of *Impatiens glandulifera* – a century of spreading reconstructed. *Biological Conservation*, **74**, 41–48.

Pysek, P., Prach, K. and Smilauer, P. (1995) Relating invasion success to plant traits: and analysis of the Czech alien flora, in *Plant Invasions*, (eds P. Pysek, K. Prach, M. Rejmánek and P.M. Wade), SPB Academic Publishing, The Hague, pp. 237–247.

Randall, J.M. (1995) Assessment of the invasive weeds problem on preserves across the United States. *Endangered Species Update*, **12**, 4–6.

Reed, C.F. (1977) *Economically Important Foreign Weeds. Potential Problems in the United States*. Agriculture Handbook No. 498. U.S. Department of Agriculture, Washington, DC.

Reichard, S.H. (1994) *Assessing the potential of invasiveness in woody plants introduced in North America*. University of Washington dissertation, Seattle, Washington.

Reichard, S.H. and Hamilton, C.W. (1997) Predicting invasions of woody plants introduced into North America. *Conservation Biology*, **11**, 193–203.

Rejmánek, M. (1989) Invasibility of plant communities, in *Biological Invasions. A Global Perspective*, (eds J.A. Drake, H.A. Mooney, F. di Castri, R.H. Groves, F.J. Kruger, M. Rejmanek and M. Williamson), John Wiley, Chichester, UK, pp. 369–388.

Rejmánek, M. (1995) What makes a species invasive?, in *Plant Invasions*, (eds P. Pysek, K. Prach, M. Rejmánek and P.M. Wade), SPB Academic Publishing, The Hague, pp. 3–13.

Rejmánek, M. (1996a) A theory of seed plant invasiveness: the first sketch. *Biological Conservation*, **78**, 171–181.

Rejmánek, M. (1996b) Species richness and resistance to invasions, in *Diversity and Processes in Tropical Forest Ecosystems*, (eds G.H. Orians, R. Dirzo. and J.H. Cushman), Springer-Verlag, Berlin, pp. 153–172.

Rejmánek, M. and Randall, J.M. (1994) Invasive alien plants in California: 1993 summary and comparison with other areas in North America. *Madroño*, **42**, 161–177.

Rejmánek, M. and Richardson, D.M. (1996) What attributes make some plant species more invasive? *Ecology*, **77**, 1655–1661.

Richardson, D.M., ed. (1998) *Ecology and Biogeography of* Pinus. Cambridge University Press, Cambridge.

Richardson, D.M., Williams, P.A. and Hobbs, R.J. (1994) Pine invasions in the Southern Hemisphere: determinants of spread and invadibility. *Journal of Biogeography*, **21**, 511–527.

Robertson, D.J., Robertson, M.C. and Tague, T. (1994) Colonization dynamics of four exotic plants in a northern Piedmont natural area. *Bulletin of the Torrey Botanical Club*, **121**, 107–118.

Robinson, G.R., Quin, J.F., and Stanton, M.L. (1995) Invasibility of experimental habitat islands in a California winter annual grassland. *Ecology*, **76**, 786–794.

Roy, J., Navas, M.L. and Sonie, L. (1991) Invasion by annual brome grasses: a case study challenging the homoclime approach to invasions, in *Biogeography of Mediterranean Invasions*, (eds R.H. Groves and F. di Castri), Cambridge University Press, Cambridge, pp. 207–224 .

Ryser, P. and Notz, R. (1996) Competitive abillity of three ecologically contrasting grass species at low nutrient supply in relation to their maximal relative growth rate and tissue density. *Bulletin of the Geobotanical Institute ETC*, **62**, 3–12.

Sanders, R.W. (1976) Distributional history and probable ultimate range of *Galium pedemontanum* (Rubiaceae) in North America. *Castanea*, **41**, 73–80.

Sarukhán, J. and Gadgil, M. (1974) Studies on plant demography: *Ranunculus repens*, L., *R. bulbosus* L. and *R. acris* L. III. A mathematical model incorporating multiple modes of reproduction. *Journal of Ecology*, **62**, 921–936.

Schiffman, P.M. (1997) Animal-mediated dispersal and disturbance: driving forces behind alien plant naturalization, in *Assessment and Management of Plant Invasions*, (eds J.O. Luken and J.W. Thieret), Springer, New York, pp. 87-94.

Selvin, S. (1995) *Practical Biostatistical Methods*. Duxbury Press, Belmont.

Sims, L.E. and Price, H.J. (1985) Nuclear DNA content variation in *Helianthus* (Asteraceae). *American Journal of Botany*, **72**, 1213–1219.

Smith, J.M.B. (1989) An example of ant-assisted plant invasion. *Australian Journal of Ecology*, **14**, 247–250.

Spencer, D.F. and Rejmánek, M. (1989) Propagule type influences competition between two submersed aquatic macrophytes. *Oecologia*, **81**, 132–137

Stohlgren, T.J., Chong, G.W., Kalkhan, M., Schell, L.D., Bull, K. and Otsuki, Y. (1997) Do invasive plant species target areas of low diversity in natural landscapes? *Bulletin of the Ecological Society of America*, **78/4**, Suppl. 191.

Strauss, S.H. and Ledig, F.T. (1985) Seedling architecture and life history evolution in pines. *American Naturalist*, **125**, 702–715.

Sutherland, S. and Vickery, R.K. (1988) Trade-offs between sexual and asexual reproduction in the genus *Mimulus*. *Oecologia*, **76**, 330–335.

Swarbrick, J.T. and Skarratt, D.B. (1994) *The Bushweed 2 Database on Environmental Weeds in Australia*. The University of Queensland Gatton College, Gatton.

Thebaud, C., Finzi, A.C., Affre, L., Debusshe, M. and Escarre, J. (1996) Assessing why two introduced *Conyza* differ in their ability to invade Mediterranean old fields. *Ecology*, **77**, 791–804.

Thomas, P.A. and Room, P.M. (1986) Taxonomy and control of *Salvinia molesta*. *Nature*, **320**, 581–584.

Thompson, K. (1990) Genome size, seed size and germination temperature in herbaceous angiosperms. *Evolutionary Trends in Plants*, **4**, 113–116.

Thompson, K., Hodgson, J.G. and Rich, C.G. (1995) Native and alien invasive plants: more of the same? *Ecography*, **18**, 390–402.

Tilman, D. (1997) Community invasibility, recruitment limitation, and grassland biodiversity. *Ecology*, **78**, 81–92.

Tucker, K.C. and Richardson, D.M. (1995) An expert system for screening potentially invasive alien plants in South African fynbos. *Journal of Environmental Management*, **44**, 309–338.

Tutin, T.G., Heywood, V.H., Burges, N.A., Moore, D.M., Valentine, D.H., Walters, S.M. and Webb, D.A., eds (1980) *Flora Europaea*. Volume 5. Cambridge University Press, Cambridge.

UCPE (1996) *Unit of Comparative Plant Ecology Annual Report*. The University, Sheffield.

USDA (1992) Conservation tree and shrub cultivars in the United States. *Soil Conservation Service Agriculture Handbook* No. **692**. U.S. Department of Agriculture, Washington, DC.

Vitousek, P.M., D'Antonio, C.M., Loope, L.L., Rejmánek, M. and Westbrooks, R. (1997a) Introduced species: a significant component of human-caused global change. *New Zealand Journal of Ecology*, **21**, 1–16.

Vitousek, P.M., Aber, J., Howarth, R.W., Likens, G.E., Matson, P.A., Schindler, D.W., Schlesinger, W.H. and Tilman G.D. (1997b) Human alteration of the global nitrogen cycle: causes and consequences. *Issues in Ecology*, **1**, 2–15.

Vitousek, P.M. and Walker, L.R. (1989) Biological invasion by *Myrica faya* in Hawaii: plant demography, nitrogen fixation, ecosystem effects. *Ecological Monographs*, **59**, 247–65.

Wakamiya I., Newton, R.J. Johnston, J.S. and Price, H.J. (1993) Genome size and environmental factors in the genus *Pinus*. *American Journal of Botany*, **80**, 1235–41.

Ware, A.B. and Compton, S.G. (1992) Breakdown of pollinator specificity in an African fig tree. *Biotropica*, **24**, 544–49.

Wells, M.J., Balsinhas, A.A. Joffe, H. Engelbrecht, V.M., Harding, G. and Stirton, C.H. (1986) A catalogue of problem plants in southern Africa. *Memoirs of the Botanical Survey of South Africa*, **53**, 1–658.

Whitmore, T. C. (1991) Invasive woody plants in perhumid tropical climates, in *Ecology of Biological Invasions in the Tropics*, (ed. P.S. Ramakrishnan), International Scientific Publications, New Delhi, India, pp. 35–40.

Whittemore, A.T. (1997) *Ranunculus*, in *Flora of North America*. Vol 3, Oxford University Press, New York, pp. 88–135.

Williams, P.A. (1996) *A Weed Risk Assessment Model for Screening Plant Imports into New Zealand*. MAF Policy, Wellington, New Zealand.

Williams, P.A. and Kral, B.J. (1996) Fleshy fruits of indigenous and adventive plants in the diet of birds in forest remnants, Nelson, New Zealand. *New Zealand Journal of Ecology*, **20**, 127–145.

Williamson, M.H. (1996) *Biological Invasions*. Chapman & Hall, London.

Williamson, M.H. and Fitter, A. (1996) The characters of successful invaders. *Biological Conservation*, **78**: 163–70.

Wilson, J.B., Rapson, G.L., Sykes, M.T., Watkins, A.J. and Williams, P.A. (1992) Distribution and climatic correlations of some exotic species along roadsides in South Island, New Zealand. *Journal of Biogeography*, **19**, 183–194.

Yeaton, R.I. (1983) The successional replacement of ponderosa pine by sugar pine in the Sierra Nevada. *Bulletin of the Torrey Botanical Club*, **110**, 292–297.

Lag times in population explosions of invasive species: Causes and implications

JEFFREY A. CROOKS and MICHAEL E. SOULÉ

Scripps Institution of Oceanography, La Jolla, California; and University of California, Santa Cruz, California, USA

Abstract

Biodiversity losses caused by invasive species may soon surpass the damage done by habitat destruction and fragmentation. Some invaders explode quickly; others have a long "lag" period. Three categories of lags in population explosions can be recognized: (1) inherent lags caused by the nature of population growth and range expansion; and prolonged lags that may be caused by (2) environmental factors related to changes (improvements) in ecological conditions that favor an alien; and (3) genetic factors related to the relative lack of fitness of the alien in a novel environment. The likelihood of overcoming a genetic lag (fitness deficit) is proportional to the population size of the alien; there is a positive feedback between population size and the rates of genetic adaptation. Some principles regarding lags include: (1) determining whether a given lag is prolonged or not is often difficult given lack of data, and can be confounded by lags in the detection of invasive species; (2) past performance of an exotic is a poor predictor of potential population growth, range expansion, and ecological impact; (3) containment can end suddenly and disastrously for both ecological and genetic reasons; and (4) the larger the size of the alien colony, the more likely it will eventually become invasive. Policy makers should understand that good surveillance and monitoring are essential, and that extirpation should be early and vigorous.

Introduction

It may not be long before invasive species surpass habitat loss and fragmentation as the major engines of ecological disintegration[1]. We make this prediction for two reasons: First, it may be soon when most of the habitat that is susceptible to destruction, modification, and fragmentation will have been so affected; there will be little more habitat to destroy for

[1] To avoid confusion, it should be noted that some invasive aliens, including feral domesticates and escaped horticultural varieties, may provide recreational and economic gains for some human beings. These benefits, however, come at the expense of costs, often catastrophic, to native species and ecosystems. It is therefore Orwellian to claim that these exotics benefit biodiversity or nature; they benefit only the hegemonic primate.

O. T. Sandlund et al. (eds.), Invasive Species and Biodiversity Management, 103–125.

urbanization, farms, clear-cutting, and water projects, at least in the tropics and temperate zones. Second, damaged, denatured lands and waters are quite vulnerable to the growing avalanche of alien species, many of which prosper in disturbed, over-grazed, over-logged, over-hunted places. This is why the basic discipline of ecology gradually may be replaced by the more applied science of "mixo-ecology" or "recombination ecology," the study of recombined biotas (Soulé, 1990; Townsend, 1991). Therefore as scientists we are compelled to understand the dynamics of invasions and to discover and promote the needed countermeasures. In this paper, we examine the critical, early stage of an alien invasion.

During these early stages, the rates of population growth and range expansion of an alien species can vary markedly. Some invasive species (e.g., Africanized bees, muskrats, and zebra mussels) have had rapid rates of local population growth and range expansion. Many other species, however, (e.g., Collared Doves and the Oxford ragwort) appear to have had long lag times between initial introduction and subsequent population explosions (Hengeveld, 1988; Cousens and Mortimer, 1995; Hobbs and Humphries, 1995; Kowarik, 1995). Despite these apparent differences in the colonizing history of invasive species, the relevant biological factors operating during the early stages of invasions are poorly understood. As such, the management and policy implications of lag times in the invasion of exotic species have been little explored (but see Hobbs and Humphries, 1995).

In order to provide examples of the lag effect and present some processes involved in the early stages of invasion, we will discuss some case histories of species that apparently had long lag times between initial invasion and subsequent population explosions. These will be used to highlight three sets of mechanisms (inherent, environmental, and genetic causes of lags) that might affect invasion dynamics. Also, presumed causes of lags presented in the case histories section will be discussed. Finally, we will speculate on the management and policy implications of lag effects.

Case histories

We group invasions into three categories: 1) invasions by exotic species, 2) ranges expansions by native species, and 3) hybridization events between native and exotic species that are followed by explosive range expansion.

Invasions by exotic species

One of the most famous examples of a lag time in the population explosion of an invasive species is the Collared Dove (*Streptopelia decaocto*) in Europe. This south Asian species spread into China and the Middle East centuries ago (Hengeveld, 1988). Probably beginning in the 16th century, the dove spread through Syria and Turkey, where it was protected by the Ottoman Turks (Hengeveld and van den Bosch, 1991). For at least two hundred years, however, the Collared Dove did not spread beyond this area. But starting suddenly in the early 1900s, the dove rapidly colonized temperate Europe and north-west Africa, covering much of this area in around 50 years (Isenmann, 1990). It has been suggested that the cause may be related to increasing urbanization in the region as well as climate change which allowed longer breeding seasons (Isenmann, 1990). Similar patterns of delayed population expansions have also been displayed by other European birds such as the Penduline Tit (*Remiz pendulinus*) and Serin Finch (*Serinus serinus*) (Hengeveld, 1989).

Several weeds in Britain have also displayed a "slow rate of spread followed by a rapid one" (Salisbury, 1953). The "Oxford ragwort" (*Senecio squalidus*) is a southern European weed that accidentally escaped from a botanical garden in Oxford, England before 1794 (Baker, 1965). For many decades the species was primarily confined to old walls in that city and Cork, Ireland (Perring, 1974). This weed slowly began to spread, however, due to the building of a railway between Oxford and London. After World War II, the species rapidly spread throughout England and Ireland, along railways and in habitat created by bombings during the war (Baker, 1965). Another weed, the gallant soldier, *Galinsoga parviflora*, from South America was purely a local resident in Kew (after its escape from the Royal Botanical Gardens) for many years before its dramatic spread during World War II. This delayed range expansion was partially attributed again to the bombings in England, which were thought to send the plant's propagules high in the air thus allowing dispersal by winds. Wild lettuce plants, *Lactuca virosa* and *L. scariola*, first reported in Britain 1570 and 1632, respectively, also displayed delayed range extensions. For example, in Surrey and Hertfordshire, both species were considered rare until the middle of the 1900s, when they became common in gravel pits (Salisbury, 1953).

The cut-leaved teasel (*Dipsacus laciniatus*) is a weed that arrived to New York prior to 1900, and in 1913 it was reported only from Albany (Solceki, 1993). However, in the last thirty years the plant, which is capable of forming monocultures that exclude most native vegetation, has spread quickly throughout much of the mid-west. This rapid spread has been attributed to dispersal via the interstate highway system, as the teasel is particularly common along highways and roads.

In Florida, two major plant invaders, the paper-bark tree or melaleuca (*Melaleuca quinquenervia*) and the Brazilian pepper (*Schinus terebinthofolius*), also were "present long before they were conspicuous elements of the landscape" (Ewel, 1986). These evergreen trees were intentionally introduced and since have become major pests in the nearly treeless Everglades. Melaleuca was first introduced into Florida in the early parts of the 1900's, although it wasn't until decades later that the population began to expand. Likewise, the Brazilian pepper, although introduced over 100 years ago, did not "explode across the landscape until the 1950s" (Ewel, 1986). The causes of these lags are unknown, although Ewel lists four possibilities: 1) Florida became more invasion-prone; 2) the species may have been undergoing rapid yet undetected expansion due to inherent lags; 3) it may have taken several decades to build up large enough populations to have significant reproductive potential ("infection pressure"); and 4) the new colonists were confined to restricted habitats until mutations favorable for further colonization became available. We will return to the issue of causation and its typology below.

In a thorough treatment of the population dynamics of introduced woody plants in Brandenburg, Germany, historical records were used to determine the length of time between the initial release for cultivation and first evidence of spontaneous spread of a large number of species (Kowarik, 1995). For the 184 species considered, there was an average of 147 years between first planting and first appearance of seedlings. Moreover, only 2% of the species became established, and 1% invaded natural vegetation. Intrinsic population factors, climatic shifts, and habitat availability were cited as possible reasons for the long lag phases displayed by many of the plant species.

The Channel Islands off southern California have been heavily affected by introduced species. Historically, the islands were almost free of large grazers, the exception being the extinct dwarf mammoths (Laughrin *et al.*, 1994). In the mid-1800s, thousands of cattle, horses, sheep, and pigs were brought to the islands. As part of a recent conservation effort on Santa Cruz Island, more than 36 000 feral sheep and 1 500 head of cattle were removed from the western 90% of the island (Brenton and Klinger, 1994). Before removal of the grazers, the European weed, fennel (*Foeniculum vulgare*), which had been present on the island for over 100 years, was not considered a dominant species except in a few small areas (Beatty and Licari, 1992). With the removal of the grazers, however, the fennel population expanded in range and density and now dominates ca. 10% of the island (Brenton and Klinger, 1994). This explosion has in turn benefited other exotics, because alien weeds (e.g. European annual grasses) are the most abundant herbaceous plants typically encountered within dense stands of fennel (Brenton and Klinger, 1994). Feral pigs (*Sus scrofa*), too, have rapidly increased due at least in part to the removal of the grazers (Crooks and Van Vuren, 1994).

The Asian mitten crab, *Eriocheir sinensis*, was first found in Europe in 1912 (Barnes, 1994). Until recently, however, it had met with considerably less success in Britain than it had in other European countries (Cohen 1995). These catadromous species spend the majority of their lives in rivers, but migrate to estuaries to reproduce. Their failure to establish in Britain has been at least partially explained by the fact that the presence of fast-flowing rivers which may inhibit the settlement and recruitment of these migratory crabs (Atrill and Thomas, 1996). From 1989 to 1992, however, severe droughts in southeast England led to reduced river flow and hordes of crabs were found upriver, some even wandering into local homes.

One particularly successful exotic species in the heavily invaded estuarine ecosystems of the Pacific coast of North America is the small, soft-sediment dwelling bivalve *Musculista senhousia* (Crooks, 1996). Although this mat forming species first appeared in San Diego in the mid-1960s, it was not until the early 1980s that the mussel could be commonly found in high density patches of around 10 000 m^2. In the summer of 1995, after a spring characterized by unusually heavy rainfall and strong red tides offshore, extraordinarily dense populations of the mussel were found carpeting thousands of square meters of Mission Bay's intertidal and shallow subtidal (Crooks, unpublished). Densities up to 170 000 per m^2 were recorded, a far higher concentration than reported for this species anywhere else in the world and among the highest ever reported for a marine bivalve.

Another southern California invader is the wood-boring gribble (Isopoda), *Limnoria tripunctata*. This small crustacean was introduced into the Long Beach–Los Angeles Harbor area before the turn of the century, probably via the hulls of wooden ships (Carlton, 1979). In portions of the harbor (such as the Dominguez Channel and East Basin), however, no isopods were found because excessive pollution from industrial, domestic, and storm wastes resulted in a nearly sterile zone. With the advent of a pollution abatement program in the late 1960s, the isopod moved into the area, underwent a population explosion, and caused the collapse of a local wharf through its extensive boring activities (Reish *et al.*, 1980).

Hybridization of native with exotic species

In the early 1800s the U.S. east coast marsh cordgrass, *Spartina alterniflora*, was accidentally introduced via ballast water into the United Kingdom (Thompson, 1991). About seventy years later, it was noticed that *S. alterniflora* had hybridized with the native European cordgrass, *S. maritima*. This hybrid, *S. townsendii*, was infertile. In the 1890s, the cordgrass began to spread out of Southampton Water and into adjacent estuaries. This spread has been attributed to the production of a new, fertile species, *S. anglica*, through chromosome doubling of *S. townsendii*. This new cordgrass species is very well adapted to the intertidal areas of estuaries, and through both natural dispersal and intentional planting for marsh reclamation, *S. anglica* is now a characteristic feature of British salt marshes (Williamson, 1996).

Range Expansions of Native Species

Range expansion of a native species may also lend insight into the processes involved in the early stages of invasion. The butterfly, *Coenonympha tullia*, is a holarctic species widely distributed throughout western North America (Wiernasz, 1989). Before the 1950s, the butterfly's distribution in eastern North America was largely restricted to Quebec and Ontario north of the Saint Lawrence and portions of the Maritime Provinces. Here, the populations were univoltine (one generation per year). In the early 1960s the species began to spread southward into much of New England and New York, following a warming trend in the region. This was associated with the development of bivoltinism (two generations per year) in the southern part of the range. Once the bivoltine populations were established, further colonization proceeded quickly, perhaps because of the rapid population growth thus afforded.

Lags in Detection of Exotics

In this consideration of the lag effect during invasions, it should be noted that many estimates of the time between initial invasion and subsequent population explosion may be conservative. This arises from yet another lag effect: our lag in determining the presence of a new invasive species. It is likely that many invaders are present in low numbers for some time before they are first recorded. Such "early stage subdetectability" was suggested to occur for the medfly (*Ceratitis capitata*) in California, which may have been present for more than 50 years prior to its discovery in 1975 (Carey, 1996). Such lags in detection of exotics will be especially likely for small or cryptic species in undersampled habitats.

The inherent lag effect

Fundamental to the examination of lag times is the definition of what in fact constitutes a lag. As can be seen from the case histories, two basic types of lags can be recognized: lags in local population increases and lags in range expansion. In order to define lags in either case, it is first necessary to explore some of the dynamics of a biological invasion.

The first thing to point out is that lags are normal; some kind of lag is built into the growth of any founder population, whether or not it is adapted to the new environment. The classic model for the early stage of an invasion is the simple exponential formula,

$$N_t = N_o e^{rt}, \quad (1)$$

where N is the number of individuals, t is time, and r is the population's intrinsic rate of increase. Inherent in this familiar model is the shallow portion early in the growth curve when the population is growing relatively slowly in absolute numbers (Figure 7.1A). Even vigorous, perfectly adapted populations follow such a trajectory.

In assessing whether an observed lag in the population growth of an invasive species is prolonged, it is necessary to determine if the observed lag is longer than the inherent lag given the r determined for the population when it is growing rapidly. Such a difference may be visualized by plotting the logarithmic equivalent of the former equation, giving

$$ln\, N_t = rt + ln\, N_o. \quad (2)$$

If plotted graphically and r is constant, the result will be a straight line with a slope of r (Figure 7.1A and 7.2A). If the slope is less steep in the early portion of the curve, however, this is evidence for a prolonged lag due to lower values of r during the early stages of the invasion (Figure 7.2A).

Models of range expansion are more complex than that for local population growth, because they include both this local numerical increase as well as emigration of individuals. There has been considerable effort devoted to describing the spatial spread of organisms, although most theory has concentrated on asymptotic rates achieved after the early stages of an invasion. The simplest model of range expansion assumes reaction-diffusion dynamics with individuals acting as random particles moving on a uniform plane (Kendall, 1948; Skellam, 1951). This model thus depends on only two processes, exponential population growth (as described above) and random diffusion of individuals. Using these, it can be determined that C, the expansion velocity, asymptotically approaches the equation:

$$C \approx 2(rD)^{0.5}, \quad (3)$$

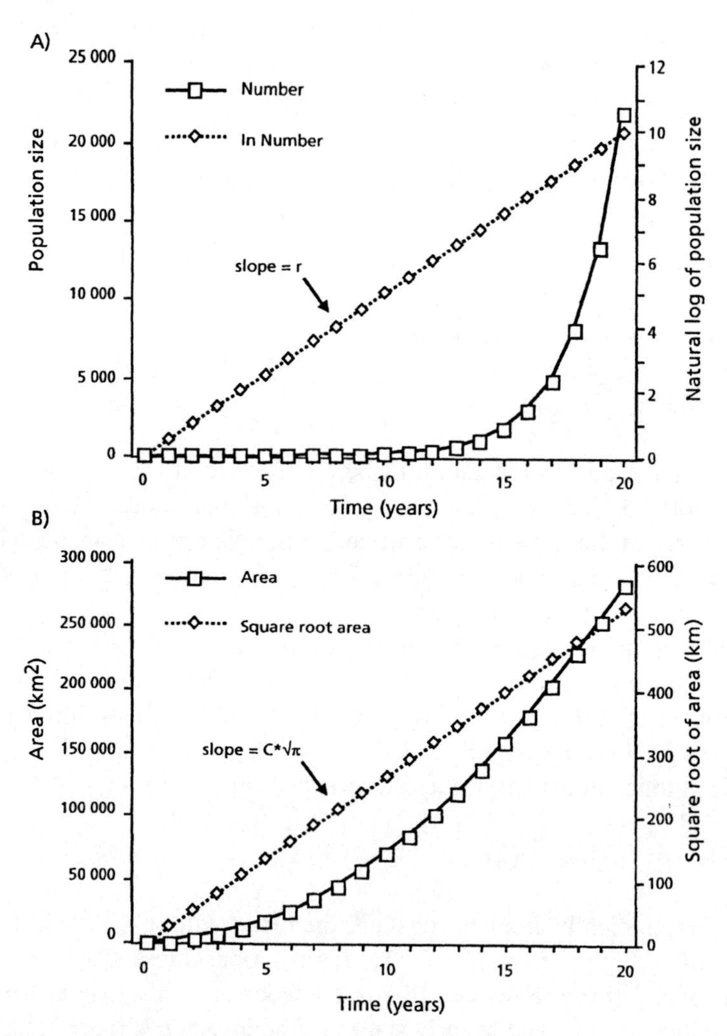

Figure 7.1 Local population growth (A) and areal expansion (B) of populations. For population growth, intrinsic increase r = 0.5/yr. For areal expansion, velocity of range expansion (C) = 15 km /yr and the correction factor (Ω) = 0.

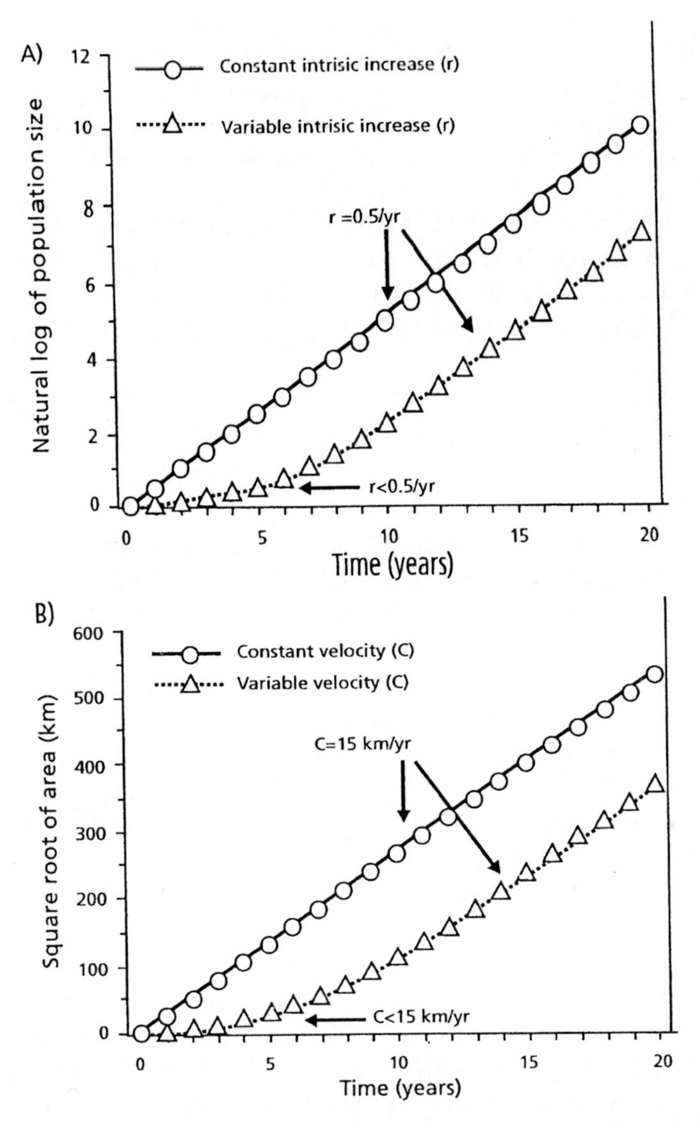

Figure 7.2 Comparisons of lines with constant versus slow early rates of population growth (A) and range expansion (B).

where D is a diffusion constant (Williamson, 1996). Thus, the velocity of the rate of spread is proportional to the square root of population growth rate (r) and the diffusion constant (D), and changes in either of these values will cause changes in expansion velocity. In many cases this model provide a very good description of the movement of organisms (Kareiva, 1983; Holmes, 1993; but see Lonsdale, 1993), although models that take into account life-history and dispersal parameters of species (e.g. Van den Bosch *et al.*, 1992; Hengeveld, 1993) may have increased predictive powers. It also should be noted that species having two different modes of dispersal, such as the cholla (*Opuntia imbricata*), which has passive dispersal by the falling of seeds or stems to the ground and active dispersal by attachment of stems to animals (Allen *et al.*, 1991), may appear to spread faster than the asymptotic rate by establishing foci beyond the range attained by passive dispersal.

A variety of related methods for quantifying the rate of spread (*C*) of an invasive species also have been developed (Andow *et al.*, 1993). In the simplest case the contours of equal population density are modelled as circles expanding at a constant velocity (van den Bosch *et al.* 1992). This gives the equation:

$$A_t = \pi p_t^2 = \pi(\Omega + Ct)^2, \quad (4)$$

where t is time since detectable spread began, A_t is the area occupied at time t, p_t is the radius of the expanding front at time t, Ω is a correction factor representing the area beyond which range expansion can be detected or the initial area occupied by the population, and C is the expansion velocity (Figure 7.3). Because area increases as the square of time, a graph of area occupied over time will be a curve with an early lag phase (Figure 7.1B). If the velocity of range expansion is constant, then plotting the square root of area versus time gives a straight line (Figure 7.1B). The expansion velocity can be easily calculated from this graph by dividing the slope by $\pi^{0.5}$. It is also possible to plot the radial equivalent of the area, $(A/\pi)^{0.5}$, versus time, in which case the slope is simply C.

Although relatively simple, this relationship explains the observation that the square root transformation typically linearizes the time course of spread of an invasive species (Williamson and Brown, 1986). This result appears to be quite robust once the invasion has "taken" in terrestrial systems (Roughgarden, 1986), although rates of spread in marine systems may be lower than predicted by the diffusion model (Grosholz, 1996). The linear relationship between the square root of area and time is exemplified by the rapid spread of the muskrat (*Odontra zibethicus*) in Europe after the release of 5 individuals near Prague in 1905 (Nowak, 1971). If the slope of the line relating the square root of area to time during the early stages of invasion is shallower than the slope later in the invasion, then early rates of range expansion are slower than that during the asymptotic spread phase (Figure 7.2B). Such early, slow spread has been witnessed for a wide variety of species (Williamson, 1996), such as the House Finch, *Carpodacus mexicanus* (Veit and Lewis, 1996) and the Starling, *Sturnus vulgaris* (Okubo, 1988). Several factors, however, make the unambiguous interpretation of early, sub-asymptotic rates difficult.

The first complicating factor is related to the ability to detect small-scale spread. In the circular growth model (eq. 4), is the scale at which expansion can be detected. Below this level, spread can be occurring but it may go unnoticed (Andow *et al.*, 1993). This would lead to an apparent lag in range

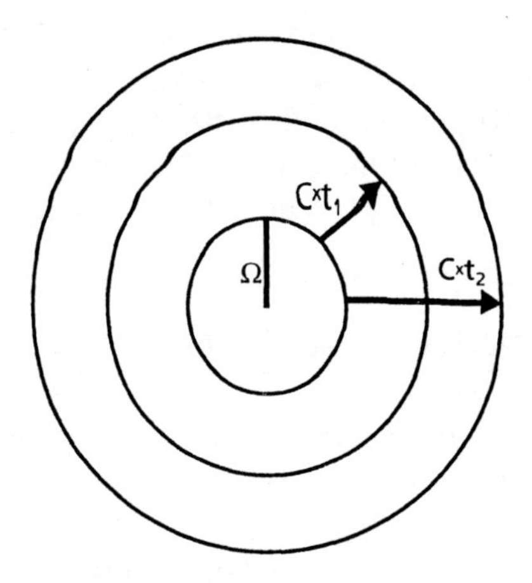

Figure 7.3 Circular areal expansion of a population, according to the equation $A_t = \pi p_t^2 = \pi(\Omega + Ct)^2$, where t is time since detectable spread began, A_t is the area occupied at time t, p_t is the radius of the expanding front at time t, Ω is a correction factor representing the area beyond which range expansion can be detected or the initial area occupied by the population, and C is the expansion velocity. See text for more details.

expansion even though the population was expanding at a constant rate. Second, equation 3, relating spreading velocity (*C*) to *r* and *D*, is for spread at an asymptotic velocity. The precise form of the equation is more complicated and actually predicts lower values of *C* early in the invasion before reaching an asymptote (Holmes *et al.*, 1994). This suggests that even given constant rates of *r* and *D*, there may be an initial, intrinsic lag. This result has also been echoed by the results of a stochastic model of areal spread, where the only parameter is the probability that an unoccupied site remains unoccupied at the next time step (Hastings, 1996). This model demonstrates that lags in areal expansion can occur early in an invasion. Like the models discussed above, however, this stochastic model also predicts that the square root of area will asymptotically increase as a linear function of time. These factors make it difficult to determine how long sub-asymptotic rates must occur before they should actually be considered prolonged. However, the length of time before an asymptotic rate of range expansion is reached would appear to provide a relative indication of the likelihood of a prolonged lag.

Even given theory like that presented above, it is possible that the examination of any given invasion will prove difficult because of lack of information to analyze critically whether prolonged lags exist. Nevertheless, it is possible to recognize two broad categories of mechanisms able to produce prolonged lags in population growth and/or range expansions beyond that of inherent lags discussed above. These categories, which may act singly or in concert, are: 1) an increase in r or D following a change in the biotic and/or abiotic environment, and 2) an increase in r or D following a change in the phenotype (and presumably, the genotype) of the invader.

Prolonged lags: environmental factors

An environmental change that enhances the fitness of an exotic will, by definition, trigger an increase, or a "release" in its growth rate, r. Therefore, if such an environmental enhancement has occurred in the history of an exotic, a plot of the colonist's population history will indicate a lag in its growth that is more profound (longer) than would be expected for the inherent lags discussed above. Such prolonged lags of this kind might be caused by any natural or anthropogenic change in a factor that limits the distribution and abundance of an invasive species.

There are several major ecological mechanisms that may result in the release of an invasive species. These include changes in habitat and food resources, climate, dispersal vectors, interspecific interactions, and intraspecific interactions (e.g., Ewel, 1985; Hengeveld, 1988; Hobbs and Humphries, 1995; Williamson, 1996). Such changes may occur either in the local environment, affecting population growth, or in potentially habitable areas, affecting range expansion.

Habitat and Food Resources

If increased quantities of a limiting resource are made available to an invasive species, rapid population growth and range expansions may result. For invasive species that are human commensals, such increased resources may be provided by expanding urbanization and human-mediated modification of natural habitats, as exemplified by the Oxford ragwort and the introduced lettuce species in the U.K. described above. In addition, the spread of the Collared Dove in Europe, which typically lives in human settlements, is probably related at least in part to increasing urbanization. Also related to increased human activities is the provision of artificial food resources, which has been suggested to account for the increase in numbers of birds such as gulls and the fulmar in Europe, both of which may feed on offal and/or garbage (Isenmann, 1990).

In addition to changing the quantity of habitat, an alteration (usually a deterioration) of habitat quality may also permit an invasive population to explode. This mechanism corresponds to Ewel's (1986) hypothesis that the delayed spread of the trees in the Everglades was caused by increasing anthropogenic disturbance which recently has allowed more sites to become invasible. Change in habitat quality has also been implicated in permitting the recent explosion of the mitten crab in the U.K. and possibly *M. senhousia* in California. In Los Angeles Harbor, the beginning of a pollution abatement program allowed the release of the wood-boring gribble populations (an interesting counter-example to the idea that increased, rather than decreased, disturbance would favor an invasive species).

Climate

The proximate mechanisms through which climate change may affect a species are numerous, but relationships to breeding and growing seasons may be important. For example, warming trends may have permitted longer breeding seasons for the Collared Dove in Europe and may have aided in the development of the bivoltinism and southern expansion in the butterfly, *C. tullia*. In species like mosquitoes with temperature-dependent reproductive thresholds, a slight increase in temperature can produce a large increase in r, facilitating both a population increase and an expansion in geographic range (Soulé, 1992). Mosquitoes such as *Aedes albopictus* and *A. aegypti*, the former already established in the southeastern United States, can spread as the climate warms. These mosquitoes can act as vectors for dengue fever, yellow fever, equine encephalitis, filariasis, and the viruses that cause hemorrhagic fevers. In the face of long-term and universal greenhouse warming, it is not unreasonable to expect concomitant changes in the ranges and densities of many invasive species, including vectors of human disease (Soulé, 1992).

Dispersal Vectors

In some instances, the delayed spread in invasive species can be attributed to the provision of a new or more efficient means of dispersal. The teasel in the eastern U.S. was provided with a means of transport via the interstate highway system, the Oxford ragwort spread along the railway lines in the U.K., and the gallant soldier may have been spread by the bombings in England. Roads, themselves, are probably the major avenue of transport of many terrestrial exotics, including many plant diseases. In Australia "dieback" caused by the fungus *Phytopthora cinnamoni* and related species affects many plant communities and is an agent of extinction; its spores are spread more rapidly along roads and where traffic and machinery disrupts the soil. Thus, the increasing volume and efficiency of local and global transportation will not only continue to introduce to new exotic species from abroad, it will also serve to spread invasive species already present in a region.

Interspecific interactions

In addition to responding to conditions such as habitat quality and climate, invasive species must interact with resident plants and animals. Changes in a host of interspecific interactions, such as competition, predation, disease, grazing, pollination, as well as indirect effects including animal-caused habitat modification, may facilitate the release of an invasive species. Such changes may be caused by natural population cycles of interacting species (e.g. Sinclair and Pech, 1996) or anthropogenic effects such as hunting or habitat degradation. For example, on the Channel Islands, the removal of the exotic grazers allowed the release of fennel. Similarly, the elimination of feral goats in the Volcanoes National Park on the island of Hawaii was followed by an explosive spread of alien, fire-conducting grasses; the result has been the near-deforestation of some Hawaiian uplands (D'Antonio and Vitousek, 1992).

Many other interspecific interactions are undoubtedly important, including the anthropogenic initiation of trophic cascades, including those triggered by the removal of predators. For instance, there is evidence (K. Crooks, pers. comm.) that the elimination of coyotes (*Canis latrans*) from remnant patches of scrub vegetation and coastal estuaries in urban areas removes a check on the distribution and activity of alien red foxes (*Vulpes vulpes*) and domestic cats (*Felis catus*), both of which harm native wildlife, including endangered species (see also Soulé *et al.*, 1988).

Intraspecific interactions

A variety of intraspecific interactions may be affected during the early stages of an invasion, when population densities are low. For example, species may have difficulty finding mates or fully utilizing a resource because of low numbers of individuals or "undercrowding" (Williamson and Brown, 1986). These "Allee effects" have been suggested to be important in causing observed lags in the spread of invasive species (Lewis and Kareiva, 1993). Models incorporating Allee dynamics (i.e., disproportionately low fecundity below a certain critical level) have successfully recounted the spatial spread, including the early lag phase, of the House Finch (Veit and Lewis, 1996).

Prolonged lags: genetic factors

Even though the general subject of the genetics of colonizing species has received considerable attention, and such fundamental concepts such as the founder effect, population bottlenecks, and genetic drift are intimately related with the problem of invasive species, genetics remains the great unknown in the biological basis of lag effects and the causation of sudden, explosive growth and expansion of exotics. The question is to what extent are lags, when they occur, caused by the lack of local genetic adaptation to the abiotic environment, the biotic environment, or both?

The possibility of the lack of genetic "fit" of a colonizing population to cause prolonged lags was widely speculated upon at a conference on the genetics of colonizing species (Baker and Stebbins, 1965). Fraser (1965) discussed situations where "migrants move into an environment to which they are not specifically adapted" and "will have an initial phase during which the specific adaptations will have to evolve". Lewontin (in Mayr, 1965) also discussed this issue of "break-out" colonizations, where "under continuous identical selection, there is a long period of stalling of increase of fitness followed by a rapid rise." Similarly, Mayr (1965) suggested that the sudden spreading of the Serin Finch and Collared Dove may be caused by genetic mutation. Baker (in Mayr, 1965; Baker, 1965) commented that the "sudden explosive spread of animals after a period when nothing very much seems to be happening is paralleled by plants," and that "if a newly introduced plant does not have appropriate 'general purpose' genotypes available, it may be confined to a restricted area until these do become available through recombination or introgression." The possibility of lags has also been recognized in the introduction of biocontrol agents, where time might be needed for post-colonization adaptation to the new environment (DeBach, 1965; Wilson, 1965; Waddington, in Wilson, 1965).

Despite these general predictions that time might be needed for some invasive populations to adapt when they are introduced into marginal environments, little empirical support has been forthcoming (Williamson, 1996). Similarly, although genetics has been suggested to play a role in the outbreak of insect species, there has been little documentation of this (Mitter and Schneider, 1987; Myers, 1987). However, possible genetic changes and development of new "biotypes" have been suggested for outbreaks of greenbugs (*Schizaphis graminum*) and Hessian flies (*Mayetiola destructor*) (Mitter and Schneider, 1987). Some experimental evidence for genetic change allowing population increases does exist, as inbred lines of *Drosophila* increased their population sizes after introduction of new genetic material (Carson, 1961, 1968; Cannon, 1963). However, we know of only one case where invasiveness of an exotic species has a demonstrable genetic cause; the cordgrass, *Spartina anglica* mentioned above. Even this case is special, in that the "mutation" was doubling of the genome, not a point mutation in a single gene. Statistically, however, it is inevitable that natural selection is a factor in the survival and fitness of any population.

From a technical standpoint, it is unlikely that particular gene mutations contributing to the success of an introduced species will be detected. Most mutations that are likely to contribute to fitness are subtle, quantitative changes in the phenotype, rather than qualitative, "Mendelian", phenotypic alterations. But the chances of researchers stumbling on such beneficial new mutations by random search are virtually nil. For example, mutations in major genes, such as those detectable by routine surveys of enzymes, are extremely rare (ca 10^7). New ones have yet to be found in all introduced rabbits in Australia, which number in the millions (Richardson *et al.*, 1980), and the expected time for such a new mutation in a major gene to be detectable is about one million years (Gorman *et al.*, 1980). On the other hand, it is quite likely that slightly advantageous mutations in the thousands of genes affecting quantitative traits (such as vigor, metabolic rate, growth rate, resistance to toxins) will occur and be selectively incorporated into the genome quite frequently in relatively large populations. But finding them is like searching for a needle in a haystack. Thus, mutations that enhance invasiveness are unlikely to be detected.

Nevertheless, population genetics theory provides some insight into the interplay between population size and genetic evolution. First, because of founder effects (Mayr, 1963), very small populations (less than 50 individuals or so) are unlikely to be able to evolve improvements in fitness (Franklin, 1980; Soulé, 1980). Although some examples exist where very small populations have successfully recovered to large, healthy populations

(Mayr, 1963), populations having gone through very small bottlenecks are more likely to decline genetically due to inbreeding (Soulé, 1980).

Calculations based on balancing total mutation rates with genetic drift suggest that until the population size increases to about one thousand, natural selection will not be a very effective force in counteracting the randomizing effects of genetic drift (chance changes in the frequencies of genes, including new mutations), and most beneficial mutations, even if they occur, will have a low probability of being incorporated into the population (Soulé, 1980). Furthermore, recent evidence suggests that near-neutral, potentially adaptive mutations may in fact occur an order of magnitude less frequently than mutations with large phenotypic effects, which tend to be highly detrimental (Culotta, 1995; Lande, 1995). This suggests that calculations based on total mutation rates may represent underestimates, and that even larger populations are required to overcome the effects of population bottlenecks. Only when populations are quite large (at least ten thousand) are slightly beneficial mutations likely to increase in frequency because of natural selection and are slightly harmful mutations likely to be weeded out efficiently.

What this implies is a positive feedback between population size and the chances that the population will improve genetically. It also implies that the longer a population exists, at least if it numbers in the thousands, the more likely is a genetic "discovery" that makes it more invasive. The larger the population, the more chance favorable mutations will arise, which in turn allows for larger populations. Moreover, mathematical models also suggest that the faster a population grows after a population bottleneck, the less the effects on average heterozygosity in the population (Nei *et al.*, 1975).

In addition to acquiring new genetic material by mutation, existing populations of invasive species can overcome potential founder effects by repeated introductions over time. This could serve to quickly increase the amount of genetic variability and allow for rapid population explosions. For example, the success of the cladoceran invader, *Bosmina coregoni*, in the Great Lakes may be related to repeated ballast water-mediated introductions (Demelo and Herbert, 1994).

Implications for policy and management

The most effective form of protection against invasive and destructive alien species would appear to be a diverse and healthy assemblage of native species (e.g., Elton, 1958; Case, 1990). Thus, alien species are often found easily penetrating zones of disturbance, particularly agricultural areas or urbanized lakes or estuaries. Furthermore, once an introduction has taken hold, there is often little that can be done to stop it, so we can expect very large losses in native biodiversity and ecological integrity of many ecosystems.

The first line of defense against invasive aliens should be vector management (i.e. the control of the means by which exotics are spread), including inspection and quarantine at ports and transportation hubs. One of the obvious goals of such practices is to prevent the introduction of new exotic species. Recognition of the lag effect also highlights at least two additional benefits of vector management. It decreases the potential for further introductions of previously established exotics, thus preventing the addition of new genetic stock which may make existing populations more aggressive invaders. Also, regional vector management will serve to slow the spread of already established species.

The lag effect also has important implications for the evaluation of the potential extent and effects of an invasive species. Recognition of both inherent and prolonged lags suggest that the past performance of an invasive species may be a poor predictor of its future potential for numerical increase, range extension, and ecological effects. It is dangerous to assume that ecological containment (mal-adaptation) will last forever, especially if numbers of individuals pass the threshold that increase the likelihood of enhancements of local adaptation by natural selection. Also, the lag phase (containment) of an exotic species can end suddenly when some aspect of the biotic or abiotic environment is altered. Often, such changes may be caused by human activities. For example, when another species, particularly a browsing or grazing mammal is removed from the system, explosive and disastrous growth of exotic weeds may result (e.g., fennel on the Channel Islands). Therefore, care in manipulations of systems where exotics are present, such as the removal of feral livestock, is recommended. Given the ever increasing human-mediated alteration of whole ecosystems and the global climate, however, it is likely that there will be corresponding changes in the dynamics of established invaders. A further consideration is the potential for lags in ecological effects of non-native species to occur even after lags in numerical increase or range expansion have ended (Moyle, 1998), or that introductions that were initially considered beneficial may have unanticipated negative effects (resulting in the "Frankenstein Effect"; Moyle *et al.*, 1986).

In the effort to control exotic species, the consideration of the lag effect also suggests an additional line of defense: the extirpation of founder colonies before the explosive growth phase has begun (Hobbs and Humphries, 1995). Experience shows that the elimination of an exotic once the lag phase is over can be virtually impossible. Therefore, careful monitoring programs that may lead to early detection and, if possible, elimination of incipient invaders should be instituted, particularly if the population can be prevented from achieving sizes in the thousands of individuals.

On a global scale, international travel and commerce will accelerate the current rates of introductions of exotic species. Assuming the impossibility of reversing globalization, only superior surveillance and the development of innovative control measures can counter the current growing momentum of ecological disintegration, depauperization, and cosmopolitanization of biodiversity. A better understanding of the biological processes at work during the early stages of invasion will contribute to effective policy development and enforcement.

Acknowledgements

We would like to thank Odd Terje Sandlund, Peter J. Schei, and Åslaug Viken for their efforts in organizing the Conference on Alien Species and preparing this volume. We would also like to thank Lisa Levin, Kevin Crooks, and Jim Enright for comments on this manuscript.

References

Allen, L.J.S., Allen, E.J., Kunst, C.R.G., and Sosebee, R.E. (1991) A diffusion model for dispersal of *Opuntia imbricata* (cholla) on rangeland. *Journal of Ecology,* **79,** 1123–1135.

Andow, D.A., Kareiva, P.M., Levin, S.A. and Okubo, A. (1993) Spread of invading organisms: patterns of spread, in *Evolution of Insect Pests,* (eds K.C. Kim and B.A McPheron), John Wiley and Sons, New York, pp. 219–242.

Atrill, M.J. and Thomas, R.M. (1996) Long-term distribution patterns of mobile estuarine invertebrates (Ctenophora, Cnidaria, Crustacea: Decapoda) in relation to hydrological parameters. *Marine Ecology Progress Series,* **143**: 25–36.

Baker, H.G. and Stebbins, G.L. (1965) *The Genetics of Colonizing Species.* Academic Press. New York.

Baker, H.G. (1965) Characteristics and modes of origins of weeds, in *The Genetics of Colonizing Species,* (eds H.G. Baker and G.L. Stebbins), Academic Press, New York, pp. 147–172.

Barnes, R.S.K. (1994) *The Brackish-water Fauna of Northwestern Europe.* Cambridge University Press, Cambridge.

Beatty, S.W. and Licari, D.L. (1992) Invasion of fennel (*Foeniculum vulgare*) into shrub communities on Santa Cruz Island, California. *Madroño,* **39**, 54–66.

Brenton, R. and Klinger, R. (1994) Modeling the expansion and control of fennel (*Foeniculum vulgare*) on the Channel Islands, in *The Fourth Channel Island Symposium,* (eds W.L. Halvorson and G.J. Maencer), Santa Barbara Museum of Natural History, Santa Barbara, California, pp. 497–504.

Cannon, G.B. (1963) The effects of heterozygosity and recombination on the relative fitness of experimental populations of *Drosophila melanogaster. Genetics*, **48**, 919–942.

Carey, J.R. (1996) The incipient Mediterranean fruit fly population in California: implications for invasion biology. *Ecology*, **77**, 1690–1697.

Carlton, J.T. (1979) *History, biogeography, and ecology of the introduced marine and estuarine invertebrates of the Pacific coast of North America*. Ph.D. Dissertation. University of California, Davis, CA, USA.

Carson, H.L. (1961) Heterosis and fitness in experimental populations of *Drosophila melanogaster*. *Evolution*, **15**, 496–509.

Carson, H.L. (1968) The population flush and its genetic consequences, in *Population Biology and Evolution*, (ed. R.C. Lewontin), Syracuse University Press. New York, pp. 123–137.

Case, T.J. (1990) Invasion resistance arises in strongly interacting species-rich communities. *Proceedings of the National Academy of Science*, **87**, 9610–9614.

Cohen, A.N. (1995) Chinese mitten crabs in North America. *Aquatic Nuisance Species Digest*, **1(2)**, 20–21.

Cousens, R, and Mortimer, M. (1995) *Dynamics of Weed Populations*. Cambridge University Press. Cambridge, UK.

Crooks, J.A. (1996) The population ecology of an exotic mussel, *Musculista senhousia*, in a southern California bay. *Estuaries*, **19**, 42–50.

Crooks, K.R. and Van Vuren, D. (1994) Conservation of the island spotted skunk and island fox in a recovering island ecosystem, in *The Fourth Channel Island Symposium*, (eds W.L. Halvorson and G.J. Maencer), Santa Barbara Museum of Natural History, Santa Barbara, California, pp. 379–385.

Culotta, E. (1995) Minimum population size grows larger. *Science*, **270**, 31–32.

D'Antonio, C.M. and Vitousek, P.M. (1992) Biological invasions by exotic grasses, the grass/fire cycle, and global change. *Annual Review of Ecology and Systematics*, **23**, 63–87.

DeBach, P. (1965) Some biological and ecological phenomena associated with colonizing entomophagous insects, in *The Genetics of Colonizing Species*, (eds H.G. Baker and G.L. Stebbins), Academic Press, New York, pp. 287–306.

Demelo, R. and Hebert, P.D.N. (1994) Founder effects and geographical variation in the invading cladoceran *Bosmina* (*Eubosmina*) *coregoni* Baird 1857 in North America. *Heredity*, **73**, 490–499.

Elton, C.S. (1958) *The Ecology of Invasions by Plants and Animals*. Methuen, London.

Ewel, J.J. (1986) Invasibility: Lessons from southern California, in *Ecology of Biological Invasions of North America and Hawaii*, (eds H.A. Mooney and J.A. Drake), Springer-Verlag, New York, pp. 214–239

Franklin, I.R. (1980) Evolutionary change in small populations, in *Conservation Biology. An Evolutionary–Ecological Perspective*, (eds M.E. Soulé and B.A. Wilcox), Sinauer Associates Inc., Sunderland, Massachusetts, pp. 135–149.

Fraser, A. (1965) Colonization and genetic drift, in *The Genetics of Colonizing Species*, (eds H.G. Baker and G.L. Stebbins), Academic Press, New York, pp. 117–125.

Gorman, G.C., Buth, D.G., Yang, S.Y. and Soulé, M.E. (1980) The relationship of the *Anolis cristatellus* species group: electrophoretic analysis. *Journal of Herpetology*, **14**, 269.

Grosholz, E.D. (1996) Contrasting rates of of spread for introduced species in terrestrial and marine systems. *Ecology*, **77(6)**, 1680–1686.

Hastings, A. (1996) Models of spatial spread: is the theory complete? *Ecology*, **77(6)**, 1675–1679.

Hengeveld, R. (1988) Mechanisms of biological invasions. *Journal of Biogeography*, **15**, 819–828.

Hengeveld, R. (1989) *Dynamics of Biological Invasions.* Chapman and Hall, New York.

Hengeveld, R. (1993) Small-step invasion research. *Trends in Ecology and Evolution,* **9(9)**, 339–342.

Hobbs, R.J. and Humphries, S.E. (1995) An integrated approach to the ecology and management of plant invasions. *Conservation Biology,* **9**, 761–770.

Hengeveld, R. and van den Bosch, F. (1991) The expansion velocity of the collared dove *Streptopelia decaocto* population in Europe. *Ardea,* **79**, 67–72.

Holmes, E.E. (1993) Are diffusion models too simple? A comparison with telegraph models of invasion. *The American Naturalist,* **142**, 779–795.

Holmes, E.E., Lewis, M.A. Banks, J.E. and Veit R.R. (1994) Partial differential equations in ecology: spatial interactions and population dynamics. *Ecology,* **75**, 17–29.

Isenmann, P. (1990) Some recent bird invasions in Europe and the Mediterranean Basin, in *Biological Invasions in Europe and the Mediterranean Basin,* (eds F. di Castri, A.J. Hansen, and M. Debussche), Kluwer Academic Publishers, The Netherlands, pp. 245–261.

Kareiva, P.M. (1983) Local movement in herbivorous insects: Applying a passive diffusion model to mark–recapture field experiments. *Oecologia,* **57**, 322–327.

Kendall, D.G. (1948) A form of wave propogation associated with the equation of heat conduction. *Proceedings of the Cambridge Philosophical Society* **44**, 591–594.

Kowarik, I. (1995) Time lags in biological invasions with regard to the success and failure of alien species, in *Plant Invasions,* (eds P. Pysek, K. Prach, M. Rejmánek, and M. Wade), SPB Academic Publishing, The Netherlands, pp. 15–38.

Lande, R. (1995) Mutation and conservation. *Conservation Biology,* **9,** 782–791.

Laughrin, L., Carroll, M., Bromfield, A. and Carroll, J. (1994) Trends in vegetation changes with removal of feral animal grazing pressures on Santa Catalina Island, in *The Fourth Channel Island Symposium,* (eds W.H. Halvorson and G.J. Maender), Santa Barbara Museum of Natural History, Santa Barbara, California, pp. 523–530.

Lewis, M.A. and Kareiva, P. (1993) Allee dynamics and the spread of invading organisms. *Theoretical Population Biology,* **43**, 141–158.

Lonsdale, W.M. (1993) Rates of spread of an invading species – *Mimosa pigra* in northern Australia. *Journal of Ecology,* **81**, 513–521.

Mayr, E. (1963) *Animal Species and Evolution.* Harvard University Press, Cambridge, Massachusetts.

Mayr, E. (1965) The nature of colonizations in birds, in *The Genetics of Colonizing Species,* (eds H.G. Baker and G.L. Stebbins), Academic Press, New York, pp. 29–47.

Mitter, C and Schneider, J.C. (1987) Genetic change and insect outbreaks, in *Insect Oubreaks,* (eds P. Barbarosa and J.C. Shultz), Academic Press, San Diego, pp. 505–528.

Moyle, P.M. (1998) Effects of invading species on freshwater and estuarine ecosystems, in *Invasive Species and Biodiversity Management,* (eds O. T. Sandlund, P. J. Schei and Å. Viken), Kluwer Academic Publishers, Dordrecht, The Netherlands.

Moyle, P.M., Li, H.W. and Barton, B.A. (1986) The Frankenstein effect: impact of introduced fishes on native fishes in North America, in *Fish Culture in Fisheries Management,* (ed. R.H. Shroud), American Fisheries Society, Bethesda, Maryland, pp. 415–426.

Myers, J.H. (1987) Population outbreaks of introduced insects: lessons from the biological control of weeds, in *Insect Oubreaks* (eds P. Barbarosa and J.C. Shultz), Academic Press, San Diego, pp. 173–193.

Nei, M., Maruyama, T. and Chakraborty, R. (1975) The bottleneck effect and genetic variability in populations. *Evolution,* **29**, 1–10.

Nowak, E. (1971) *The Range Expansion of Animals and its Causes.* Translated from Polish 1975. Foreign Scientific Publications Department, U.S. Department of Commerce, Washington, DC.

Okubo, A. (1988) Diffusion-type models in avian range expansion. *Acta XIX Congressus Internationalis Ornithologici,* **1**, 1038–1049.

Perring, F.H. (1974) Changes in our native vascular plant flora, in *The Changing Flora and Fauna of Britain*, (ed. D.L. Hawksworth), Academic Press, London, pp. 7–25.

Reish, D.J., Soule, D.F. and Soule, J.D. (1980) The benthic biological conditions of Los Angeles – Long Beach Harbors: results of 28 years of investigations and monitoring. *Helgolander Meeresuntersuchungen,* **34**, 193–205.

Richardson, B.J., Rogers, P.M. and Hewitt, G.M (1980) Ecological genetics of the wild rabbit in Australia. II. Protein variation in British, French, and Australian rabbits and the geographical distribution of the variation in Australia. *Australian Journal of Biological Science,* **33,** 371–383.

Roughgarden, J. (1986) Predicting invasions and rates of spread, in *Ecology of Biological Invasions of North America and Hawaii,* (eds H.A. Mooney and J.A. Drake), Springer-Verlag, New York, pp. 179–188.

Salisbury, E.J. (1953) A changing flora as shown in the study of weeds of arable land and waste places, in *The Changing Flora of Britain*, (ed. J.E. Lousley), Botanical Society of the British Isles, Oxford, England, pp. 130–139.

Sinclair, A.R.E. and Pech, R.P. (1996) Density dependence, stochasticity, compensation, and predator regulation. *Oikos,* **75**, 164–173.

Skellam, J.G. (1951) Random dispersal in theoretical populations. *Biometrika,* **38**: 196–218.

Solecki, M.K. (1993) Cut-leaved teasel and common teasel (*Dipsacus laciniatus* L. and *D. sylvestris* Huds.): profile of two invasive aliens, in *Biological Pollution: the Control and Impact of Invasive Exotic Secies* (ed. B.N. McKnight), Indiana Academy of Natural Sciences, Indianapolis, pp. 85–92.

Soulé, M.E. (1980) Thresholds for survival: maintaining fitness and evolutionary potential, in *Conservation Biology. An Evolutionary–Ecological Perspective*, (eds M.E. Soulé and B.A. Wilcox*)*, Sinauer Associates Inc., Sunderland, Massachusetts, pp. 151–169.

Soulé, M.E. (1990) The onslaught of alien species and other challenges in the coming decades. *Conservation Biology,* **4**: 233–240.

Soulé, M.E. (1992) The social and public health implications of global warming and the onslaught of alien species. *Journal of Wilderness Medicine,* **3,** 118–127.

Soulé, M.E., Bolger, D.T., Alberts, A.C., Sauvajot, Wright, R.J., Sorice, M. and Hill, S. (1988) Reconstructed dynamics of rapid extinctions of chaparral-requring birds in urban habitat islands. *Conservation Biology,* **2**, 75–92.

Thompson, J.D. (1991) The biology of an invasive plant. *Bioscience,* **41(6)**, 393–401.

Townsend, C.R. (1991) Exotic species management and the need for a theory of invasion ecology. *New Zealand Journal of Ecology,* **15**, 1–3.

Van den Bosch, F., Hengeveld, R. and Metz, J.A.J. (1992) Analysing the velocity of animal range expansion. *Journal of Biogeography,* **19**, 135-150.

Veit, R.R. and Lewis, M.A. (1996) Dispersal, population growth, and the Allee effect: Dynamics of the house finch invasion of eastern North America. *The American Naturalist,* **148**, 255–274.

Wiernasz, D.C. (1989) Ecological and genetic correlates of range expansion in *Coenonympha tullia. Biological Journal of the Linnean Society,* **38**, 197–214.

Williamson, M. and Brown, K.C. (1986) The analysis and modelling of British invasions. *Philosophical Transactions of the Royal Society of London, B,* **304**, 505-522.

Williamson, M. (1996) *Biological Invasions.* Chapman and Hall, London.
Wilson, F. (1965) Biological control and the genetics of colonizing species, in *The Genetics of Colonizing Species*, (eds H.G. Baker and G.L. Stebbins), Academic Press, New York, pp. 307–329.

8 Modelling the impact of biological invasions

ROB HENGEVELD
Institute of Forest and Nature Research, Wageningen, The Netherlands

Abstract

This paper gives an overview of principal problems encountered when dealing with invasions. These problems are both theoretical and practical. They arise from difficulties with predicting what will happen to a species after its transfer into an area alien to it. Many of the variables are known, qualitatively, and they can be put into context by analytical modelling. The difficulty of prediction using these models is primarily technical; we cannot make adequate estimates before a potential invader has been introduced and actually spreads. Despite this impossibility of prediction, the equations do show how the variables interrelate, which practical difficulties of the measurements should be solved, and how to stop an invasion during its progression. Ecological science can also explain why species of particular environments might be favoured whereas other species run increasingly greater risks of extinction. The reshuffling of species according to the spatially dynamic potential has genetic consequences such as altering the likelihood of hybridisation.

Introduction

It is becoming increasingly important that scientists build the capacity to predict which species will invade an area alien to them and to estimate what harm they can do. This harm can be both economical and ecological. Economic harm results when an invader reduces crop yields, for example, or when it incurs disease. Ecological harm is done when its effect is against the native wild species. In this paper, I consider whether or not we can predict the fate of such species introduced into an alien, continental area. The answer is: "No, we cannot predict this". Equations describing the spatial invasion process may, however, be helpful in selecting the most sensitive variables which determine an invader's spatial progression, in order to slow down or stop further progression. These models may also help us to evaluate the risk of hybridisation based on the spatial dynamics of the species. Thus, it is possible to make qualitative inferences about various aspects of invasions, and about the possible effects of various countermeasures. But we cannot predict in quantitative terms what will happen to species which are introduced into an area alien to it.

O. T. Sandlund et al. (eds.), Invasive Species and Biodiversity Management, 127–138.

Three ways of predicting invasions

The first and most common way of prediction is a statistical approach. Based on analyses of previous or still progressing invasions, principal process components may be identified. Components common in many invasions are assumed to be predictive driving forces of this process. For example, seed size, seed number, number of eggs or offspring, dispersal capacity, range size, ecological tolerance, etc., are all considered important variables. As these specific properties are intrinsic to a species, the resulting invasiveness is often considered an intrinsic property as well. The weakness of this approach is that one can only estimate the probability with which some variable will be involved, implying that it may not always operate. To improve predictability, one therefore estimates sets of variables recognising that all of them add equally to the predictability of a particular invasion. However, one can still argue that other sets of variables may contribute signficantly. A further weakness of the statistical approach is that the environmental conditions are likely to differ between the areas of origin and introduction of the species.

The second way of predicting species invasiveness is by identifying the species' physiological requirements as inferred from its preferred climatic conditions in the area of origin. The future range can be predicted according to climatic conditions in the new area. This approach may generate a more distinct, testable hypothesis (cf. Bartlein *et al.*, 1986; Huntley *et al.*, 1995). Yet, this approach does not include the demographic components of the statistical approach as the driving forces of invasions. Neither does it consider the possible interactions between an invader and other species.

In the third, analytical approach, one postulates the operation of several variables, and models the spatial progression of an invasion. This provides predictive and testable models, though they are valid only within the framework of the postulated variables and model structure. This approach is more developed than the previous two because it studies the combined function of the identified variables. This is important since the variables or parameter values commonly counteract or compensate each other. For example, small numbers of offspring per clutch may be compensated by a great number of subsequent clutches in longer-lived individuals, or by reduced juvenile mortality.

Of course, these three approaches are interdependent; the variables postulated in the analytical approach are often obtained from statistical or physiological studies.

Some analytical models

Invading ecologically uniform areas

Van den Bosch *et al.* (1990) formulated an analytical invasion model based on a reaction/diffusion process. This assumes an invasion to progress from randomly Brownian moving propagules diffusing from an occupied area into the surrounding, unoccupied space. After settlement, the propagules would reproduce, thus maintaining the process. This represents the reaction component. Reproduction capacity is expressed by the net reproduction rate, which is estimated from data on the individual survival and fertility rates per age class. The diffusion component is expressed by the frequency distribution of the distances between the place of birth and reproduction of the individuals. This model has been applied to several bird species, as well as to the muskrat, *Ondatra zibethicus*, invading Europe and North America (Van den Bosch *et al.*, 1992). It estimates a species' expected expansion rate, which may be compared with the actual expansion rate as calculated from observed range expansion. Figure 8.1 shows that the data points representing the various species follow the 45 degree line, implying that the model structure is correct and contains the relevant process variables.

Thus, the invasion of these species prove to be predictable on this spatial scale. This model may be validated by estimating the sensitivity of the results by varying the values of the process components. Slight changes in the component values result in large changes in invasion rate when the process is indeed sensitive. When it is robust, hardly any change occurs, even with considerable changes in the component values. It appears that the expected invasion rate is least sensitive to changes in survival rate by short-distance dispersal and most sensitive to those in dispersal over long distances (Hengeveld, 1992). As these variable values are under environmental control, invasiveness cannot be an intrinsic character in species. For example, as soon as juvenile mortality decreases to give net reproduction rate values exceeding 1, the population starts to increase in numbers and expand in space. Also, long-distance dispersal increases when the rate of interception of young, potential reproducers decreases due to a decreasing number of nesting sites in homogenised agricultural landscapes. In both cases changes in ecological conditions may turn a stable population into a spatially expanding one. In other words, the species turns into an invader. More generally, climatic variability affects parameter values, causing range contractions or expansions, and turning stable ranges into mobile or expanding ones. Invasiveness is not an intrinsic species property, but largely an expression of the match between species properties and requirements on the one hand and environmental conditions on the other.

This is an important conclusion when considering introduction of species into other regions or continents. Under different conditions, stable species may easily turn into noxious, rapidly expanding invaders (Forcella, 1985).

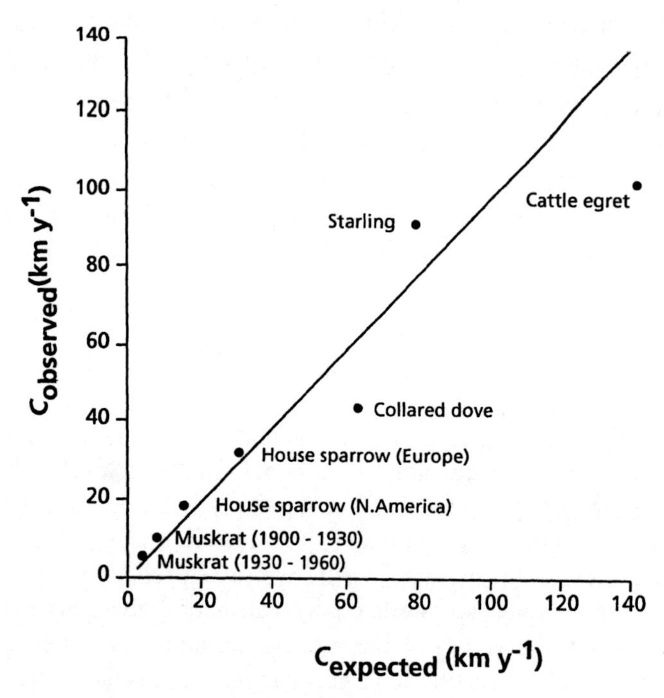

Figure 8.1 Relationship between the expected, calculated invasion rate and the observed rate for several species (after data in Van den Bosch *et al.*, 1992).

The values of various process components may counteract or reinforce each other. For example, great numbers of seeds or eggs are usually compensated for by high mortality rates. Thus, the great tit, *Parus major*, with up to 15 eggs per clutch, is not considered and intrinsic invader, whereas the collared dove, *Streptopelia decaocto*, with only two eggs per clutch, has become a hobby horse for invasion research in Europe. The net reproduction rate, which is a combination of several variables, may be similar in species differing widely in their fertility rates. Moreover, it is not known how a species will respond to the conditions in an alien environment. For example, the house sparrow, *Passer domesticus*, is a sedentary bird in Europe. Yet, outside Europe it has conquered many parts of the world at high spreading rates after introduction (e.g., Long, 1981). In Europe, the number of fledgelings produced annually per pair is 3-5. However, when the species was spreading in North America and New Zealand, the average numbers of fledgelings were recorded at 24 (Barrows, 1889) and 31 per year (Kirk, 1890), respectively! The house sparrow thus proved to be capable of continuous breeding. This had never been recorded in Europe, and could consequently not be predicted. One reason for the difference is that the male sparrows under low population densities experience few interferences, resulting in changes in their hormonal balance. Their reproductive drive is not suppressed as is normally the case under denser conditions. To reach the maximum number of fledgelings, the females must be able to produce eggs throughout a large part of the year, and they must be able to leave brooding to the young of the previous clutch while feeding, etc.

Other species simultaneously introduced into North America, such as the starling, *Sturnus vulgaris*, do not show this kind of change in behaviour. Thus, the response of species to new conditions is often unpredictable and specific. Species are ecologically idiosyncratic, showing individual reaction patterns (Hengeveld, 1990; Huntley, 1988). Consequently, the impact of the invader on the local, alien biota also remains unpredictable.

Invading ecologically non-uniform areas

Changes in the penetrability or interception rate of landscapes can considerably alter the rate of invasion. Landscapes are typically non-uniform in an ecological sense; they offer variable living conditions to any species. Over the past few hundred years, however, landscapes in many parts of the world become more uniform, due to new agricultural practices. This may benefit some species, but not others. For instance, mortality rates due to dispersal, as well as dispersal rates will change. These two variables may counteract or reinforce each other.

The model has therefore been extended to include variables related to spatial non-uniformity. These variables are the fraction of suitable area, the increase of spatial variance, and the dispersal mortality. We assume that the randomly moving individuals follow a Gaussian distribution with variance ω increasing over time τ. The fraction of suitable area is δ, and the mortality rate during dispersal ψ. The rate of settlement becomes the product $\psi\delta$, and the risk of dispersal in ecologically non-uniform areas ψ/ω. This risk is small when ω is large, when ψ is small, or both (e.g. Hengeveld and Van den Bosch, 1997). Thus, in a mathematical sense, the invasion process can be understood for non-uniform conditions. The problem of prediction is not so much due to inadequate modelling, but rather to the lack of adequate data. Technical difficulties in obtaining data were already apparent for the model pertaining to ecologically uniform areas, as sufficient data were found for only five species in the literature (Figure 8.1). The additional information needed for non-uniform areas concerns the area to be invaded. This means an area with ecological conditions to which the invading species has not yet been subjected. Consequently, however realistic the model may be, its application for quantitative prediction remains virtually impossible.

Practical use of the equations

The models may still be used to improve understanding of the invasion process. The behaviour of the model demonstrates that invasion of uniform areas is a linear function, whereas invasion of non-uniform areas is non-linear (Figure 8.2). In the model for uniform areas, increasing values of the determining variables results in increasing expected invasion rates. In the model for non-uniform areas expected invasion rates initially increase with the fraction of suitable area (δ), reache a maximum, and subsequently decline with higher values of δ. The position of the maximum is determined by dispersal risk (ψ/ω) as a function of the fraction of suitable area. The fraction of suitable area most favourable for the invader's progress depends on its dispersal and mortality rates, which determine the dispersal risk. Under these model conditions, the fractions of suitable area must be higher than approximately 20%, as a lower fraction makes the area nearly impenetrable (Figure 8.3). Thus, both the variables and their interaction are important during the invasion process.

The statistical analyses indicate which variables might be relevant, whereas analytical modelling demonstrates the impact of the process structure within which these variables operate. Yet, the result concerning e.g. the value of minimum suitable area fraction for invader penetration is determined by which variables are included or excluded in the model. For example, the model for invasion of non-uniform areas assumes that the suitable sites are uniformly distributed in space, which is not realistic. When we assume that suitable sites are spatially clumped, the area will sooner become impenetrable, depending on the degree of clumping. Moreover, the spatial scale is not universally defined for all species. An area which is penetrable for one species, can easily prove impenetrable to another.

The model for the invasion process structure indicates the sensitivity of the expected invasion rate to values of one or more ecological variables. This has important practical implications. Not only is invasiveness an expression of the match between ecological requirements, competitiveness, and environmental conditions, but it can also be manipulated. As validation studies indicate the sensitivity of the various process components, we may in principle manipulate the invasion rate efficiently. For example, spatial outliers have often been observed to be foci for hearth formation, which are important to the rate of progression of invasions. The validation results of the model for invasion of uniform areas show how sensitive the expected rates are to hearth formation and long-distance dispersal. To reduce invasion rates, it is therefore important to control these hearths. The results of validation experiments by Collingham (1995) show that in areas with few suitable sites, long-distance dispersal does not necessarily increase the invasion rates. Rather, spatial progression may in those cases be particularly dependent on short-distance dispersal.

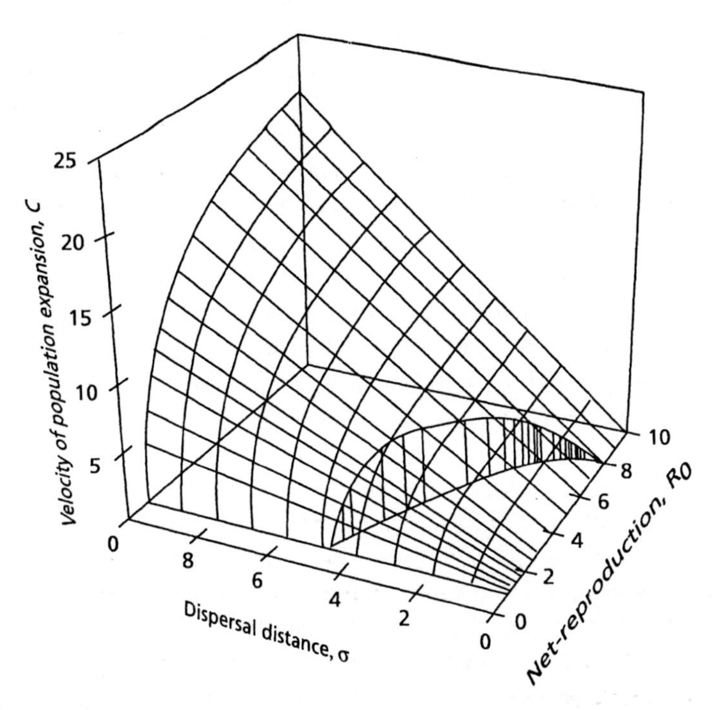

Figure 8.2 Invasion rate (C) plotted as a function of both the net rate of reproduction (R_0) and dispersal distance (σ).

Thus, despite their inapplicability in quantitative prediction of invasion rates, extended models can be of great help in taking practical measures. They indicate possible process structures with their expected outcomes and their sensitive components. They show whether the process can be manipulated at a certain stage of development, and how this can be done most efficiently. Finally, they show explicitly under what assumptions this can be done or not done. What holds under the assumptions of one model expressing one particular process may not hold for other models based on other assumptions and therefore representing other processes.

Hybridisation

In dynamic environments, species must track their preferred habitat conditions by moving (Hengeveld, 1997). To survive, they must be able to 1) reach sites or habitat patches with these conditions on time, and 2) recognise the sites. Moreover, the habitat patches should be large enough for the species to build up a new population large enough for sending sufficient propagules for settling elsewhere in the future. The easier the individuals can recognise their preferred conditions and the sharper these stand out against other conditions, the better species separate out (Anderson, 1948; Paterson, 1993; Hengeveld, 1996). However, under the present regime of habitat degradation, individuals will be less and less able to discriminate between good and bad habitats, with the result that they hybridise (Levin *et al.,* 1996; Simberloff, 1996). Firstly, habitats become fragmented and too small to contain large populations. This may affect the possibility to find a mate, leading to hybridisation between different species. Moreover, when habitat differences are reduced or altogether obliterated, individuals of different species find themselves together more often, which increases the chance of hybridisation. Finally, when alien species are introduced, they can also hybridise with native ones when they have insufficient discriminative recognition systems (Crawford *et al.,* 1987). Hybridisation can have the direct effect of loss of well-adapted genotypes. Also, genotypic variation and integrity can be reduced, be lost by introgression, or gametes can be wasted. Simberloff (1996), analysing many cases of recent instances of observed hybridisation, therefore fears that this effect of species introduction and invasion poses a major threat to species conservation. In many cases, little can be done, if anything at all. Often, it is difficult to control the hybrids, particularly when these can only be distinguished by molecular techniques (Simberloff, 1996). Moreover, once a species has been introduced or a hybrid recognised, it will usually be exceedingly difficult to eradicate it. Only species with large individual body sizes can be controlled or eradicated; in mammals, the lower limit of body size will be roughly that of the muskrat. The possibility of spotting and eradicating the species also depends on the mobility and behaviour of the animals, their number of offspring, the character of their habitat, etc.

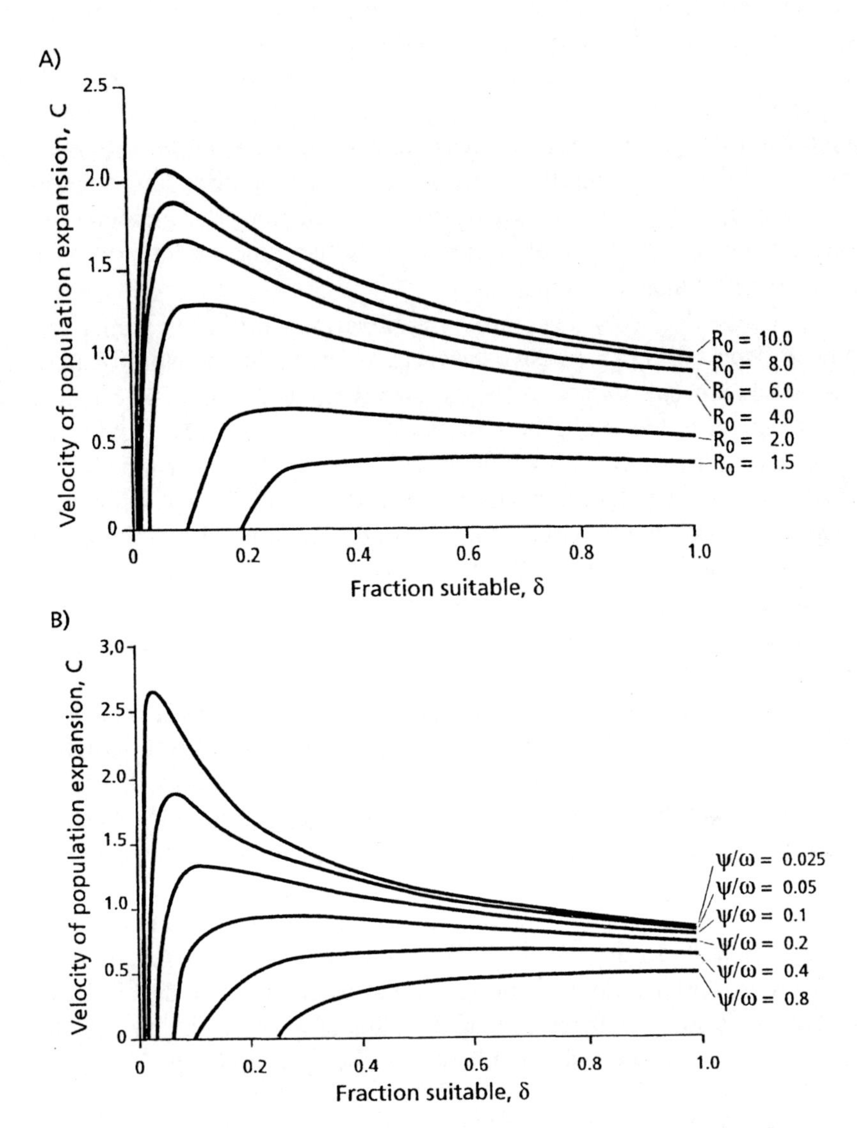

Figure 8.3 Invasion rate (C) as a function of the fraction of suitable habitat (δ) in a non-uniform area for various values of (A) net reproduction rate (R_0) and (B) dispersal risk (ψ/ω).

Modes of dispersal

It seems profitable to look at the mode of dispersal as a sensitive process component in invasions. For example, when the invasion progresses mainly through long-distance dispersal, it is important to locate and eradicate the initial foci, however small they may be. On the other hand, the advance of species progressing through short-distance dispersal is mainly controlled by net reproduction rates being larger than one (Mollison, 1977). Thus, to reduce invasion rates in this case, reproductive output should be reduced. This can be done in several ways, depending on the most sensitive component of this subprocess of invasion, e.g., by lowering fertility rate, by increasing juvenile mortality, or by reducing the number of generations. When propagule production depends on reproduction, the measures to be taken to reduce the advance of an invasion are the same for long distance as well as for short distance dispersal.

The practical efforts to control invasions may differ according to the mode of dispersal in the invasive species. For example, the North American ruddy duck, *Oxyura jamaicensis*, is spreading in Europe; in Spain also hybridising with the white-headed duck, *O. leucocephala* (Anonymous, 1993). The expansion of this duck will mainly be determined by the diffusion component through long-distance dispersal. Its expansion should preferably be stopped by eradicating the outlying "bridgehead" populations or groups of birds. At the same time, eggs should be destroyed to prevent new propagules from spreading. Hunting may in this case have the adverse effect of making the ducks restless, causing migration. In contrast to this, the expansion of the Egyptian Goose, *Alopochen aegyptiaca*, within and from the Netherlands is caused mainly by density-dependent reproductive overflow (Lensink, in press). This expansion progresses very slowly and could be retarded or stopped or the species eradicated by hunting the reproductive adults, or preferably by that of the pre-adults when they congregate for moulting. Thus, according to the biology of the species, our measures should be different.

Another way of reducing species invasiveness is to increase the amount of area unsuitable to the invader. Thus, vast stretches of monocultures will be open to certain invaders, enabling them to reach outbreak densities and to produce many long distance propagules. Spatially more varied and mixed habitats run a smaller risk in relation to this process (e.g. Vandermeer, 1989).

Conclusions

Invasions cannot be predicted quantitatively because parameter estimation is technically close to impossible for conditions still alien to a species. Also, any species can turn into a noxious invader as long as its net reproduction rate exceeds one; even a notoriously sedentary species like the house sparrow can rapidly invade large parts of the world in only part of a century, subsequently dominating the local bird faunas. Recent models can, however, indicate which process components are most sensitive and therefore easiest to manipulate to yield the greatest effect. In this sense, they have to be further extended and studied.

References

Anonymous (1993) *UK Ruddy Duck Working Group. Information*. Joint Nature Conservation Committee, Peterborough, UK.

Anderson, E. (1948) Hybridisation of the habitat. *Evolution,* **2,** 1–9.

Barrows, W.B. (1889) *The English Sparrow in North America, Especially in its Relations to Agriculture.* U.S. Department of Agriculture, Division of Economics Ornithology and Mammology. Bulletin 1.

Bartlein, P.J., Prentice, I.C. and Webb, T. (1986) Climatic response surfaces from pollen data for some eastern North America taxa. *Journal of Biogeography*, **13,** 35–57.

Collingham, Y.C. (1995) *The development of a spatially explicit landscape-scale model of migration and its application to investigate the response of trees to climatic change.* PhD. Thesis, University of Durham, UK.

Crawford, D.J., Witkus, R. and Stuessay, T.F. (1987) Plant evolution and speciation on oceanic islands, in *Differentiation Patterns in Higher Plants,* (ed. K.M. Urbanska), Academic Press, London, pp. 183–199.

Forcella, F. (1985) Final distribution is related to rate of spread in alien weeds. *Weed Research*, **25,** 181–191.

Hengeveld, R. (1998) *Dynamic Biogeography*. Cambridge University Press, Cambridge.

Hengeveld, R. (1992) Potential and limitations of invasion research. *Florida Entomologist,* **75,** 60-72.

Hengeveld, R. (1997) Impact of biogeography on a population-biological paradigm shift. *Journal of Biogeography,* **24,** 541-548.

Hengeveld, R. and Van den Bosch, F. (1997) Invading into an ecologically non-uniform area, in *Past and Future Rapid Environmental Changes*, (eds B. Huntley, W. Cramer, A.V. Morgan, H.C. Prentice and J.R.M. Allen), Springer, Berlin, pp. 217-225.

Huntley, B. (1988) Europe, in *Vegetation History,* (eds B. Huntley and T. Webb), Junk, The Hague, pp. 341-383.

Huntley, B., Berry, T.M., Cramer, W. and MacDonald, A.P. (1995) Modelling present and potential future ranges of some European higher plants using climatic response surfaces. *Journal of Biogeography*, **22,** 967-1002.

Kirk, T.W. (1890) Note on the breeding habits of the European Sparrow (*Passer domesticus*) in New Zealand. *Transactions of the Proceedings of New Zealand Institute of Zoology*, **23,** 108-110.

Lensink, R. (1998) Temporal and spatial expansion of the Egyptian Goose, *Alopochen aegyptiaca*, in the Netherlands, 1967-1994. *Journal of Biogeography,* **25,** 251-263.

Levin, D.A., Francisco-Ortega, J. and Jansen, R.K. (1996) Hybridisation and the extinction of rare plant species. *Conservation Biology*, **10,** 10-16.

Long, J.L. (1981) *Introduced Birds of the World*. David and Charles, Newton Abbot, UK.

Mollison, D. (1977) Spatial contact models for autecological an epidemic spread. *Journal of the Royal Statistical Society, Series B,* **39,** 283-326.

Paterson, H.E.H. (1993) *Evolution and the Recognition Concept of Species: Collected Writings*. (ed. by S.F. McEvey). John Hopkins University Press, Baltimore.

Simberloff, D. (1996) Hybridisation between native and introduced a wild life species: importance for conservation. *Wildlife Biology*, **2**, 143-150.

Van den Bosch, F., Hengeveld, R. and Metz, J.A.J. (1992) Analysing the velocity of range expansion. *Journal of Biogeography*, **19**, 135-150.

Van den Bosch, F., Metz, J.A.J. and Diekmann, O. (1998) The velocity of spatial population expansion. *Journal of Mathematical Biology*, **28**, 529-565.

Vandermeer, J. (1989) *The Ecology of Intercropping*. Cambridge University Press, Cambridge.

9 Biological invasions and global change

HAROLD A. MOONEY and ANNIKA HOFGAARD

Stanford University, California, USA; and Climate Impacts Research Centre, Abisko, Sweden

Abstract

Global change, in its various dimensions, has the capacity to exacerbate the invasive species problem. Climate change, land use change, changing commercial trade intensity and pathways, changes in atmospheric composition, including CO_2 and nitrogen deposition are all predicted to aid the establishment of alien invasive species. We need to stem the tide of invaders under present conditions since future conditions may make it more difficult to accomplish this task. We should also work to reduce the potential for global change to head off this scenario of increasing invasions, as well as for other obvious reasons.

Introduction

There is no question that we are currently witnessing a massive exchange of biotic material among the nations of the world as has been so forcefully chronicled by Elton (1958). This exchange has been going on through geological time, and has increased in tempo as commerce among nations has ever expanded and as landscapes have become increasingly impacted by human activities. These activities have altered the composition of flora and fauna world-wide through their deliberate or accidental dispersal of species beyond their native ranges or through range expansions of existing species (Mack, 1997).

A brief look at the last millennia indicates both small-scale and large-scale interference with the natural landscape due to the use of alien species by humans. During the Middle Ages, a fairly warm period predating the Little Ice Age (i.e.c. 1450–1880 AD; Grove, 1988), the small scale use of "useful plants and animals" (e.g. for production of food, fodder and medicine) increased as a result of increasing monastery activities that subsequently was adopted by local people. The Middle Ages was also the starting period for the European imperialism (Lodge, 1993). Exchanges of biota have occurred constantly as species accompany humans into non-native habitats. Movements of diseases, weeds, and agricultural plants and animals with Europeans during this epoch was apparently vital to the

O. T. Sandlund et al. (eds.), Invasive Species and Biodiversity Management, 139–148.

success of European imperialism (Lodge, 1993). This biological expansion of the biota of Europe caused profound and large scale changes where the species assemblages of entire landscapes were changed (Clout, 1998; Pech, 1998).

The 14th to 19th centuries involved establishment of the feudal system accompanied by an ever increasing use of alien species for ornamental purposes. Gradually this use of alien species spread among common people and represents today a world-wide industry. Within the boreal region this exchange of biotic material became more large scale after the end of the Little Ice Age as climate improved and growing conditions became more suitable.

The history of world-wide exchange of biotic material has brought us into the present large scale use of alien species in modern farming and modern forestry. As reported by many in this volume consequences of introductions are sometimes great, and devastating to the native environment, and can additionally be directly related to global climate change (Lodge, 1993). However, some of these historic activities only involved "short" distance displacements – causing extended distribution limits of species or changed distribution patterns of the species within its natural range. These changes of the biota are more or less impossible to separate from what would have been the natural "setting". Nevertheless, they represent an important part of the historic human-caused alternations of the world's native habitats.

Through time the increased volume and efficiency of local and global transportation have continued to introduce new exotic species and spread those already present (McNeely, 1998). This has been a continuing process. It might be seen locally as an increase in biodiversity but at a larger scale it is a process of homogenising of the world's biota.

As the human population continues to grow it is quite clear that we will be required to manage, to an ever greater extent, the landscapes of the Earth in order to sustain them so that they will continue to provide us with the goods and services upon which we depend.

This is a daunting task considering that we do not, even now, have the tools and information that is required to accomplish this goal. Our knowledge of how to manage ecosystems is, in general, fairly primitive as is our capacity to reconstruct, or rehabilitate, damaged ones.

An increasingly challenging problem is the management and control of invasive species, including weeds, pests, and diseases, not only in agricultural and forestry systems, but also in nature preserves. Our tools and the understanding necessary to accomplish these objectives are generally poorly developed. The objective of the following is to indicate that the problem will, in fact, become more difficult as we witness an increasingly changing Earth.

Global change elements

Most discussion of global change has focused on climate change. In a sense this is unfortunate since there are many elements of global change that are a much greater immediate threat to biological systems (Vitousek, 1994). These include changes in atmospheric composition caused by human activities, such as increases in concentrations of ozone, methane, nitrous oxide, tropospheric ozone, and so forth. These increases are substantial. Between 1959 to 1989 the annual mean CO_2 concentration at Mauna Loa went from 316 ppm to 353 ppm, a 12% increase in 30 years, which potentially should have had a very high impact on biotic systems. Many of these gases have a direct metabolic effect on biota as well as a potential indirect effect through greenhouse warming as noted below.

Another important and increasingly greater impact is land use change, which obviously has direct effects on biota through habitat destruction and alteration of competitive relationships, but also indirect effects through alterations of fire frequency and other processes that affect alterations of trace gas emissions, and nutrient and water balance.

The other element of global change is alterations of the biotic structure of the earth including the severing of biotic connections and the intercontinental exchanges of biotic material – i.e, invasions. Here we discuss the effects of some global change elements on potential invasions, which in turn are, as stated, a global change component.

Interacting mechanisms

Invasive species may fail to become dominating due to an inappropriate climate for one or more stages in life (e.g., fertility, survival, dispersal). This means that seemingly benign invading species might turn into "aggressive" species as environmental prerequisites change in a favourable direction to species expansion. A long-term view suggests that at all spatial scales, species ranges are changing constantly and are an important structuring force of natural communities (Lodge, 1993), and long term environmental trends and episodic events will govern both the invading capacity of a species and the resistance of receiving communities. Additionally, all areas are unique in time and space, which means that knowledge from introductions in one space and time compartment can not unconditionally be applied to other points in space and time (cf. Sprugel, 1991; Hofgaard, 1997).

Global changes in climate and land use will both cause and respond to range changes of species that alter ecosystem processes like gas and water vapour fluxes (Lodge, 1993). Consequently, future changes in species distributions will result from all these interactions and feedback mechanisms. The dominating two forces, climate change and land use change will be operating simultaneously on both the volume and character of species dispersal and subsequent establishment. As Mack (1997) points out, "the challenge will be to predict how all these forces will operate together, and that it is likely that their effect will be synergistic, i.e. climate change expanding the potential ranges for some species, while an ever-increasing level of human activity enhances the opportunities for species immigration to reach these new ranges".

Up to the present the research and discussion of global change have concentrated primarily on climate change. It is for this reason that there has been little interaction between global change research efforts and those in biodiversity – and that there are entirely different Conventions dealing with each. It is with invasive species that the interrelationships between global change and biodiversity are most strongly seen. The information provided in this book should promote much stronger interactions between the biodiversity and climate change research and policy communities.

Potential impacts of atmospheric change on ecosystems

Since CO_2 is a biologically active gas we have every reason to believe that the changes in atmospheric concentrations that we have seen since the beginning of the industrial age, have already influenced biological systems and that we will see even greater impacts in the years ahead.

Increasing CO_2 not only increases photosynthetic rates but also increases the amount of carbon fixed to water lost. This general relationship has important habitat consequences since it can result in a greater savings of water in the soil as has been seen in a California grassland under elevated CO_2 (Field *et al.*, 1996). This water savings in turn has the potential to provide habitat for late-season annuals, including invasive species that are in this group such as star thistle, already a very serious pest in California.

Increasing CO_2 also has the potential for altering plant and habitat biogeochemistry thereby changing plant/pest relationships as well as decomposition. These changes could, in turn, alter competitive balances affecting the success of invasive species.

Thus increasing CO_2 directly effect ecosystems through a number of mechanisms that in turn modify habitat conditions in ways that may be favorable for invasive species. Further these short and long term responses can feed back to atmospheric processes through both the hydrological and biogeochemical cycles to influence potential climate change, forcing again ecosystem adjustments.

A very direct effect of CO_2 on plants is to increase photosynthesis. The effect differs depending on whether plants have the C3 or C4 photosynthetic pathways. C4 plants saturate their photosynthetic capacity at a relatively low CO_2 concentration. In contrast C3 plants benefit from increasing CO_2 concentrations up to fairly high levels. The predictions from these basic responses is that C3 plants will prevail in the high CO_2 world to come. However, in natural systems, considering all components that limit growth and reproduction, these simple predictions might not be realized.

What does this have to do with invasives? Most of all of the world's worst weeds are C4 species, and the principal crops are C3 species (Patterson, 1995) – thus according to the above, the weeds would begin to lose out in a competitive battle. However, C4 plants are more water and nutrient use efficient, as well as temperature tolerant. Thus it is not so clear whether they will be the losers or the winners in the future world which will not only be higher in CO_2 but also be warmer.

Even among a given photosynthetic type, such as C3 plants, there is variation in response to CO_2. In a study of a number of grass species of the Great Basin it was found that the aggressive weed, *Bromus tectorum*, benefited the most from enhanced CO_2 (Smith *et al.*, 1987). This weedy species is known to promote fire once established, which results in habitat degradation which, in turn, favors *Bromus*. Thus CO_2 increase could enhance this positive feedback process of increasing an invasive species.

Of course, CO_2 is not the only gas that is increasing in the atmosphere. We are also seeing increases in tropospheric ozone, which is detrimental to the growth of most plants. Increasing CO_2 may decrease the sensitivity of plants to ozone since CO_2 decreases stomatal conductance which is why water use efficiency increases, thereby decreasing the potential uptake of ozone. As with all factors of the environment, species will differ in the response to these changes, perhaps favoring certain invasives.

We are also seeing changes in the deposition of nitrates from the atmosphere due to fossil fuel combustion. Large changes in the vegetation have been noted due to this phenomenon in western Europe (Berendse *et al.*, 1993). In California S. Weiss (pers. comm.) has noted, that in areas of high N deposition the native herbs are being overtopped by taller invasive grasses.

We obviously have a lot of experimental work ahead in order to gain the information necessary to make predictions of potential responses of organisms of the changing atmosphere as well as to a changing climate. As John Lawton (1995) has said it will "be an exceptionally long, complicated and difficult journey".

Climate change

There is widespread awareness that we will be experiencing substantial changes (increase in most areas) in temperatures in the future due to the accumulation of greenhouse gases (Houghton *et al.*, 1990; Watson *et al.*, 1995) as well as a change in precipitation patterns. The predictions available are not very locally specific nor is there good agreement on expected precipitation changes. Nevertheless we can begin to make some predictions of the influence of climate change on biota and on invasive species.

We can get some indication of the influence of climate change on the distribution of invasives from examing population responses to normal climate incursions. In the California annual grasslands, which are developed on serpentine soils, the resident plants are principally native endemic species. It has been observed by R. Hobbs (pers. comm.) that in years of high annual precipitation this grassland is invaded by the invasive annual, *Bromus mollis*. In dry years it retreats to neighboring more fertile soils. It can be predicted then that as precipitation increases with global warming this invasive species may overcome the native vegetation.

The development of insects is temperature dependent. It has been predicted by many that increasing temperatures will enhance the winter survival of insects and increase the numbers of generations with time, hence changing ranges. These enhancements in the sizes of populations could lead to an increase in pesticide uses (Smith and Tirpak, 1989).

For many pest organisms we have good models of the environmental controls of the growth of organisms and populations. Many of these are strongly dependent on temperature, using degree days, for example. Williams and Liebhold (1995) have made quantitative predictions of the changing distributions of spruce budworm in Oregon with differing degrees of temperature and precipitation changes. They indicate a very substantial increase in distribution of this pest under certain climate scenarios for a doubling of atmospheric CO_2 concentrations. We have the capacity, through the use of growth models, to also predict the change in distribution of pest host species, such as crop species (Rosenzweig and Parry, 1994).

The variability of climate is expected to change in addition to the general increase in mean temperatures (Watson *et al.*, 1995). Ecosystem resilience to future swings in climate will be buffered in complex systems (Walker and Steffen, 1997). However the increasing simplification and homogenization of ecosystems due to the spread of invasive species may decrease ecosystem resilience to climatic perturbation. At the same time climatic variability will favor the invasion process by creating ecosystem stress and hence avenues for opportunistic species.

In addition to recent climate change and variability, history, chance and determinism together interact to shape ever-changing communities and community invasibility. Accordingly some guidelines about the potential of introduced species to spread over the landscape, and time lags involved, can be obtained from paleoecological studies. Such data on e.g. – past range shifts in response to changing climate (cf. Davis and Sugita, 1997); – presence of an invading species several millennia before climate conditions favoured its development into a landscape-dominating species (Kullman, 1995); and – disappearance of native species due to interacting mechanisms between climate change and human interference (cf. Zimov *et al.*, 1995; Kullman, 1998), provides information on potential future responses to present changes of the environment. Consequently, there is a need to take into account and discuss the potential invasibility both in a long-term perspective and with a shorter perspective.

Land use change

Many contributions in this book address the profound influence of land use change on invasive species. D'Antonio and Vitousek (1992) indicate the kinds of major shifts that can occur after land clearing which favor the invasions of grasses, that in turn are fire prone, may lead to a permanent loss of forests. Land use change and associated habitat loss has been cited as the prime destroyer of biodiversity. It is also the case that land use change and habitat modification is the principal cause of the spread of invasive species which in turn are the second most important threat to biodiversity. The spread of native species will become more restricted as they move through an increasingly impassable landscape due to human activities. Many species will get caught in and suffer from "the double bind of climate change and habitat degradation" (Pitelka *et al.,* 1997). These conditions will promote the success of invasive species.

Can we protect habitats from global change and invaders?

Obviously we cannot protect organisms from climate change or atmospheric change. Can we even protect them from human interference? Probably not. For one, we are seeing increasing invasions of species even into protected areas due to pervasive impacts of humans. For example in parks there is a linear relationship between the numbers of invaders and the numbers of park visitors (Macdonald *et al.*, 1988)

Even in protected areas that are not heavily impacted by visitors there may be large additions of exotics. A research habitat that was fenced and was free from major human-based disturbances for nearly a hundred years accumulated over 52 exotic species, a number of these becoming well established in natural communities (Burgess *et al.*, 1991). The cause of these invasions is presumably in large part due to the proximity of large seed sources from the surrounding increasingly developed areas.

Ecologists can make some powerful and wide-ranging predictions about invasions under changing environmental conditions, e.g. by the use of models explaining extent of invasions (cf. Rejmanek, 1998). Nevertheless, such predictions will always suffer uncertainty, and it will be impossible to accurately predict the results of a single invasion or introduction event (Lodge, 1993). Additionally, in the light of the discussion above it might be hazardous to encourage the use of alien species that up to present have shown no sign of invasive spread for controlling the further spread of some, at present, highly invasive alien species (cf. Crooks and Soulé, 1998; Oduor, 1998).

References

Berendse, F., Aerts, R. and Bobbink, R. (1993) Atmospheric nitrogen deposition and its impact on terrestrial ecosystems, in *Landscape Ecology of A Stressed Environment*, (eds C.C. Vos and P. Opdam), Chapman and Hall, London.

Burgess, T.L., Bowers, J.E. and Turner, R.M. (1991) Exotic plants at the Desert Laboratory, Tucson, Arizona. *Madroño*, **38**, 96–114.

Clout, M.N. (1998) Biodiversity conservation and the management of invasive animals in New Zealand, in *Invasive Species and Biodiversity Management*, (eds O.T. Sandlund, P.J. Schei and Å. Viken), Kluwer Academic Publishers, Dordrecht, The Netherlands.

Crooks, J. and Soulé, M.E. (1998) Lag times in population explosions of invasive species: causes and implications, in *Invasive Species and Biodiversity Management*, (eds O.T. Sandlund, P.J. Schei and Å. Viken),. Kluwer Academic Publishers, Dordrecht, The Netherlands.

D'Antonio, C.M. and Vitousek, P.M. (1992) Biological invasions of exotic grasses, the grass/fire cycle and global change. *Annual Review of Ecology and Systematics,* **23**, 63–87.

Davis, M.B. and Sugita, S. (1997) Reinterpreting the fossil pollen record of Holocene tree migration, in *Past and Ffuture Rapid Environmental Changes: the Spatial and Evolutinary Responses of Terrestrial Biota,* (eds B. Huntley, W. Cramer, A.V. Morgan, H.C. Prentice and J.R.M. Allen), NATO ASI Series, Vol. I 47, Springer Verlag, pp. 181–193.

Elton, C. (1958) *The Ecology of Invasions by Plants and Animals.* Methuen, London.

Field, C.B., Chapin, III, F.D. Chiariello, N.R., Holland, E.A. and Mooney, H.A. (1996) The Jasper Ridge CO_2 experiment: design and motivation, in *Carbon Dioxide and Terrestrial Ecosystems,* (eds G.W. Koch and H.A. Mooney), Academic Press, San Diego, pp. 121–145.

Grove, J.M. (1988) *The Little Ice Age.* Methuen, London.

Hobbs, R.J. and Mooney, H.A. (1995) Spatial and temporal variability in California annual grassland: results from a long-term study. *Journal of Vegetation Science,* **6**, 43–57.

Hofgaard, A. (1997) Structural changes in the forest-tundra ecotone: A dynamic process, in *Past and Future Rapid Evironmental Changes: The Spatial and Evolutionary Responses of Terrestrial Biota,* (eds B. Huntley, W. Cramer, A.V. Morgan, H.C. Prentice, and J.R.M. Allen), NATO ASI Series, Vol. I 47, Springer Verlag, pp. 255–263.

Houghton, J.T., Jenkins, G.J. and Ephraums, J.J., eds (1990) *Climate Change: The IPCC Scientific Assessment.* Cambridge Press, Cambridge.

Kullman, L. (1995) New and firm evidence for Mid-Holocene appearance of *Picea abies* in the Scandes Mountains, Sweden. *Journal of Ecology,* **83**, 439–447.

Kullman, L. (1998) Paleoecological and biogeographic implications of early Holocene immigration of *Larix sibirica* Ledeb. into the Scandes, Sweden. *Global Ecology & Biogeography Letters*, **7**, in press.

Lawton, J.H. (1995) The response of insects to environmental change, in *Insects in a Changing Environment*, (eds R. Harrington and N.E. Stork), Academic Press, London, pp.3–26.

Lodge, D.M. (1993) Biological invasions: Lessons for ecology. *Trends in Ecology and Evolution,* **8**, 133–137.

Macdonald, I.A.W., Loope, L.L. Usher, M.B. and Hamann, O. (1988) Wildlife conservation and the invasion of nature reserves by introduced species: A global perspective. in *Biological Invasions. A Global Perspective*, (eds J.A. Drake, H.A. Mooney, F. di Castri, R.H. Groves, F.J. Kruger, M. Rejmanek and M. Williamson), John Wiley, Chichester, UK, pp. 215–255.

Mack, R.N. (1997) Plant invasions: Early and continuing expressions of global change, in *Past and Future Rapid Environmental Changes: The Spatial and Evolutionary Responses of Terrestrial Biota,* (eds B. Huntley, W. Cramer, A.V. Morgan, H.C. Prentice, and J.R.M. Allen), NATO ASI Series, Vol. I 47, Springer Verlag, pp 205–216.

Oduor, G. (1998) Biological pest control and invasives, in *Invasive Species and Biodiversity Management*, (eds O.T. Sandlund, P.J. Schei and Å. Viken), Kluwer Academic Publishers, Dordrecht, The Netherlands.

Patterson, D.T. (1995) Weeds in a changing climate. *Weed Science,* **43**, 685–701.

Pech, R.P. (1998) Managing alien species: the Australian experience, in *Invasive Species and Biodiversity Management*, (eds O.T. Sandlund, P.J. Schei and Å. Viken), Kluwer Academic Publishers, Dordrecht, The Netherlands.

Pitelka, L.F. and the Plant Migration Workshop Group (1997) Plant migration and climate change. *American Scientist,* **85**, 464–473.

Rejmánek, M. (1998) Invasive plant species and invasible ecosystems, in *Invasive Species and Biodiversity Management*, (eds O.T. Sandlund, P.J. Schei and Å. Viken), Kluwer Academic Publishers, Dordrecht, The Netherlands.

Rosenzweig, C. and Parry, M.L. (1994) Potential impact of climate change on world food supply. *Nature,* **367**, 133–138.

Smith, J.B. and Tirpak, D., eds (1989) *The Potential Effects of Global Climate Change on the United States.* U. S. Environmental Protection Agency, Washington, DC.

Smith, S.D., Strain, B.R. and Sharkey, T.D. (1987) Effects of CO_2 enrichment on four Great Basin grasses. *Functional Ecology,* **1**, 139–143.

Sprugel, D.G. (1991) Disturbance, equilibrium, and environmental variability: What is "natural" vegetation in a changing environment? *Biological Conservation,* **58**, 1–18.

Vitousek, P.M. (1994) Beyond global warming: ecology and global change. *Ecology,* **75**, 1861–1876.

Walker, B. and Steffen, W. (1997) An overview of the implications of global change for natural and managed terrestrial ecosystems. *Conservation Ecology [online],* 1(2), 2. (available through http://www.consecol.org/vol1/iss2/art2).

Watson, R.T., Zinyowera, M.C. and Moss, R.H. (1995) *Climate Change 1995: Impacts, Adaptations, and Mitigation of Climate Change.* The IPCC Second Assessment Report. Cambridge University Press, New York.

Williams, D.W. and Liebhold, A.M. (1995) Potential changes in spatial distribution of outbreaks of forest defoliators under climate change, in *Insects in a Changing Environment*, (eds R. Harrington and N.E. Stork), Academic Press, London, pp. 509–513.

Zimov, S.A., Chuprynin, A.P., Oreshko, F.S., Chapin III, F.S., Chapin, M.C. and Reynolds, J.F. (1995) Effects of mammals on ecosystem change at the Pleistocene–Holocene boundary, in *Arctic and Alpine Biodiversity. Patterns, Causes and Ecosystem Consequences,* (eds F.S. Chapin III and C. Körner), Ecological Studies 113, Springer Verlag, pp. 127–135.

10 Introductions at the level of genes and populations

KJETIL HINDAR
Norwegian Institute for Nature Research (NINA), Trondheim, Norway

Abstract

The diversity provided by genetically different and locally adapted populations, both in the wild and in domesticated species, is crucial for their continued productivity and evolution. This diversity is threatened by introductions of non-native populations. Loss of subspecific biodiversity caused by introductions includes loss of adapted genes or gene complexes through interbreeding, loss of entire populations through competitive displacement or disease introduction, and homogenization of the genetic structure of a species. Such losses have followed introductions even when the goal has been to maintain biodiversity. Here I review some of the genetic lessons from introductions within the fields of wildlife and fisheries management, and agriculture. They show that interbreeding between differently adapted populations may have detrimental, and immediate, consequences. Moreover, they suggest that it is extremely difficult to predict the outcome of a particular introduction. I recommend that future introductions at the population level be carefully assessed along the same procedures as have been proposed for genetically modified organisms, and that genetic analyses, such as estimates of the genetic differentiation among populations, can be used to provide guidelines for introductions.

Introduction

In its definition of "biodiversity", the Convention on Biological Diversity lists diversity within species as well as that between species and of ecosystems. In Article 8h on prevention of harmful introductions, the Convention stresses explicitly that ecosystems, habitats and species must be protected from introductions, but ignores the fact that introductions may also be harmful to the diversity within species. Here I present examples of within-species diversity (i.e. genetic diversity) being threatened by introductions, even when the introductions were aimed at maintaining biodiversity. I discuss the genetic mechanisms involved and show how genetic techniques can aid in formulating guidelines for assessing the genetic impact of introductions.

O. T. Sandlund et al. (eds.), Invasive Species and Biodiversity Management, 149–161.

Concerns about loss of genetic diversity

Thirty years ago it became evident that erosion of animal and plant genetic resources, the material basis for development of breeds and varieties, was occurring at an alarmingly high rate. One third of the more than 3000 livestock breeds worldwide are now considered to be in danger of extinction (Hall and Bradley, 1995). In the developed world, extreme concentration of use occurs on a single or very few, successful breeds. In the developing world, breeds are not always as clearly defined and recognized, but may nevertheless exist as local populations which are adapted to existing production systems (Barker, 1994). Each breed is likely to represent a unique combination of genes because of different evolutionary histories, perhaps over many centuries. Many breeds in the developing world are threatened primarily by crossbreeding with introduced breeds from the developed world. For example, the native, small-bodied Somba cattle in West Africa are notable for their tolerance to trypanosomiasis. It is however threatened by genetic introgression with the more recently arrived, large-bodied humped breeds from further north, which are not trypanotolerant. Various genetic techniques are now employed in order to protect the valuable genetic resource represented by inherited trypanotolerance (Hall and Bradley, 1995).

Plant geneticists have expressed similar concerns following the disappearance of primitive or traditional crop varieties and their replacement by modern, genetically uniform, cultivars (Soulé and Mills, 1992). Traditional crop varieties represent the products of thousands of years of selection to local variation in habitat quality, microclimate, diseases and pests. Discarding this diversity within a single human generation is quite short-sighted. Even when modern varieties are favoured for market reasons, the abandonment of traditional varieties means losing genetic variation whose value may be crucial in the future. Today, there is considerable international activity for the conservation of breeds of livestock and varieties of crops by organisations within the Food and Agriculture Organisation.

For wild species, Ehrlich (1988) has noted that the loss of genetically distinct populations within a species is at least as important a problem as the loss of entire species. A species goes extinct when the last population goes extinct; in most species, many populations have gone extinct before this point is reached. Whereas habitat destruction may be the most important driving force for population extinction in most species, introductions of species and populations may be equally important for others. Many commercially important species are bred for release into and harvest from environments inhabited by wild relatives. These include several game species and, at considerably larger scales, forest trees and fishes where releasing cultivated individuals is a common practice in resource management. While these releases may not threaten biodiversity at the species level, between-population genetic diversity may be lost as cultivated or captive-bred populations displace or interbreed with wild ones. Even when captive breeding is used in attempts to protect threatened species from extinction, ignorance of genetic differences between captive and remaining natural populations may prove detrimental to the success of the reintroduction. Attention to population differences, and to genetic adaptation to captivity, is now being recognized as an important part of reintroduction programmes (May, 1991; Frankham, 1995).

Genetic threats from introductions

Genetic variation within species is the foundation of evolution and hence the foundation of the diversity of species. Genetic variation is distributed among individuals within populations and between populations. In widespread species, substantial genetic differences can evolve among local populations even if there is some dispersal of individuals among them (Slatkin, 1987). The genetic differences between populations accumulate partly as a response to different selection pressures and partly because of random events. Forced interbreeding between such semi-isolated and genetically different populations may have a number of undesirable consequences such as (Ryman *et al.*, 1995):

- loss of adapted genes or gene complexes through interbreeding,
- loss of entire populations as a result of competitive displacement or eradication through disease introduction,
- homogenization of a previously structured population through swamping a region with a common gene pool, and
- no readaptation to local conditions if the introductions continue.

When interbreeding of genetically different populations results in a loss of fitness, it is often referred to as "outbreeding depression" irrespective of the mechanism which caused this fitness loss. The mechanisms fall into two different categories: (1) local adaptation, where the hybrid population lacks adaptations to its environment, and (2) coadaptation, where the hybrid population contains combinations of genes not adapted to each other (Templeton, 1986). Outbreeding depression may occur in the first hybrid generation, or among their offspring. The degree of fitness loss seems to depend on how wide a cross is; crosses between species usually produce severe outbreeding depression, whereas lesser effects are found between different sub-species, and more minor but still significant effects can be detected between crosses of genetically different populations within a species. Quantitative data are still lacking on the frequency and severity of outbreeding depression, and on the situations where it is found (Frankham, 1995).

One might argue that interbreeding itself does not lead to loss of alleles, and that it therefore cannot lead to loss of genetic diversity. However, the original genotypes of the two parent populations will occur with very low probabilities, and may not be found in a single individual of the hybrid population. This loss of genotypic variation may be as irreversible as the loss of alleles from a population (Barker, 1994).

The threats to intraspecific diversity caused by introductions extend to interspecific diversity as well. At the species level, introductions may lead to hybridization between species which otherwise do not meet. This interspecific hybridization may lead to loss of entire species, especially when rare species are swamped by introduced, closely related ones (Ebenhard, 1988; Levin *et al.*, 1996).

It was pointed out already by Haldane (1964), one of the founders of theoretical population genetics, that theory alone cannot provide precise predictions about the outcome of mixing genetically different populations. A similar position has also been taken when considering releases of genetically modified organisms, where the internationally agreed curriculum is that careful experimentation in confined situations must precede releases to the wild (Tiedje *et al.*, 1989). This does not suggest that we should eschew theoretical considerations. Rather, we must acknowledge that doing basic biological research with relevance to introductions is an important task which is different from trying to predict the outcome of a particular introduction. At the same time, we must not forget that the knowledge needed to make precise predictions about a particular release is enormous, and probably not reached for a single species (Ryman *et al.*, 1995). For some species, releases of exogenous or captive populations have been carried out for such a long time, that a considerable body of empirical evidence now exists to aid in predicting the effects of future releases. Some lessons from this experience are presented below.

Case studies

The Tatra mountain ibex

When the Tatra mountain ibex (*Capra ibex ibex*) in Slovakia became extinct through overhunting, ibex were successfully introduced from nearby Austria. Later additions to the Tatra herd of bezoars (*C. ibex aegagrus*) from

Turkey and of Nubian ibex (*C. ibex nubiana*) from Sinai, resulted in hybrids which gave birth to young during the middle of winter when no young could survive, and the population went extinct (Templeton, 1986). Presumably, this outbreeding depression was caused by different climatic adaptations in the donor and recipient populations.

Salmonid fishes

Salmonid fishes provide an excellent model system for evaluating the genetic consequences of releasing exogenous populations (including cultured fish) into natural environments. They are well known genetically and ecologically, they are characterized by numerous local populations which are genetically different, and they have been artificially reared and released for more than a century. A review of the genetic effects on native populations, following releases of non-native salmonid populations (Hindar *et al.*, 1991), provided two broad conclusions:

- the genetic effects of (intentionally or accidentally) released salmonids on natural populations are typically unpredictable; they vary from no detectable effect to complete introgression or displacement.
- where genetic effects on performance traits have been detected, they appear always to be negative in comparison with the unaffected native populations. For example, reduced total population size has been observed following introductions of exogenous populations, and also reduced performance in a number of traits which can explain such population declines (e.g. lower survival in fresh and sea water).

The typically negative effects of releases of non-native salmonids are not unexpected. Salmonid populations are believed to be adapted to their local environments (Taylor, 1991), and introduced populations or crosses involving introduced populations should be expected to perform worse than the native ones.

Loss of salmonid populations has resulted from disease introductions. Transfer of salmonid fishes between drainages has led to pathogens and parasites coming in contact with fish populations to which they were not adapted. These new host-pathogen confrontations have led to unnaturally high fish mortality. In Norway, Atlantic salmon (*Salmo salar*) populations in 35 rivers have been severely reduced since 1975 following introductions of fish from hatcheries infected by the monogenean parasite *Gyrodactylus salaris*. This parasite does not appear to be native to Norway, and was probably imported with salmonid eggs or juveniles from infected localities in the Baltic (Johnsen and Jensen, 1991).

The recent growth of Atlantic salmon aquaculture has been accompanied by large numbers of escaped fish which eventually enter rivers to spawn in competition with local, wild salmon. On average, 20-30% of Atlantic salmon spawning in Norwegian rivers are escapes from fish farms, and in some rivers, they outnumber the native fish (Sægrov *et al.*, 1997). Spawning experiments show that farmed fish are able to reproduce in nature. Their spawning success is 20 to 50% of wild fish, dependent on whether they escape as adults or smolts (Fleming *et al.*, 1996; 1997). Most farmed salmon in Norway originate from a breeding program which was initiated in the early 1970s, and which consists of four selected strains based on parental fish collected from 40 wild populations (Gjedrem *et al.*, 1991). The farmed salmon show lower genetic variability than wild fish in allozymes and DNA fingerprints, and significant genetic divergence from wild fish including one of its principal founding populations (Mjølnerød *et al.*, 1997). Genetic divergence of farmed from wild fish has also been demonstrated in fitness-related traits such as aggression, behavioural response to predators, and growth rate (Einum and Fleming, 1997; Fleming and Einum, 1997). Experiments in Ireland and Norway show that offspring of escaped farmed salmon, including crosses with wild fish, can complete the life cycle in the wild. Continued escape of cultured salmon will therefore lead to an increase in culture-adapted traits in wild populations and an associated loss in population fitness in the wild.

One experiment with temporally isolated populations of pink salmon (*Oncorhynchus gorbuscha*) from the same stream provides an instructive case of outbreeding depression. This species has a rigid 2-year life cycle and breeds only once, followed by death. Thus, two populations exist in the same river, one spawning in odd-numbered years and the other in even-numbered years. Gharrett and Smoker (1991) produced hybrids between two such temporally isolated populations by fertilizing eggs of one population with cryopreserved sperm from the other. They observed good return rates among first-generation hybrids, and increased variances in body size suggesting that hybrids were more genetically variable than pure-bred groups. Second-generation hybrids, however, produced extremely low return rates, suggesting that outbreeding depression had occurred. This outbreeding depression, however, was not likely caused by the two populations being adapted to different environments; in the long run, the environments in odd- and even-numbered years in the same river should be alike. Rather, the outbreeding depression seems to have been caused by disrupting coadapted gene complexes evolved independently in the two populations.

Discussion

Introductions may result in a reduction of biodiversity, even at sub-specific levels. Some important lessons from man-mediated introductions of populations are firstly, that interbreeding between differently adapted populations may have detrimental, and immediate, consequences, and secondly, that unwanted genetic effects can occur in the absence of interbreeding, for example through the introduction of a parasite which is lethal to individuals of the recipient population. Finally, as is true for introductions of species, it is extremely difficult to predict the outcome of a particular introduction.

Different motivations for introductions

Introductions of exogenous populations are used for a variety of reasons. Possible motivations include (1) supporting a small natural population in danger of extinction, (2) supporting an intensive harvest regime by releasing captive-bred individuals, and (3) increasing the productivity of livestock and cultured plants. In addition, modern gene technology allows the transfer of single genes into the genome of a host organism, thus producing exogenous populations which may differ from the recipient, natural populations by a single gene. Finally, the introduction may not at all be deliberate, but may result from accidental release of captive-bred populations and of genetically modified organisms.

With such variable motivations for introductions of populations, the justified genetic concerns will vary as well (Ryman, 1991; Ryman *et al.*, 1995). In the case of supporting an endangered population, the chief motivation may be to increase the total population size even at the expense of the genetic integrity of the endangered population. This increase in population size may be necessary in order to prevent extinction from demographic reasons, and incidentally, the endangered population may also gain from genetic input if inbreeding depression has accompanied the population decline. On the other hand, setting up large-scale release programs for captive-bred fishes or cultivated trees to be harvested from natural environments, increases the risk of diluting the genetic variation among natural populations which may be essential for future productivity of these species.

Whatever motivation exists for a particular introduction, it should be designed as an experiment and followed by a monitoring program, so that scientific knowledge is accumulated alongside the various management actions. Also, the introductions must be carried out such that the genetic and ecological consequences for native populations are minimized. Two procedures for minimizing these consequences are presented below; one motivated by the planned release of genetically modified organisms (GMOs), the other based on population genetics theory.

Risk assessment of transgenic organisms

In the past decade, we have witnessed an increasing number of releases of transgenic varieties of crop plants. In the future we may see that even forestry and aquaculture turns to using transgenic varieties (MacKenzie, 1996; Moffat, 1996), i.e. organisms which are meant for use in close connection with natural populations of the same species. This raises a number of environmental concerns that need careful consideration before any release can be contemplated. Tiedje *et al.* (1989) provide a checklist of questions relating to the invasion potential of the GMO, the ecological role it plays, the potential for gene flow to wild relatives, the vulnerability of the release environment, and the possibilities for monitoring the effects and controlling unwanted ones. This checklist is instructive, because it can aid in assessing the impact of any introduction, be it an exotic species or a non-native population.

The checklist can be illustrated by reference to the possible environmental effects of transgenic cultured salmon, should they escape to the wild (Figure 10.1 and 10.2). Two types of genetic modifications are considered: (1) insertion of genes for growth hormone (increased growth rate), and (2) insertion of anti-freeze protein genes for increased cold tolerance (cf. MacKenzie, 1996). In the former case, the inserted genes may come from the species being the target for genetic modification (salmon genes inserted in salmon, but with a "read-me" gene [promotor] attached to it), in the latter, the anti-freeze gene is from a quite different species (e.g. polar marine fish living in salt water at below-zero temperatures).

According to Tiedje *et al.*'s (1989) checklist for classifying target species and release environments, Atlantic salmon would be classified as belonging to the high-risk end of the spectrum (Fig. 10.1). First, transgenic Atlantic salmon have conspecific individuals in the wild, so that the transgene would easily spread to a number of populations, should the transgenic fish escape. Second, the release environment, being the North Atlantic Ocean, is uncontrollable with respect to call back unwanted effects. Making a high-risk organism more environment-tolerant (cold tolerance) or more fast-growing (growth hormone) by gene technology can make things worse (Figure 10.2). The most hazardous genetic modification seems to be increased cold tolerance, because it would invite a northward expansion of salmon aquaculture and could also lead to a range expansion of the species in the wild. My conclusion from this checklist example (see also Hindar, 1995), is that genetic modification is interesting only if it can make fish farming less hostile to the environment, for example, if it makes the fish sterile or if it makes the fish unable to live outside the fish farm. Another example involving transgenic Norway spruce (*Picea abies*) is discussed by Tømmerås *et al.* (1996).

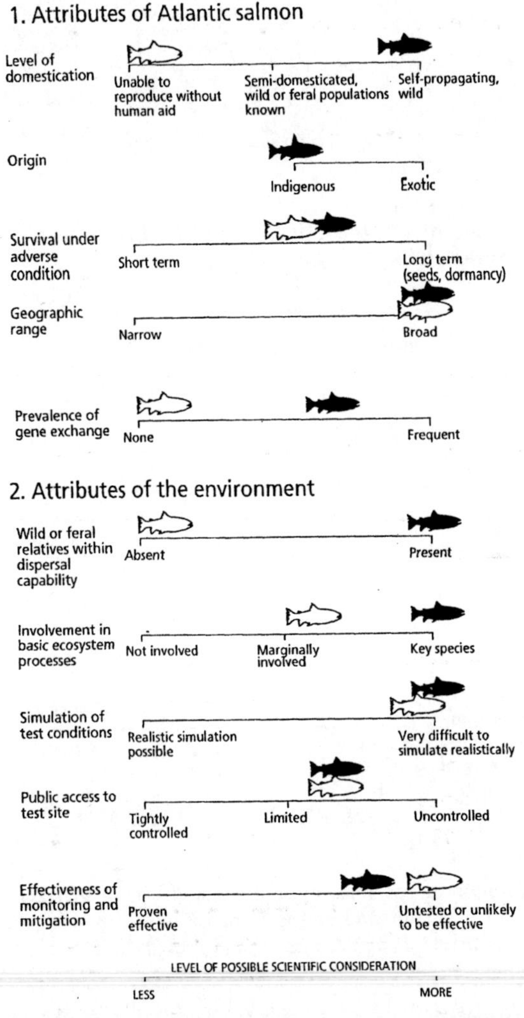

Figure 10.1 Consideration of possible environmental effects of transgenic salmon, based on attributes of the target species, and of the release environment. Filled silhouettes designate fertile, transgenic salmon; open silhouettes designate sterile, transgenic salmon. The positions of the silhouettes are my own suggestions.

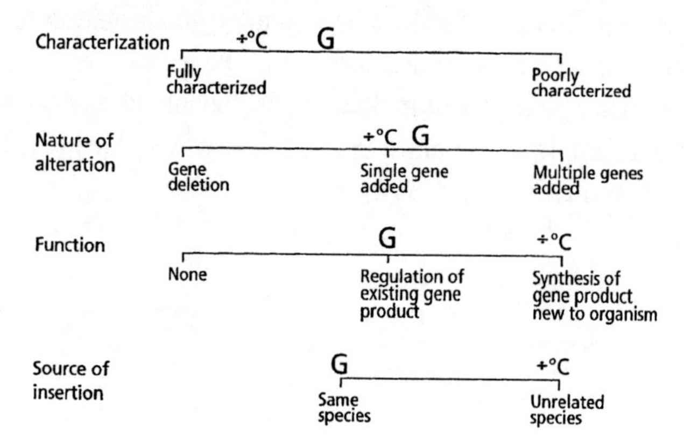

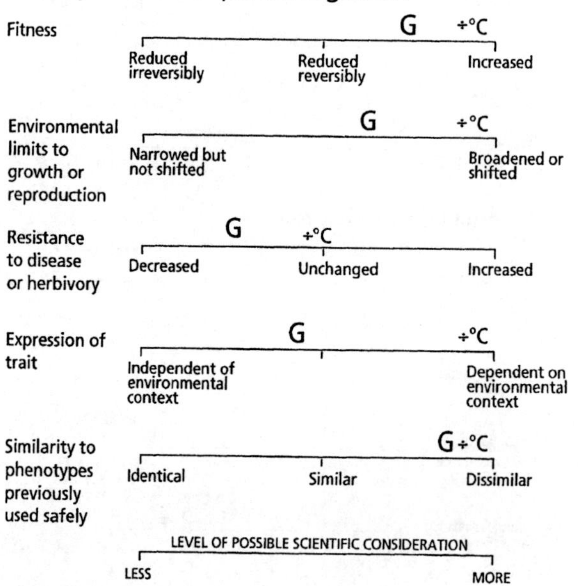

Figure 10.2 Consideration of possible environmental effects of transgenic salmon, based on attributes of the genetic alteration and on the comparison of the transgenic organism with its non transgenic counterpart. G designates additional growth hormone genes with allospecific promotors; –C designates genes for increased cold tolerance.

Some have lamented that GMOs have received more attention than they deserve (e.g. relative to exotic species). My own view is that we should take advantage of the current focus on risk assessment of GMOs and suggest similar procedures for risk assessment of introductions of non-modified organisms.

Permissible level of introgression

How much genetic input can be allowed without the recipient population losing its genetic integrity? To answer this question, Ryman (1991; Ryman *et al.*, 1995) suggested to use the amount of genetic differentiation (F_{ST}) between populations as a guideline for permissible levels of man-mediated gene flow between them. Moreover, he suggested that Wright's (1969) island model of genetic differentiation be used where F_{ST} relates to the effective number of migrants per generation (*Nm*) by the equation $F_{ST} = 1/(4Nm + 1)$. F_{ST} is the standardized variance of allele frequencies among populations and is readily estimated from various biochemical genetic data sets (e.g. Brown and Schoen, 1992; Ward *et al.*, 1992). For example, data show that an F_{ST} of 0.10 is commonly observed between conspecific populations. The corresponding level of permissible gene flow is only about 2 (genetically effective) individuals per generation. Even with an F_{ST} of 0.01, which is smaller than that observed in most species, the recommended input is not more than 25 immigrants per generation. Thus, it is clear that if our goal is to preserve naturally occurring genetic diversity, very limited man-mediated introgression can be accepted.

There are problems associated with such a simplified guideline, discussed by Ryman *et al.* (1995). It may nevertheless serve as a starting-point for more detailed analyses, for example along those procedures suggested by Tiedje *et al.* (1989) for transgenic organisms.

Recommendations

This essay provides a basis for the following recommendations regarding the potential release of exogenous populations:

1. Strive to maintain genetic variation *between* as well as *within* populations.
2. Accept the shortcomings of making *a priori* predictions of releasing exogenous populations.
3. Base introductions on careful genetic (and other biological) analyses, including estimates of the genetic differentiation among populations of the species in question.

References

Barker, J.S.F. (1994) Animal breeding and conservation genetics, in *Conservation Genetics,* (eds V. Loeschcke, J. Tomiuk and S.K. Jain), Birkhäuser Verlag, Basel, pp. 381–395.

Brown, A.H.D. and Schoen, D.J. (1992) Plant population genetic structure and biological conservation, in *Conservation of Biodiversity for Sustainable Development,* (eds O.T. Sandlund, K. Hindar and A.H.D. Brown), Scandinavian University Press, Oslo, pp. 88–104.

Ebenhard, T. (1988) Introduced birds and mammals and their ecological effects. *Swedish Wildlife Review,* **13,** 5–107.

Ehrlich, P. (1988) The loss of diversity: causes and consequences, in *Biodiversity,* (ed. E.O. Wilson), National Academy Press, Washington DC, pp. 21–27.

Einum, S. and Fleming, I.A. (1997) Genetic divergence and interaction in the wild among native, farmed and hybrid Atlantic salmon. *Journal of Fish Biology,* **50,** 634–651.

Fleming, I.A. and Einum, S. (1997) Experimental tests of genetic divergence of farmed from wild Atlantic salmon due to domestication. *ICES Journal of Marine Science,* **54,** 1051–1063.

Fleming, I.A., Jonsson, B., Gross, M.R. and Lamberg, A. (1996) An experimental study of the reproductive behaviour and success of farmed and wild Atlantic salmon (*Salmo salar*). *Journal of Applied Ecology,* **33,** 893–905.

Fleming, I.A., Lamberg, A. and Jonsson, B. (1997) Effects of early experience on the reproductive performance of Atlantic salmon. *Behavioural Ecology,* **8,** 470–480.

Frankham, R. (1995) Conservation genetics. *Annual Review of Genetics,* **29,** 305–327.

Gharrett, A.J. and Smoker, W.W. (1991) Two generations of hybrids between even- and odd-year pink salmon (*Oncorhynchus gorbuscha*): a test for outbreeding depression? *Canadian Journal of Fisheries and Aquatic Sciences,* **48,** 1744–1749.

Gjedrem, T., Gjøen, H.M. and Gjerde, B. (1991) Genetic origin of Norwegian farmed Atlantic salmon. *Aquaculture,* **98,** 41–50.

Haldane, J.B.S. (1964) A defense of beanbag genetics. *Perspectives in Biology and Medicine,* **7,** 343–359.

Hall, S.J.G. and Bradley, D.G. (1995) Conserving livestock breed biodiversity. *Trends in Ecology and Evolution,* **7,** 267–270.

Hindar, K. (1995) Ecological and genetic effects of transgenic fish, in *Pan-European Conference on the Potential Long-Term Ecological Impact of Genetically Modified Organisms,* (Council of Europe), Council of Europe Press, Strasbourg, pp. 233–244.

Hindar, K., Ryman, N. and Utter, F. (1991) Genetic effects of cultured fish on natural fish populations. *Canadian Journal of Fisheries and Aquatic Sciences,* **48,** 945–957.

Johnsen, B.O. and Jensen, A.J. (1991) The *Gyrodactylus* story in Norway. *Aquaculture,* **98,** 289–302.

Levin, D.A., Francisco-Ortega, J. and Janzen, R.K. (1996) Hybridization and the extinction of rare plant species. *Conservation Biology,* **10,** 10–16.

MacKenzie, D. (1996) Can we make supersalmon safe? *New Scientist,* **149(2014),** 14–15.

May, R.M. (1991) The role of ecological theory in planning re-introduction of endangered species. *Symposia of the Zoological Society, London,* **62,** 145–163.

Mjølnerød, I.B., Refseth, U.H., Karlsen, E., Balstad, T., Jakobsen, K.S. and Hindar, K. (1997) Genetic differences between two wild and one farmed population of Atlantic salmon (*Salmo salar*) revealed by three classes of genetic markers. *Hereditas,* **127,** 239–248.

Moffat, A.S. (1996) Moving forest trees into the modern genetics era. *Science,* **271,** 760–761.

Ryman, N. (1991) Conservation genetics considerations in fishery management. *Journal of Fish Biology,* **39 (Suppl. A),** 211–224.

Ryman, N., Utter, F. and Hindar, K. (1995) Introgression, supportive breeding, and genetic conservation, in *Population Management for Survival and Recovery: Analytical Methods and Strategies in Small Population Conservation*, (eds J.D. Ballou, M. Gilpin and T.J. Foose), Columbia University Press, New York, pp. 341–365.

Sægrov, H., Hindar, K., Kålås, S. and Lura, H. (1997) Escaped farmed Atlantic salmon replace the original salmon stocks in the River Vosso, western Norway. *ICES Journal of Marine Science,* **54,** 1166–1172.

Slatkin, M. (1987) Gene flow and the geographic structure of natural populations. *Science,* **236,** 787–792.

Soulé, M.E. and Mills, L.S. (1992) Conservation genetics and conservation biology: a troubled marriage. in *Conservation of Biodiversity for Sustainable Development*, (eds O. T. Sandlund, K. Hindar and A.H.D. Brown), Scandinavian University Press, Oslo, pp. 55–69.

Taylor, E.B. (1991) A review of local adaptation in Salmonidae, with particular reference to Pacific and Atlantic salmon. *Aquaculture,* **98,** 185–207.

Templeton, A.R. (1986) Coadaptation and outbreeding depression, in *Conservation Biology: the Science of Scarcity and Diversity,* (ed. M.E. Soulé), Sinauer Associates, Sunderland, MA, pp. 105–116.

Tiedje, J.M., Colwell, R.K., Grossman, Y.L., Hodson, R.E., Lenski, R.E., Mack, R.N. and Regal, P.J. (1989) The planned introduction of genetically engineered organisms: ecological considerations and recommendations. *Ecology,* **70,** 298–315.

Tømmerås, B.Å., Johnsen, Ø., Skrøppa, T., Hindar, K., Holten, J. and Tufto, J. (1996) Long-term environmental impacts of release of transgenic Norway spruce (*Picea abies*). *NINA•NIKU Project Report,* **3,** 1–48.

Ward, R.D., Skibinski, D.O.F. and Woodmark, M. (1992) Protein heterozygosity, protein structure, and taxonomic differentiation. *Evolutionary Biology,* **26,** 73–159.

Wright, S. (1969) *Evolution and the Genetics of Populations, vol. 2. The Theory of Gene Frequencies.* Chicago University Press, Chicago.

11 Alien species and emerging infectious diseases: past lessons and future implications

RALPH T. BRYAN
Centers for Disease Control and Prevention, National Center for Infectious Diseases, Albuquerque, New Mexico, USA

Abstract

Traditional approaches to public health rarely address human health threats from the perspective of alien or exotic species introductions and, when alien species are discussed, they are usually not identified or referenced as such. The cumulative impact of alien species introductions on human health, however, has been profound. Such introductions have played a pivotal role in fostering the emergence of "new" infectious diseases or the re-emergence or translocation of "old", or previously known, diseases. In addition to well-known historical examples such as small pox, yellow fever, and plague, modern introductions of alien species continue to impact the public's health. Ranging from the introduction of malaria-transmitting mosquitoes to the translocation of snail vectors of schistosomiasis, alien species introductions are an ongoing threat to global health. Global changes, such as major population shifts, expanding international tourism and commerce, deforestation, and water development projects, could increase the likelihood of future introductions of vectors or reservoirs of infectious diseases, or the infectious agents themselves. In order to be better prepared for such situations, early warning systems should be developed. Surveillance for conditions that may foster alien species introductions requires a multi-disciplinary approach involving professionals from many fields of biology, ecology, and climatology, in addition to medicine and public health. These professionals should take advantage of multiple opportunities for scientific exchange and cooperation to address the threats presented by alien species introductions to biodiversity and to human health.

Introduction

Traditional approaches to public health rarely address human health threats from the perspective of alien or exotic species introductions and, when alien species are discussed, they are usually not identified or referenced as such. Nevertheless, a careful review of both historical and current medical literature, particularly that emphasizing new, emerging, or re-emerging diseases, suggests that the cumulative impact of alien species introductions on human health has been profound. The focus of this paper will be on the

O. T. Sandlund et al. (eds.), Invasive Species and Biodiversity Management, 163–175.

role of alien species introductions in fostering the emergence of "new" infectious diseases or the re-emergence or translocation of "old," or previously known, diseases. Other important public health problems related to alien species introductions, such as those arising from Africanized bees, fire ants, and toxic dinoflagellates, are not covered herein, but will be discussed in subsequent publications.

Emerging Infectious Diseases

As we approach the next millennium, it is important to note that infectious diseases remain the leading cause of death worldwide (Murray and Lopez, 1996; WHO, 1996). Even in the United States, infectious disease mortality remains a serious threat – having actually increased by nearly 60% between 1980 and 1992 – an increase that is only partially accounted for by human immunodeficiency virus (HIV). Infectious diseases are now the third leading cause of death in the U.S. (after cardiovascular diseases and malignancies) – surpassing accidents, cerebrovascular disease (e.g., strokes), and chronic obstructive pulmonary disease (e.g., emphysema) (Pinner *et al.,* 1996).

The Institute of Medicine (IOM) is an arm of the U.S. National Academy of Sciences that reports directly to the U.S. Congress. In 1992, IOM issued the report, "Emerging Infections, Microbial Threats to Health in the United States," that effectively sounded the alarm over emerging infections in North America (IOM, 1992). In this landmark publication, IOM defined emerging infectious diseases as those infections whose incidence in humans has increased in the past two decades or whose incidence threatens to increase in the near future. Included in this definition are newly recognized syndromes associated with novel infectious agents, previously well-known diseases that have resurged, and infections associated with organisms resistant to antimicrobial therapy. Emergent zoonotic, vectorborne, and waterborne or foodborne infectious diseases have been particularly problematic in recent years. IOM also developed the concept of "factors in emergence" – factors that promote the emergence or re-emergence of infectious diseases. These factors are: human demographics and behavior; technology and industry; economic development and land use; international travel and commerce; microbial adaptation and change; and breakdown of public health measures. They represent convenient conceptual categories that are by no means mutually exclusive, but serve to illustrate the intricate relationship between our rapidly changing world and emerging infectious diseases.

Historical examples of alien species introductions and emerging infections

Medical historians tell us that approximately 56 million persons died in association with European exploration of the New World. Although combat and social disruption undoubtedly contributed, by far most deaths were attributed to introduced infectious diseases. Even as recently as the 1960s and 1970s, previously uncontacted Amazonian tribes experienced mortality rates of up to 75% due to their exposure to, what to them, were new and novel infectious agents (Black, 1992). Hence, we may view various microbes (the viruses, bacteria, protozoa, helminths, and fungi) as alien species themselves – agents of disease and death that travel in our food, our animals, our commercial goods, as well as in ourselves.

The Aztecs described a pestilence that they called *hueyzahuatl*, and gave vivid accounts of the agony, disfigurement, and death brought to Mexico in the 16th century. The disease was smallpox (Crosby, 1991). Likely introduced by the Spanish army in 1520, smallpox killed over 3.5 million people, over half of the population, in only two years time (Horse Capture, 1991; Crosby, 1991; Black, 1992).

In 1648, the first ships bearing Africans for the slave trade arrived in the Caribbean. As if the indignity and injustices of the slave trade were not enough, the slave traders also unwittingly carried with them (in the ships' water casks) what was to become one of the most notorious arthropod vectors of the Western Hemisphere – the mosquito, *Aedes aegypti* (Weil, 1981). It was the slaves themselves, however, that carried the serious viral infections transmitted by *Ae. aegypti*: yellow fever and dengue fever. Despite intermittently effective mosquito control programs during the 20th century, *Ae. aegypti* continues to pose a serious threat to human health, even in this decade. Because of breakdowns in public health infrastructures for the elimination of *Ae. aegypti* breeding sites, in 1995 the geographic distribution of this mosquito is similar to its distribution prior to the successful eradication campaigns of the 1950s and 1960s (IOM, 1992; Gubler and Clark, 1995).

As an example of the continuing impact of this centuries-old alien species introduction, yellow fever re-emerged with alarming force in Peru in 1995. Affecting approximately 500 persons in 9 departments, the case fatality rate approached 40% (PAHO, 1995). Yellow fever currently occurs only in sub-Saharan Africa and South America, and varies in severity from a flu-like illness to severe hepatitis and hemorrhagic fever. Many non-human primates maintain the infection in sylvatic (jungle) transmission cycles. Humans contract the disease when bitten by mosquitoes that have been infected by other primates. Because *Ae. aegypti*, the principal vector for *urban* yellow fever, has reinfested most urban centers of tropical America, the region is at the highest risk in over 50 years for urban epidemics. Similarly, dengue fever and dengue hemorrhagic fever have resurged throughout Latin America. The number of cases of dengue and dengue hemorrhagic fever continues to rise, with the epidemic extending to new areas in the Americas. The Pan American Health Organization (PAHO) reported recently that by late 1995 there were almost 200 000 cases of dengue and more than 5 500 cases of dengue hemorrhagic fever reported from the Western Hemisphere (F. Pinheiro, pers. comm.).

Livestock also accompanied the early Europeans to the New World. Swine, brought along to satisfy European dietary preferences, often escaped and readily adapted to their new environment, where their subsequent impact on native flora and fauna is well-described (Crosby, 1991). Perhaps less well appreciated, however, is the role played by introduced swine in the New World emergence of the human brain infection known as neurocysticercosis. Unknown in the Western Hemisphere prior to the introduction of swine, this disease emerged as swine became commonplace in and around homes in rural Latin America. Its transmission and life cycle are closely linked to husbandry practices and human hygiene. Taeniasis – human intestinal infection with adult *Taenia solium* tapeworms – develops when people eat inadequately cooked pork that is infested with cysticerci, the larval stage of *T. solium*. Humans pass ova, or eggs, of *T. solium* into the environment with feces, and the cycle is completed when pigs consume fecally contaminated food. Human neurocysticercosis is also a result of fecal–oral contamination, and develops when people inadvertently ingest *Taenia* ova passed by persons with taeniasis. Today, neurocysticercosis is the most common parasitic infection of the central nervous system and the most common cause of adult-onset seizures in certain areas of Latin America. It is a major source of morbidity that consumes huge amounts of resources every year (Bryan, 1992).

Also travelling aboard ships during the 16th and 17th century periods of European exploration and colonization were rats, particularly *Rattus norvegicus*, also known as the brown or Norway rat. *Rattus* dispersal and invasion was a global event, but we will focus here on its impact in the New World. Although well-established elsewhere in North America (e.g., eastern seaboard) perhaps a century earlier, *Rattus* did not reach the west coast until 1899. Its arrival heralded the spread of the Third, or Modern, plague pandemic to the shores of western North America. Plague is a serious condition due to infection with the plague bacillus, *Yersinia pestis*. Of its three primary clinical forms, bubonic, septicemic, and pneumonic, bubonic is the most common and pneumonic the most deadly and most feared. Although the pneumonic form of plague is highly contagious from person to person, in most cases, plague is transmitted from rodent reservoir hosts to humans via the bite of infected fleas – often in epidemic conditions associated with urban rats (Campbell and Dennis, in press).

The interesting occurrence in North America was the gradual shift from *Rattus* to indigenous ground-dwelling rodents as primary plague reservoirs. This shift created, for the first time in North America, permanent, endemic, rural foci of plague. Plague epidemiology then evolved from urban, *Rattus*-based **epidemics** to rural or sylvatic, **endemic** plague which is maintained in species such as prairie dogs and rock squirrels. After 1920, the international spread of plague was interrupted by regulations for rat control and inspections in harbors and ships, and the last reported urban outbreak in the United States occurred in Los Angeles in 1924–1925 (Campbell and Dennis, in press).

Today, plague is maintained in enzootic cycles in rural areas of Africa, Asia, South America, and the western United States. The impact of its introduction, however, is still felt as evidenced by a recent large outbreak in Peru. In Peru, sporadic cases of human plague have been reported for the past 40–50 years. However, in October 1992, a plague epidemic emerged. By the end of 1995, a total of 1 299 cases were diagnosed, with 62 deaths or a case fatality rate of 4.8% (PAHO, 1995). Human cases were all associated with plague enzootics in wild rats and domestic guinea pigs – the latter are commonly raised as food sources in the Andean regions of South America. Even in the United States, approximately 15 cases per year are reported and, despite its susceptibility to common antibiotics, plague mortality in U.S. cases is 14% (Campbell and Dennis, in press).

Modern examples of alien species introductions and emerging infections

Malaria is a severe febrile illness due to infection with protozoa of the genus *Plasmodium* that is transmitted to humans via the bite of infected female Anopheline mosquitoes. Malaria infects some 300-500 million persons worldwide causing approximately 800,000 deaths in sub-Saharan Africa alone in 1990 (Olliaro *et al.*, 1996). The *Plasmodium* life cycle has obligatory developmental stages in the mosquito gut and in human liver and red blood cells; humans are the primary reservoirs for human disease. Although competent malaria vectors exist throughout the tropics, there have been at least two well-documented instances where a particularly anthropophilic and efficient mosquito vector of *P. falciparum* – the most deadly malaria species – has been introduced as an alien species with subsequent disastrous public health impact.

In 1930, *Anopheles gambiae* travelled aboard a ship from west Africa to northeastern Brazil. Less than a year after its introduction, approximately 10 000 cases of malaria occurred within a six square mile area inhabited by only 12 000 people. Over the next several years, epidemics recurred and the vector's geographic range substantially widened. Toward the end of that decade, an intense and unprecedented mosquito eradication program, involving thousands of workers and millions of dollars, was implemented. The last *A. gambiae* was collected there in 1940 (Diggs, 1992). A similar invasion occurred in upper Egypt's Nile Valley in association with water development projects in 1942–1943. This introduction, thought to have originated from Sudan, resulted in a malaria outbreak with over 130 000 deaths (Hunter *et al.,* 1993).

In addition to humans, their livestock, rodents, and various arthropods, mollusks as alien species have also had a negative impact on human public health. Schistosomiasis (bilharzia) is a trematode, or fluke, infection caused primarily by three species of the genus *Schistosoma*. This disease afflicts about 200 million persons worldwide causing severe, chronic disease of the gastrointestinal, hepatic, and genitourinary organ systems. Humans are infected by contacting fresh water wherein infected snails have released the free-living cercarial phase of the schistosome, which penetrates intact human skin. As mature flukes develop in human hosts, they begin to produce eggs that are shed into the environment in urine and feces. Miracidia emerge from eggs in fresh water where they seek out snail intermediate hosts in which to complete the cycle with their development into cercariae (Bryan and Michelson, 1995).

There are two relatively recent examples – one well-documented as to source, the other less so – wherein the unintentional introductions of competent *Schistosoma* snail vectors of the genus *Bulinus* occurred. The first example took place in Ghana in western Africa after the completion of a dam across the Volta River and the creation of Lake Volta in 1964. As early as 1969, the prevalence of schistosomiasis in local inhabitants above the dam had increased from 5-10% to near 90%. Instrumental in this increase was the introduction of *Bulinus truncatus* by tribal fishermen coming to the new lake from distant coastal lagoons. These fishermen carried with them both the snail itself and the particular strain of *S. hematobium* specifically adapted to that snail (Hunter *et al.,* 1993). *B. truncatus* also found its way to Jordan in the 1970s. Prior to that, neither the snail nor the disease were known to occur there. The route of its introduction is speculative, but some regional experts contend that the snails were transported there inside Iraqi military water carriers (Wurzinger and Saliba, 1979; Burch and Bruce, 1989; J. Sullivan, pers. comm.).

Ship ballast water is a well-documented source of alien species introductions (OTA, 1993). Its role in the introduction of human infectious diseases is less certain, but recent experience in South America suggests that microorganisms contained in ballast water may have resulted in a devastating epidemic. Pandemic cholera returned to the Western Hemisphere in 1991 infecting over 1 million persons and causing some 10 000 deaths. In the United States more imported cases occurred in 1992 than in any other year since cholera surveillance began. The explanation for its sudden return after an absence of over a century is uncertain, but many experts have suggested that ship ballast water discharged in a Peruvian harbor may have seeded the epidemic (IOM, 1992; Tauxe *et al.,* 1994, 1995).

Perhaps the most cogent and telling example of modern alien species introductions with potentially significant public health impact is that of the Asian tiger mosquito, *Aedes albopictus*. Until its discovery in Houston, Texas in 1985, *Ae. albopictus* was unknown in the Western Hemisphere. In addition to the United States, it has been recently introduced into the Dominican Republic, Brazil, Nigeria, Italy, and New Zealand. The vehicle of introduction into the United States was scrap tires imported from Japan. Capable of holding water in any position in which they are stored, discarded tires provide excellent breeding habitat. Each year in the United States, one quarter billion tires are discarded and several million scrap tires are imported. To further exacerbate the problem, Asian tiger mosquitoes also breed well in many forms of discarded containers such as cans, flower pots, buckets, etc. This fact, plus limited interstate transport regulations and inspections, have no doubt contributed to the expanded range of this mosquito in the United States. *Ae. albopictus* is now well-established throughout much of the southeastern United States, with extensions in range as far west as western Texas and as far north as Chicago (IOM, 1992; OTA, 1993; Serufo *et al.,* 1993; Moore *et al.,*1988).

In addition to being a significant nuisance species, *Ae. albopictus* has also been shown to be a competent vector for mosquito-borne encephalitis viruses such as eastern equine and LaCrosse encephalitis viruses, as well as for dengue virus. Because of its ability to transmit dengue transovarially (mother to offspring via eggs), there is concern that *Ae. albopictus* in the southeastern United States, the Caribbean, and Brazil could result in the establishment of new enzootic foci of dengue viruses. Perhaps most concerning, however, is the potential, particularly in areas such as Brazil and Nigeria, for *Ae. albopictus* to facilitate the transfer of jungle yellow fever to urban cycles involving *Ae. aegypti* (IOM, 1992; Moore *et al.,* 1988; Craven *et al.,* 1988; Mitchell *et al.,* 1992; C. Moore, pers. comm.)

Global change, alien species, and emerging diseases

Most of the factors listed in Table 11.1 can be said to apply to alien species introductions as well as to emerging infectious diseases. This observation is no coincidence – the same global changes that foster the emergence or re-emergence of infectious diseases often play a similar role in alien species introductions. Indeed, in their 1992 report the U.S. Institute of Medicine suggested that most emerging infectious diseases could be attributed to environmental changes (IOM, 1992). Careful examination of some of these factors in emergence may help to identify those situations in which global changes are likely to foster alien species introductions with potential impact on human health – introductions that could involve vectors or reservoirs of infectious diseases or the infectious agents themselves.

Table 11.1 Factors in emergence of infectious diseases.

- Human demographics and behavior
- Technology and industry
- Economic development and land use
- International travel and commerce
- Microbial adaptation and change
- Breakdown of public health measures

World population growth has taken a precipitous rise that is projected to approach 12 billion by the year 2050. Furthermore, the distribution of the world's population is shifting toward greater numbers and greater urban densities in developing countries (Horton, 1996). As more and more of the rural poor migrate to urban environments (i.e., the cities or "mega-cities"), crowding and poverty increase, and adequate sanitation decreases. This combination creates settings in which insect and/or rodent vectors of human disease may be readily introduced, setting the stage for the emergence of new diseases or the resurgence of well-known scourges such as plague, malaria, dengue, yellow fever, and typhus. Evidence for this process already exists in some areas of Latin America where hematophagous insect vectors (triatomine species) of American trypanosomiasis have been carried among the belongings of rural residents as they migrate to urban centers, creating a unique epidemiologic situation for a disease that is characteristically rural (Mott *et al.*, 1990; Rorattini, 1989).

International tourism and commerce can transport new infectious diseases or alien species across one or several international frontiers in a matter of hours. Worldwide tourist traffic from 1980 to 1994 increased dramatically to the point that total international arrivals exceeded 500 million in 1994 (data from World Tourism Organization). In addition to tourism, other forms of population movements can readily introduce "exotic" infectious agents to new geographic areas or population groups. In the United States, for example, the presence of infected visitors or immigrants from endemic regions has resulted in the autochthonous transmission (transmission in the place where the patient is) of diseases usually considered exotic to the U.S. such as dengue, cholera, malaria, neurocysticercosis, and transfusion-associated trypanosomiasis (IOM, 1992; Gubler and Clark, 1995; Finelli *et al.*, 1992; CDC, 1995a; Zucker, 1996; Schantz et al., 1992; Grant *et al.*, 1989; Nickerson *et al.*, 1989).

International trade agreements such as the North American Free Trade Agreement (NAFTA) and the General Agreement on Tariffs and Trade (GATT) may be fostering broader opportunities for both alien species and infectious disease introductions. One area of particular concern is current global traffic in food items, especially fresh produce. For instance, consumer demand for fresh fruits and vegetables is increasing. In the United States, the seasonal importation of produce from Mexico, Central America, and other tropical areas has increased to the point that over 75% of out of season fresh fruits and vegetables consumed here are harvested abroad (Hedberg *et al.*, 1994).

As noted above in several of the historical and modern examples of alien species introductions with human health impact, numerous other environmental, demographic, and sociocultural conditions can facilitate these kinds of introductions. Other such conditions include commercial trade in exotic game animals and pets; deforestation and other human land use modifications; and water development projects such as dams, irrigation, and aquaculture (Mott *et al.*, 1990; IOM, 1992; Hunter *et al.*, 1993; CDC, 1995b).

Early warning systems and cross-disciplinary cooperation

The preceding examples highlight situations or circumstances where we might anticipate alien species introductions with a significant potential for human health impact. In order to be better prepared for such situations in the future, early warning systems should be developed. The need for such systems has been emphasized by experts in the fields of ecology and biodiversity, as well as those in public health (Berkelman *et al.*, 1994; Bryan *et al.*, 1994; Satcher, 1995; Morse, 1995). Strategies for surveillance that are underway or under development within these fields provide ample opportunity for cooperation and mutual support (CDC, 1994; PAHO, 1995).

Catalyzed in large part by the 1992 IOM report, the Centers for Disease Control and Prevention (CDC) has taken a lead role in the United States and internationally in developing and implementing strategies to address threats to health from emerging infectious diseases. The CDC plan, "Addressing Emerging Infectious Disease Threats – A Prevention Strategy for the United States," contains four main goals that emphasize surveillance, applied research, prevention and control, and infrastructure (CDC, 1994). Particularly relevant to the issue of biodiversity and alien species introductions is the first goal and some of its specific components. Goal I of the CDC plan is to "detect, promptly investigate, and monitor emerging pathogens, the diseases they cause, and the **factors influencing their emergence**." The latter is emphasized because surveillance for factors in emergence requires a multi-disciplinary approach involving professionals from many fields of biology, ecology, and climatology, in addition to medicine and public health. As illustrated by many of the examples presented herein, vectorborne and zoonotic infectious diseases may present excellent opportunities for bridging the gap between ecologic and epidemiologic sciences. An important objective of Goal I of the CDC plan is to "strengthen and integrate programs to monitor, control, and prevent emerging vectorborne and zoonotic diseases." A collaborative effort to develop mutual strategies to address this objective may be a reasonable first step toward reaching common goals for early warning systems.

In addition to collaborative projects to develop early warning systems, another means to enhance interaction between ecologists and public health professionals is through the scientific literature, particularly that accessible by electronic means. As part of its efforts to disseminate important information regarding emerging infectious diseases, CDC's National Center for Infectious Diseases now publishes a quarterly, peer-reviewed journal, "Emerging Infectious Diseases" (EID), that seeks to promote the recognition of new and reemerging infectious diseases and to improve the understanding of factors involved in disease emergence, prevention, and elimination. EID is international in scope and publishes articles from professionals in multiple disciplines whose studies elucidate the factors influencing the emergence of infectious diseases. EID is available electronically on the World Wide Web (WWW), through file transfer protocol (FTP), or by electronic mail. Access the journal at http://www.cdc.gov/ncidod/EID/eid.htm, from the CDC home page (http://www.cdc.gov), or download it through anonymous FTP at ftp.cdc.gov.

Conclusions

The purpose of this paper has been to highlight the role of alien species introductions in fostering the emergence or re-emergence of infectious diseases, and to help identify areas in which closer cooperation between professionals in ecology and biodiversity and those in public health and epidemiology will enhance efforts to attain common goals. It is clear that alien species introductions have had, and will undoubtedly continue to have, significant impact on human health. The cases presented here are just a sampling of many other citable examples of alien species introductions with human health implications, both infectious and non-infectious. Hopefully, the information presented and issues raised in this paper will prompt ecologists and public health professionals to take advantage of multiple opportunities for scientific exchange and cooperation to address the threats presented by alien species introductions to biodiversity and to human health.

Acknowledgements

The author gratefully acknowledges the assistance and contributions of Drs. James Sullivan and Raymond Beach (Division of Parasitic Diseases, NCID, CDC, Atlanta, GA), Drs. Chester Moore and David Dennis (Division of Vectorborne Infectious Diseases, NCID, Ft. Collins, CO), Dr. Martin Cetron (Division of Bacterial and Mycotic Diseases, NCID, CDC, Atlanta, GA), Dr. Paul Ettestad (Division of Epidemiology, Evaluation, and Planning, New Mexico Department of Health, Santa Fe, NM), Dr. C.J. Peters (Division of Viral and Rickettsial Diseases, NCID, CDC, Atlanta, GA), and Dr. Stephen Ostroff (Office of the Director, NCID, CDC, Atlanta, GA).

References

Berkelman, R.L., Bryan, R.T., Osterholm, M.T., LeDuc, J.W. and Hughes, J.M. (1994) Infectious disease surveillance: a crumbling foundation. *Science*, **264**, 368–370.

Black, F.L. (1992) Why did they die? *Science,* **258**, 1739–1740.

Bryan, R.T. (1992) Current issues in cysticercosis: proteins, proglottids, pigs, and privie, in *Global Medicine: Current Status and Directions for Control in the Twenty-first Century,* (ed. D.H. Walker), Springer-Verlag, Vienna.

Bryan, R.T. and Michelson, M.K. (1995) Parasitic infections of the liver and biliary tree, in *Gastrointestinal and Hepatic Infections*, (eds C. Surawicz and R.L. Owen), W.B. Saunders, Philadelphia, pp. 405–454.

Bryan, R.T., Pinner, R.W. and Berkelman, R.L. (1994) Emerging infections in the United States: improved surveillance, a requisite for prevention, in *Disease in Evolution*, (eds R. Levins, M.E. Wilson, and A. Spielman), Annals of the New York Academy of Sciences, 740, New York, pp. 346–361.

Burch, J.B., Bruce, J.I., and Amr, Z. (1989) Schistosomiasis and malacology in Jordan. *Journal of Medical & Applied Malacology*, **1**, 139–163.

Campbell, G.L. and Dennis, D.T. (in press) Plague, in *Harrison's Principals of Internal Medicine, 14th ed.*, (eds A.S. Fauci *et al.*), McGraw Hill, New York.

CDC (1994) *Addressing Emerging Infectious Disease Threats: a Prevention Strategy for the United States.* U.S. Department of Health and Human Services, Public Health Service, Atlanta, Georgia.

CDC (1995a) Cholera associated with food transported from El Salvador – Indiana, 1994. *MMWR*, **44**, 385–387.

CDC (1995b) Reptile-associated salmonellosis – selected states, 1994–1995. *MMWR*, **44**, 347–350.

Craven, R.B., Eliason, D.A. and Francy, P. *et al.* (1988) Importation of *Aedes albopictus* and other exotic mosquito species into the United States in used tires from Asia. *Journal of the American Mosquito Control Association,* **4**, 138–142.

Crosby, A.W. (1991) Metamorphosis of the Americas, in *Seeds of Change – Five Hundred Years Since Columbus*, (eds H.J. Viola and C. Margolis), Smithsonian Institution Press, Washington, DC., pp. 70–89.

Diggs, C.L. (1992) Malaria control in the twenty-first century, in *Global Medicine: Current Status and Directions for Control in the Twenty-first Century*, (ed. D.H Walker), Springer-Verlag, Vienna.

Finelli, L., Swerdlow, D., Mertz, K., Ragazzoni, H. and Spitalny, K. (1992) Outbreak of cholera associated with crab brought from an area with epidemic disease. *Journal of Infectious Diseases,* **166**, 1433–1435.

Forattini, O.P. (1989) Chagas' disease and human behavior, in *Demography of Vector-Borne Diseases*, (ed. M.W. Service), CRC Press, Boca Raton, pp. 107–118.

Grant, I.H,, Gold, J.W.M. and Wittner, M. *et al.* (1989) Transfusion-associated acute Chagas disease acquired in the United States. *Annals of Internal Medicine*, **111**, 849–851.

Gubler, D., and Jand Clark, G.G. (1995) Dengue/dengue hemorrhagic fever: the emergence of a global health problem. *Emerging Infectious Diseases*, **1**, 55–57.

Hedberg, C.W., MacDonald, K.L. and Osterholm, M.T. (1994) The changing epidemiology of foodborne disease: a Minnesota perspective. *Clinical and Infectious Diseases*, **18**, 671–682.

Horse Capture, G.P. (1991) An American Indian perspective, in *Seeds of Change – Five Hundred Years Since Columbus,* (eds H.J. Viola and C. Margolis), Smithsonian Institution Press, Washington, D.C., pp. 186–207.

Horton, R. (1996) The infected metropolis. *Lancet*, **347**, 134–135.

Hunter, J.M., Rey, L., Chu, K.Y., Adekolu-John, E.O. and Mott, K.E. (1993) *Parasitic Diseases in Water Resources Development.* World Health Organization, Geneva.

IOM (1992) *Emerging Infections: Microbial Threats to Health in the United States.* Institute of Medicine, National Academy Press, Washington, DC.

Mitchell, C.J., Niebylski, M.L. and Smith, G.C. *et al.* (1992) Isolation of eastern equine encephalitis virus from *Aedes albopictus* in Florida. *Science*, **257**, 526–527.

Moore, C.G., Francy, D.B., Eliason, D.A. and Monath, T.A. (1988) *Aedes albopictus* in the United States: rapid spread of a potential disease vector. *Journal of the American Mosquito Control Association,* **4**, 356–361.

Morse, S.S. (1995) Factors in the emergence of infectious diseases. *Emerging Infectious Diseases*, **1**, 7–15.

Mott, K.E., Desjeux, P., Moncayo, A., Ranque, P. and de Raad , P. (1990) Parasitic diseases and urban development. *Bulletin of the WHO*, **68**, 691–698.

Murray, C.J.L. and Lopez, A.D. (1996) *Global Health Statistics.* Harvard University Press, Cambridge, MA.

Nickerson, P., Orr, P. and Schroeder, M-L. *et al.* (1989) Transfusion-associated *Trypanosoma cruzi* infection in a non-endemic area. *Annals of Internal Medicine*, **111**, 851–853.

Olliaro, P., Cattani, J. and Wirth, D. (1996) Malaria, the submerged disease. *Journal of the American Medical Association,* **275**, 230–233.

OTA (1993) *Harmful Non-indigenous Species in the United States.* Office of Technology Assessment, U.S. Government Printing Office, Washington, DC.

PAHO (1995) *Regional Plan of Action for Combatting New, Emerging, and Re-emerging Infectious Diseases in the Americas.* Pan American Health Organization, PAHO/HCP/95.060, Washington, DC.

Pinner, R.W., Teutsch, S.M. and Simonsen, L. *et al.* (1996) Trends in infectious diseases mortality in the United States. *Journal of the American Medical Association*, **275**, 189–93.

Satcher, D. (1995) Emerging infections: getting ahead of the curve. *Emerging Infectious Diseases*, **1**, 1–6.

Schantz, P.M., Moore, A.C., and Muñoz, J.L. *et al.* (1992) Neurocysticercosis in an orthodox Jewish community in New York City. *New England Journal of Medicine*, **327**, 692–695.

Serufo, J.C., Montes, de Oca H. and Tavares, V.A. *et al.* (1993) Isolation of dengue virus type 1 from larvae of *Aedes albopictus* in Campos Altos City, State of Minas Gerais, Brazil. *Memoires Instituto de Oswaldo Cruz*; **88**, 503-504.

Tauxe, R., Seminario, L., Tapio, R. and Libel, M. (1994) The Latin American epidemic, in Vibrio cholerae *and Cholera: Molecular to Global Perspectives*, (eds I.K. Wachsmuth, P.A. Blake and O. Olsvik), American Society for Microbiology, Washington, DC, pp. 326–327.

Tauxe, R.V., Mintz, E.D. and Quick, R.E. (1995) Epidemic cholera in the New World: translating epidemiology into new prevention strategies. *Emerging Infectious Diseases*, **1**, 141-146.

Weil, C. (1981) Health problems associated with agricultural colonization in Latin America. *Science and Medicine*, **15D**, 449–461.

WHO (1996) *The World Health Report 1996: Fighting Disease, Fostering Development.* World Health Organization, Geneva.

Wurzinger, K.H., and Saliba, E.K. (1979) A cytological and electrophoretic comparison of Jordanian *Bulinus* with three other tetraploid *Bulinus* populations. *Malacological Review*; 12, 59–65.

Zucker, J.R. (1996) Changing patterns of autochthonous malaria transmission in the United States: a review of recent outbreaks. *Emerging Infectious Diseases*, 2, 37–43.

12 Effects of invading species on freshwater and estuarine ecosystems

PETER B. MOYLE
University of California – Davis, California, USA

Abstract

Freshwater and estuarine environments are among the most altered ecosystems worldwide, because they are subject to water diversions and intense human use of both the aquatic systems and the surrounding watersheds. They are also among the most invaded ecosystems worldwide, particularly in temperate regions of the world. While invading species unquestionably have been responsible for major changes in aquatic ecosystems, including extinction of native species, the invaded environments are typically already severely altered by human activity. Because invasion is a natural process (although not at the rates observed today), most aquatic communities seem to have considerable capacity to adjust to most invaders through niche shifts and other adaptive mechanisms. Extinctions are most likely to occur when the successful invader is a top carnivore, when the invader carries with it novel disease organisms, when the invaded ecosystem has naturally low diversity, and/or when the invaded ecosystem has been highly disturbed by human or natural factors. While our increased understanding of how ecosystems work has increased our ability to predict the general effects of potential invaders, our knowledge of both individual ecosystems and species is limited enough so that many species introductions are likely to have unexpected, usually harmful, consequences (the "Frankenstein Effect"). Such consequences are likely to occur with increasing frequency in the future as humans continue to degrade aquatic environments and move aquatic organisms about with little regard for long-term environmental effects. The rate of invasion into aquatic environments can be greatly slowed if governments cease to sanction introductions that are not justified by a detailed risk analysis and if industries that benefit from introduced species (especially the aquaculture, shipping, and aquarium fish industries) are no longer allowed to externalize the costs of harmful invasions they cause.

Introduction

Freshwater and estuarine ecosystems are among the most altered ecosystems worldwide (Moyle and Leidy, 1992; Allan and Flecker, 1993). The reasons for this are multiple and interactive:

- They are the ultimate sumps for watersheds, concentrating pollutants and the effects of watershed abuse;

O. T. Sandlund et al. (eds.), Invasive Species and Biodiversity Management, 177–191.

- Their water is increasingly diverted for human use (Postel *et al.*, 1996), often with dams that change the character of the ecosystem by impounding water, blocking movement of organisms, and changing flow regimes;
- They are the focus of most human activity, especially large cities and intensive agriculture, resulting in direct changes to the aquatic and riparian environments;
- They are strongly affected by human-induced climate change, further stressing the native organisms;
- The fish and other ecologically important organisms are subject to intense exploitation.

On top of these massive environmental changes, freshwater and estuarine ecosystems are among the most invaded ecosystems in the world, especially in temperate regions, and invasions of exotic organisms are still occurring at a high rate. Human mediated invasions of fish, invertebrates, plants, and pathogens have caused further, usually irreversible, alterations of already stressed aquatic ecosystems. As a result of this combination of events, Moyle and Leidy (1992) estimated that 20% of all freshwater fish species, the best monitored group of aquatic organisms, are at risk of extinction in the near future if present trends continue. Percentages of at-risk species are highest in temperate regions, especially in regions with Mediterranean climate, where competition between humans and aquatic organisms for water is most severe. Percentages of many freshwater invertebrates, especially bivalves (Bogan, 1993) and crayfishes (Taylor *et al.*, 1996), are probably as high or higher. While introduced species unquestionably have been a major factor causing the declines and extinctions of some native species, their overall role is less clear because of their close association with ecosystems heavily altered by other human actions.

The purpose of this paper is to briefly answer the following questions:

1. Why are freshwater and estuarine ecosystems so subject to invasion?
2. Are most aquatic invasions harmful?
3. What are the sources of aquatic invaders?
4. Can we predict the effects of aquatic invasions?
5. How can we minimize future invasions?

Why are freshwater and estuarine ecosystems so subject to invasion?

Those of us who live in areas with arid or Mediterranean climates tend to see freshwater and estuarine ecosystems as being exceptionally invasible because we deal with faunas dominated by introduced organisms. In the Sacramento-San Joaquin estuary (California), for example, over 212 exotic species have been established and exotic invertebrates almost completely dominate the benthos and plankton (Cohen and Carlton, 1995). In the freshwater and brackish portions of the estuary, about 60% of the fish species are non-native species and they tend to dominate in terms of numbers and biomass (Meng *et al.*, 1995). Likewise, of the freshwater fishes of California, 52 species are non-native and 55 are native; the non-native species dominate the highly altered lowland environments and reservoirs, as well as high elevation waters that were originally fishless. In addition, some of the native fishes have been moved to new drainages, where they are often quite successful (e.g., Brown and Moyle, 1997). In Italy, massive introductions of native fishes into drainages to which they are not native have completely obscured the original distribution patterns; in addition 26 species (37% of the fauna) have been introduced from outside Italy (Bianco, 1995). In contrast, successful marine invasions are comparatively few; for example, only five species of fish have invaded marine environments in California, even in bay systems that are dominated by exotic invertebrates.

Freshwater systems outside of arid and Mediterranean regions have also been extensively invaded. The Laurentian Great Lakes of North America are now dominated by introduced piscivores and planktivores among the fishes (Moyle, 1986) and exotic mussels (*Dreissena* spp.) in the epibenthos (Ludyanskiy *et al.*, 1993). In fact, most streams and lakes in the eastern United States contain one or more introduced species, often unauthorized introductions from nearby drainages (Moyle, 1986). Lake Victoria in East Africa is now dominated by introduced Nile perch (*Lates* sp.) and tilapia (*Oreochromis niloticus*) (Kaufman, 1992; Ogutu-Ohwayo, 1998). African tilapias (*Oreochromis* spp.) are rapidly becoming among the most abundant species in Lake Nicaragua in Central America (McKaye *et al.*, 1995). In Sri Lanka, four introduced species quickly became abundant in tropical streams (Wikramanayake, 1990). On the Hawaiian islands, every stream has been invaded by the Tahitian prawn (*Macrobrachium lar*) and often by exotic fishes and other invertebrates as well (A. Brasher, personal communication). Most suitable alpine lakes and streams worldwide contain one or more species of introduced trout or charr (Salmonidae). With the possible exceptions of the upper reaches of large tropical river systems and most arctic and subarctic waters, it is unusual to find a lake or stream without one or more introduced species.

Does the fact that most freshwater and estuarine ecosystems have been invaded mean that they are exceptionally invasible? I would argue this is not the case. Instead, the high frequency of successful invasions indicates (1) most aquatic environments have been altered by human activity, (2) there has been a high frequency of introductions into aquatic systems, many of them secondary effects of human activity (e.g., canal building, ballast water discharge, aquaculture operations) and (3) the high success of people introducing aquatic organisms in matching the organism to the local environment (Li and Moyle, 1993; Moyle and Light, 1996b). Indeed, the massive invasion of the Mediterranean Sea by fishes and invertebrates from the Red Sea, following the construction of the Suez Canal, indicates that marine systems can also suffer major invasions (Baltz, 1991; Boudouresque, 1998).

Are most aquatic invasions harmful?

The answer to this question is not as obvious as it might seem because the effects of most invading species have been poorly documented (Li and Moyle, 1993; Crivelli, 1995). There is a tendency to think of all invading species as being harmful because there are so many examples of spectacular negative effects, such as the elimination of endemic cichlids in Lake Victoria by Nile perch, the disruption of the Laurentian Great Lakes ecosystems by the invasion of sea lamprey (*Petromyzon marinus*) and alewife (*Alosa pseudoharengeus*), and the extirpation of native crayfish throughout Europe following the introduction of a North American species and its accompanying crayfish plague. However, invasion is a natural process (Vermeij, 1991) so most aquatic systems should be capable of either absorbing or resisting invaders to some extent (Elton, 1958; Moyle and Light, 1996a, b). Of the 52 exotic fishes established in California, there is evidence that (1) 20 have had or have the potential to have significant negative effects on native fishes; (2) seven are so localized in their distribution that they are unlikely to have any effects; and (3) the effects of 25 are not known (Moyle, unpublished data). Moyle and Williams (1990) list introduced species as a serious problem in the conservation of 19 of the 39 native fishes of California that are extinct or in serious decline, although, overall, water diversions and other habitat changes were more important as causes of declines. Indeed, relatively undisturbed streams and lakes in California show a high resistance to invasion by non-native species (Moyle and Light, 1996a).

On the other hand, just looking at the effects of an introduced organism on similar species (e.g., fish on fish) is likely to provide a rather limited view of its effects. The cascading effects of introduced species, especially top predators and planktivores, on ecosystems have been well documented (Li and Moyle, 1993). Such effects can range from the collapse of fisheries for native species to subtle changes in the composition of invertebrate and algal assemblages in the benthos (Power, 1992). Similar effects have been noted from the introduction of predatory invertebrates (e.g., Spencer *et al.*, 1991; Lodge, 1993). For the most part, it is difficult to anticipate all the effects of an introduction at the time of introduction. For example, wakasagi (*Hypomesus nipponensis*) were brought from Japan to California in 1952 and introduced into a few mountain reservoirs. This small fish was regarded as innocuous until 1994, when they became noticeable in the Sacramento-San Joaquin estuary and began hybridizing with delta smelt (*H. transpacificus*), a species already endangered from the effects of water diversions (Trenham *et al.*, 1998). A further unexpected threat to the delta smelt is the inland silverside (*Menidia beryllina*) which invaded the estuary in the 1970s and may be a major predator on smelt eggs and larvae (Bennett and Moyle, 1996). Various trout species (*Salmo, Oncorhynchus, Salvelinus*) were introduced into thousands of alpine lakes in California 50–100 years ago but only recently has the collapse of the amphibian fauna of these lakes, largely from trout predation or diseases, been recognized, in part because the effects were not instantaneous (Bradford *et al.*, 1993). Likewise the effects of diseases or parasites brought in with exotic organisms can have unanticipated effects. For example, whirling disease was introduced into the USA with European brown trout (*Salmo trutta*); it is lethal to native trouts and appears to have been a major factor in increasing the success of brown trout in North American streams (Vincent, 1996). The likelihood of unanticipated negative effects of introductions regarded as beneficial has been termed the "Frankenstein Effect" by Moyle *et al.* (1986).

Despite the catalogue of invasion horror stories, a surprising number of invasions seem to have been accommodated by integrating the species into established assemblages through niche shifts and other mechanisms (Moyle and Light, 1996b) or by having the species live mainly in altered environments poorly suited for native species. In the Sacramento–San Joaquin estuary, assemblages of mixed native and introduced organisms behave in many respects like co-evolved assemblages (Meng *et al.*, 1994). All species of fish have declined together in response to poor environmental conditions and only a new invader, the shimofuri goby (*Tridentiger bifasciatus*) increased in numbers. In depauperate Sri Lankan streams, the addition of four species of fish from nearby drainages appeared to quickly create an integrated assemblage of fishes (Wikramanayake, 1990) similar to that in the "home" streams. Fernando (1991) argues that, in tropical Asia and America, introduced fishes, except piscivores, have had little impact on natural aquatic systems. He points out that introduced omnivores, detritivores, and herbivores are typically most abundant in reservoirs and other artificial or altered environments, where they contribute substantially to local fisheries. Unfortunately, introduced fishes do not stay just where they are planted. In the Eel River, California, nine species of fish have been flushed into downstream areas from the only large reservoir in the system, although only two have become established in large numbers. Both are native to neighboring drainages and have changed the structure of the pre-existing ecosystem, but with no known extinctions of native species so far (Brown and Moyle, 1997).

The harm that introduced species do is a major cost of introducing species, often not considered when an introduction is made. There are, however, many benefits (real and imagined) to introductions as well, which, of course, is why they are made. In California, 42 of the 52 introduced species would still be regarded as having been beneficial introductions, because they support sport and commercial fisheries, are used in biological control of pests, or serve as forage for introduced predators in reservoirs and other altered waters (Moyle, unpublished). In general, impoundments worldwide tend to be dominated by introduced fishes and other organisms, which are important in local fisheries. In the Laurentian Great Lakes, introduction of Pacific salmon (*Oncorhynchus* spp.) resulted in restoration of a salmonid fishery, which had been largely destroyed by previous introductions (Moyle, 1986). Stow *et al.* (1995) go so far as to suggest that restoration of a fishery for the native lake charr (*Salvelinus namaycush*) may not even be desirable because the charr have much higher contaminant loads than the non-native salmon and are therefore less suitable for human consumption.

What are the sources of aquatic invaders?

Any attempt to understand the extent and causes of invasions of aquatic systems in order to reduce problems must take into account the diverse means by which invading species are introduced. Introductions are of two main types: deliberate introductions and by-product introductions.

Deliberate introductions are the result of agencies or individuals deliberately introducing species for social or personal gain. **Official introductions** are usually made to enhance fisheries, such global introductions of various trout (Salmonidae) and tilapia (Cichlidae) species, or for biological control, such as the widespread introductions of mosquitofish (*Gambusia* spp.) and grass carp (*Ctenopharyngodon idella*). While fisheries and environmental agencies increasingly recognize that introductions can create more problems than they solve, officially sponsored introductions still commonly occur. For example, massive introductions of mixed lots of fish from northern Italy to other parts of the peninsula are still frequent (Bianco, 1995). However, most introductions today do not come through official sources.

Unofficial introductions are made by individuals independent of government agencies, usually out of ignorance of long-term consequences or of not caring about them because of perceived needs to solve local problems. Such introductions reflect the fact that official attitudes (at least in the USA) towards introductions 20 or more years ago were rather careless about consequences and it will probably take at least that long to change the thinking of anglers and others (Rahel, 1997). Thus in California, there have been no officially sponsored introductions since 1969 (except sterile triploid grass carp, *Ctenopharyngodon idella,* to a limited area) but anglers have introduced two species of predatory fish (*Esox lucius* and *Morone chrysops*) in recent years that may pose a major threat to salmonid fisheries in the near future. Similarly, piscivorous Sacramento pikeminnow (*Ptychocheilus grandis*) were moved by persons unknown to the Eel River drainage, where they quite likely are having a negative impact on the anadromous fishes (Brown and Moyle, 1997).

By-product introductions are a secondary effect of some human activity. Such introductions have often been labelled as "accidental" in the past, a term which de-emphasizes the fact that, given our present level of knowledge, most are highly preventable. By-product introductions are the major source of aquatic invaders today, a direct result of the extent and volume of international trade and the efficiency of modern transportation and water transport systems. Major sources of by-product introductions are (1) ballast water, (2) the aquarium trade, (3) aquaculture, and (4) water projects.

The introduction of fish and invertebrates from the dumping of ballast water of ships is the biggest source of aquatic introductions in the world today, as millions of organisms and thousands of species are introduced daily into estuaries around the world (Carlton and Geller, 1994, Carlton, 1998). As the invasion of the zebra mussel into the Laurentian Great Lakes demonstrates, such organisms introduced via ballast water can cause enormous ecological and economic damage. Such invasions are at least partially preventable with contemporary technology (Locke *et al.*, 1991).

Another common and increasing source of introductions of fish, invertebrates, and plants is the aquarium trade. A by-product of this trade has been the release of aquarium fishes into the wild by owners tired of their charges and unwilling to kill them directly (Courtenay and Stauffer, 1990). One consequence of such releases is that in many tropical areas (especially urban streams and ponds) around the world, guppies (*Poecilia reticulata*) and swordtails (*Xiphophorus* spp.), both native to central America, are abundant. In some highly disturbed areas, such as the canal systems of central Florida and the streams of Oahu (Hawaii), a bizarre collection of exotic aquarium fishes dominate the fauna, often swimming among beds of equally exotic aquatic plants covered with exotic snails. Because a substantial portion of the world's freshwater fish fauna has probably been found at one time or another in the aquarium fish trade, the trade is potentially a major source of new introductions. Indeed, it is now common for individuals of aquarium fishes to be collected in the wild even where there is little likelihood that populations are likely to become established (Courtenay and Stauffer, 1990). There is also considerable potential for microinvertebrates to hitch-hike rides in the water transporting aquarium species or on aquatic plants, resulting in their spread.

Increasingly, many species of aquarium fishes are being reared on fish farms and these farms themselves have become major sources of releases into the wild (Courtenay and Stauffer, 1990). This industry is just one small portion of the aquaculture industry, which is growing rapidly in suitable areas all over the globe. Not surprisingly, aquaculture is a growing source of introductions into the wild, such as the releases of 1.2 million Atlantic salmon (*Salmo salar*) from Norwegian netpens in a two month period (1988–89) (Beveridge *et al.*, 1994). Many other aquaculture operations leak a steady stream of escapees into the local environment because few (if any) aquaculture operations are completely escape-proof. In still other operations (mainly for salmonids), millions of fish are deliberately released into the wild to augment wild populations or to engage in "ocean ranching." The massive and/or frequent nature of such introductions can have severe ecological and genetic implications (Kreuger and May, 1991), not the least of which is the spread of diseases from aquaculture operations into wild populations and vice versa (Stewart, 1991).

Often aquatic organisms initially established by one means are carried to new locations by canals and other water projects that carry water across watershed boundaries. Such canals can often cause faunas of neighboring watersheds to mix. This is a common means of introduction of non-native fishes in the United States. For example, at least six species of fishes native to the Central Valley of California and two species of non-native fishes have been introduced into southern California through the California aqueduct, which brings northern California water (and its fauna) to the south (Swift *et al.*, 1993). Likewise, the construction of canal systems resulted in the invasion of alewife and sea lamprey into the Great Lakes (Moyle, 1986).

Can we predict the effects of aquatic invasions?

It is clear that many invasions have unpredictable consequences. On the other hand, we are rapidly gaining experience with introduced species in aquatic systems and have a greater understanding of their effects. In addition ecological theory is developing rapidly and is providing increasing promise of predictive tools, such as assembly theory (Pimm, 1989; Moyle and Light, 1996b). Moyle and Light (1996b) develop some potential "rules" for predicting the effects of invasions (Table 12.1). They suggest that virtually any species can invade and that any ecosystem can be invaded given the right circumstances. They note that most dramatic effects of

Table 12.1 Some proposed rules for biotic invasions into freshwater and estuarine systems during two major phases of the invasions (Vermeij, 1991), initial establishment and long-term integration into the existing fish community. From Moyle and Light (1996b).

Establishment	Integration
1A. Most invasions fail.	1B. Most successful invasions are accommodated without major community effects.
2A. All aquatic systems are invasible.	2B. Major community effects are most often observed where species richness is low.
3A. Piscivores and detritivore/omnivores are most likely to be successful in systems with low levels of human disturbance. (In lakes, zooplanktivores also show a high rate of invasion success and ability to alter the ecosystems they invade. Likewise insectivores easily invade and alter historically fishless streams or low gradient streams highly altered by human activity.)	3B. Piscivores are most likely to alter invaded communities; omnivores/detritivores least likely do so.
4A. Any species with the right physiological and morphological traits can invade, given the opportunity.	4B. Long-term success depends on a close physiological match between the invader and the system being invaded.
5A. Successful invasions are most likely when native assemblages are depleted or disrupted.	5B. Long-term success is most likely in aquatic systems highly altered by human activity.
6A. Invasibility of aquatic systems is related to interactions among environmental variability, predictability, and severity.	6B. Invaders are much more likely to extirpate native species in aquatic systems with either extremely high or extremely low variability or severity.

invasions usually occur when the invader is a piscivore or planktivore and/or when the invaded ecosystem has relatively low natural diversity. Furthermore, the effects of an invasion are more likely to be reduced by environmental resistance than by the biotic resistance of the established community. However, they caution that both invading species and invaded ecosystems have idiosyncracies that often defy our ability to predict actual effects, so that it is best to expect the unexpected.

How can we minimize the effects of aquatic invasions?

Because successful invasions are largely irreversible phenomena, the only way to minimize their effects is to avoid as many new invasions as possible. Reducing invasions requires a combination of (1) developing risk analyses for proposed introductions, (2) educating potential sources of introductions, (3) developing international agreements on policies towards invasions, and (4) developing economic policies that discourage new invasions.

Risk analysis

The best ways to minimize the effect of official introductions is to adopt a very conservative approach and permit only those that have gone through a careful, independent, risk analysis (Li and Moyle, 1981, 1993; Kohler, 1992; Ruesink *et al.*, 1995) or by following a protocol for evaluating introductions such as the one devised for the introduction of fish for pest control of Ahmed *et al.* (1988) or for marine organisms by ICES (1994) . Such analyses have halted the introduction of channel catfish (*Ictalurus punctatus*) into New Zealand (Townsend and Winterbourn, 1992) and Nile perch (*Lates nilotica*) into Australia, while justifying the introduction of several detritivorous fishes into the Sepik River of New Guinea (Coates, 1993). Allendorf (1991) provides a synthesis of other recommendations for regulating such introductions.

Education

Education on a broad scale is needed, in order to make a conservative approach to introductions part of a worldwide code of environmental ethics. This is a slow process, however, so that initial efforts need to be focussed on the major sources of by-product introductions today, such as the aquaculture, aquarium fish, and shipping industries. In the past introductions from such sources have conveniently been regarded as "accidental" and therefore uncontrollable. With education, all introductions into aquatic environments no matter what the source can be regarded as deliberate introductions and therefore preventable.

International agreements

The spread of exotic species is an international problem that requires international agreements to halt or reduce, one of the ultimate purposes of this conference. In most industrialized countries, national laws are inadequate to prevent most introductions (e.g., Dentler, 1994). In the absence of strong international agreements, major source countries for trade (e.g., USA, Japan, western European countries, Australia) have an obligation to act unilaterally to force the shipping industry to modify their operations in order to reduce ballast water introductions. Strong laws in only a few countries could have a major impact worldwide.

Economic incentives

While official introductions are increasingly under control, at least in western countries, most introductions are unofficial or take place in areas where there is little concern for their potential effects. Increasingly, introductions are the by-products of other human activity, a special, self-perpetuating form of pollution. There is little hope for controlling such introductions unless there are strong economic incentives *not* to make them. For example, if the enormous economic (and environmental) costs of introductions from ballast water and aquaculture operations (e.g., fouling of water intakes, declines in wild salmon fisheries) were even partially accounted for in the costs of farming fish or transoceanic shipping, there would be considerable incentive to reduce or eliminate introductions. Because the introductions by such sources can now be regarded as deliberate (even if not predictable in time and space), the sources of the introductions should be held responsible for their effects and the various industries responsible should no longer be allowed to externalize the costs incurred when introduced species cause environmental damage. On a more sweeping scale, there is a need for economic policies that go beyond short-term market values and place value on long-term intergenerational obligations (Moyle and Moyle, 1995).

Conclusions

Unfortunately for the world's aquatic biota, sweeping new policies that would reduce introductions into aquatic systems are very difficult to implement. Exploding human populations and the exploding demand for goods and services make effective regulation of introductions unlikely in many, if not most, parts of the world. As a consequence, we are likely to see a continued reduction in aquatic biodiversity, an increased homogenization of the world's freshwater, estuarine, and marine fish faunas, and an increased loss of the many goods and services that intact aquatic ecosystems can provide to humanity (Dailey, 1997). Despite this bleak outlook, efforts at reducing biological invasions into aquatic systems are still worthwhile, even if only a few major invasions are prevented. More than ever before we have the knowledge and the resources to control invasions. The question is: do we have the will to undertake invasion prevention measures in order to protect our aquatic environments and remaining native species and to make the Earth a better place for humanity?

References

Ahmed, S.S., Linden, A.L. and Cech Jr., J.J. (1988) A rating system and annotated bibliography for the selection of appropriate indigenous fish species for mosquito and weed control. *Bulletin of the Society for Vector Ecology,* **13,** 1–59.

Allan, J.D. and Flecker, A. S. (1993) Biodiversity conservation in running waters: identifying the major factors that affect destruction of riverine species and ecosystems. *Bioscience,* **43,** 32–43.

Allendorf, F.W. (1991) Ecological and genetic effects of fish introductions: synthesis and recommendations. *Canadian Journal of Fisheries and Aquatic Sciences,* **48 (Suppl. 1),** 178–191.

Baltz, D.M. (1991) Introduced fishes in marine ecosystems and seas. *Biological Conservation,* **56,** 151–178.

Bennett, W. A. and Moyle, P. B. (1996) Where have all the fishes gone? Interactive factors producing fish declines in the Sacramento–San Joaquin estuary, in *San Francisco Bay: the Ecosystem,* (ed. J. T. Hollibaugh), Pacific Division, AAAS, pp. 519–541.

Beveridge, M.C.M., Lindsay, G.R. and Kelly, L.A. (1994) Aquaculture and biodiversity. *Ambio,* **23,** 497–502.

Bianco, P.G. (1995) Mediterranean endemic freshwater fishes of Italy. *Biological Conservation,* **72,**159–170.

Boudouresque, C.F. (1998) The Red Sea – Mediterranean link: unwanted effects of canals, in *Invasive Species and Biodiversity Management,* (eds O.T. Sandlund, P.J. Schei and Å. Viken), Kluwer Academic Publishers, Dordrecht, The Netherlands.

Bogan, A.E. (1993) Freshwater bivalve extinctions (Mollusca: Unionoida): a search for causes. *American Zoologist,* **33,** 599–609.

Bradford, D.F., Tabatabai, F. and Graber, D.M. (1993) Isolation of remaining populations of the native frog, *Rana muscosa*, by introduced fishes in Sequoia and Kings Canyon National Parks, California. *Conservation Biology*, **7**, 882–888.

Brown, L.R. and Moyle, P.B. (1997) Invading species in the Eel River, California: successes, failures, and relationships with resident species. *Environmental Biology of Fishes*, **49**, 271–291.

Carlton, J.T. (1998) The scale and ecological consequences of biological invasions in the World's oceans, in *Invasive Species and Biodiversity Management*, (eds O.T. Sandlund, P.J. Schei and Å. Viken), Kluwer Academic Publishers, Dordrecht, The Netherlands.

Carlton, J.T. and Geller, J. (1993) Ecological roulette: the global transport and invasion of nonindiginous marine organisms. *Science*, **261**, 239–266.

Coates, D. (1993) Fisheries ecology and management of a large tropical Australasian river basin, the Sepik–Ramu, New Guinea. *Environmental Biology of Fishes*, **38**, 345–368.

Cohen, A.N. and Carlton, J.T. (1995) *Nonindigenous Aquatic Species in a United States Estuary: a Case Study of the Biological Invasions of the San Francisco Bay and Delta.* Report for U.S. Fish and Wildlife Service, 245 pp.

Crivelli, A.J. (1995) Are fish introductions a threat to endemic freshwater fishes in the northern Mediterranean region? *Biological Conservation*, **72**, 311–320.

Courtenay, W.R.Jr. and. Stauffer, J.R.Jr. (1990) The introduced fish problem and the aquarium industry. *Journal of the World Mariculture Society*, **21**, 145–159.

Dailey, G.C. (1997) *Nature's Services: Societal Dependence on Natural Ecosystems.* Island Press, Washington, DC.

Dentler, J.L. (1993) Noah's farce: the regulation and control of exotic fish and wildlife. *University of Puget Sound Law Review*, **17**, 191–242.

Elton, C.S. (1958) *The Ecology of Invasions of Plants and Animals*. Methuen, London.

Fernando, C.H. (1991) Impacts of fish introductions in tropical Asia and America. *Canadian Journal of Fisheries and Aquatic Science*, **48 (Suppl. 1)**, 24–32.

ICES (1995) *ICES Code of Practice on the Introductions and Transfers of Marine Organisms 1994*. International Council for the Exploration of the Sea, Copenhagen, 12 pp.

Kaufman, L. (1992) Catastrophic change in species-rich freshwater ecosystems: the lessons of Lake Victoria. *Bioscience*, **42**, 846–858.

Kohler, C.C. (1992) Environmental risk management of introduced aquatic organisms in aquaculture. *ICES Marine Science Symposium*, **194**, 15–20.

Craggier, C.C. and May, B. (1991) Ecological and genetic effects of salmonid introductions in North America. *Canadian Journal of Fisheries and Aquatic Sciences*, **48 (Suppl. 1)**, 66–77.

Li, H.W. and Moyle, P.B. (1981) Ecological analysis of species introductions into aquatic systems. *Transactions of the American Fisheries Society*, **110**, 772–782.

Li, H.W. and Moyle, P.B. (1993) Management of introduced fishes, in *Inland Fisheries Management in North America*, (eds C. Kohler and W. Hubert), American Fisheries Society, Bethesda, Maryland, pp. 282–307.

Locke, A., Reid, D.M., van Leeuwen, H.C., Sprules, W.G. and Carlton, J.T. (1993) Ballast water exchange as a means of controlling dispersal of freshwater organisms by ships. *Canadian Journal of Fisheries and Aquatic Sciences*, **50**, 2086–2093.

Lodge, D.M. (1993) Species invasions and deletions: community effects and responses to climate and habitat change, in *Biotic Interactions and Global Change*, (eds P.M. Kareiva, J.G. Kingsolver, and R.B. Huey), Sinauer Associates, Sunderland, Massachusetts, pp. 367–387.

Ludyanskiy, M.L, Mcdonald, D. and MacNeill, D. (1993) Impact of the zebra mussel, a bivalve invader. *Bioscience,* **43**, 533–544.

McKaye, K.R., Rayan, J.D., Stauffer, J.R.Jr.,. Lopez Perez, L.J., Vega, G.I. and Van den Berghe, E.P. (1995). African tilapia in Lake Nicaragua. *Bioscience,* **45**, 406–411.

Meng, L., Moyle, P.B. Herbold, B. (1994) Changes in abundance and distribution of native and introduced fishes of Suisun Marsh. *Transactions of the American Fisheries Society,* **123,** 498–507.

Moyle, P.B. (1986) Fish introductions into North America: patterns and ecological impact, in *Ecology of Biological Invasions of North America and Hawaii,* (eds H.A. Mooney and J.A. Drake), Springer Verlag, New York, pp. 27–43.

Moyle, P.B. and Leidy, R.L. (1991) Loss of biodiversity in aquatic ecosystems: evidence from fish faunas, in *Conservation Biology: the Theory and Practice of Nature Conservation, Preservation, and Management,* (eds P.L. Fiedler and S.K. Jain), Chapman and Hall, New York, pp. 127–170.

Moyle, P.B., Li, H.W. and Barton, B.A. (1986) The Frankenstein effect: impact of introduced fishes on native fishes in North America. in *Fish Culture in Fisheries Management,* (ed. R.H. Shroud), American Fisheries Society, Bethesda, Maryland, pp. 415–426.

Moyle, P.B. and Light, T. (1996a) Fish invasions in California: do abiotic factors determine success? *Ecology,* **77**, 1666–1670.

Moyle, P.B. and Light, T. (1996b) Biological invasions of fresh water: empirical rules and assembly theory. *Biological Conservation,* **78**, 149–162.

Moyle, P.B., and Moyle, P.R. (1995) Endangered fishes and economics: intergenerational obligations. *Environmental Biology of Fishes,* **43**, 29–37.

Moyle, P.B. and Vondracek, B. (1985). Persistence and structure of the fish assemblage of a small California stream. *Ecology,* **66**, 1–13.

Moyle, P.B. and Williams, J.A. (1990) Biodiversity loss in the temperate zone: decline of the native fish fauna of California. *Conservation Biology,* **4**, 275–284.

Ogutu-Ohwayo, R. (1998) Nile perch in Lake Victoria: balancing the costs and benefits of aliens, in *Invasive Species and Biodiversity Management*, (eds O.T. Sandlund, P.J. Schei and Å. Viken), Kluwer Academic Publishers, Dordrecht, The Netherlands.

Pimm, S.L. (1989) Theories of predicting success and impact of introduced species, in *Biological Invasions: a Global Perspective,* (eds J.A. Drake, H.A. Mooney, F. di Castri, R.H. Groves, F.J. Kruger, M. Rejmánek and M. Williamson), SCOPE 37, John Wiley and Sons, New York, pp. 351–367.

Postel, S.L., Dailey, G.C. and Ehrlich, P.R. (1996) Human appropriation of renewable fresh water. *Science,* **271**, 785–788.

Power, M.E. (1992) Habitat heterogeneity and the functional significance of fish in river food webs. *Ecology,* **73**, 1675–1688.

Rahel, F.J. (1997) From Johnny Appleseed to Dr. Frankenstein: changing values and the legacy of fisheries management. *Fisheries (Bethesda),* **22**, 8–9.

Ruesink, J.L., Parker, I.M., Groom, M.J. and Kareiva, P.M. (1995) Reducing the risks of nonindigenous species introductions. *Bioscience,* **45**, 465–477.

Spencer, C.N., McClelland, B.R. and Stanford, J.A.. (1991) Shrimp stocking, salmon collapse, and eagle displacement. *Bioscience,* **41**, 14–21.

Stewart, J.E. (1991) Introductions as factors in diseases of fish and aquatic invertebrates. *Canadian Journal of Fisheries and Aquatic Sciences,* **48 (Suppl. 1)**, 110–117.

Stowe, C.A., Carpenter, S.R., Madenjian, C.P., Eby, L.A. and Jackson, L.J. (1995) Fisheries management to reduce contaminant consumption. *Bioscience,* **45**, 752–758.

Swift, C.C., Haglund, T.R., Ruiz, M. and Fisher, R.N. (1993) The status and distribution of freshwater fishes of Southern California. *Bulletin of South California Academy of Science,* **92**, 101–167.

Taylor, C.A., Warren, M.L., Fitzpatrick, J.F., Hobbs III, H.H., Jezerinac, R.F., Pfleiger, W.F. and Robison, H.W. (1996) Conservation status of crayfishes of the United States and Canada. *Fisheries (Bethesda),* **21**, 25–38.

Townsend, C.R. and Winterbourn, M.J. (1992) Assessment of the environmental risk posed by an exotic fish: the proposed introduction of channel catfish (*Ictalurus catus*) to New Zealand. *Conservation Biology,* **6**, 273–282.

Trenham, P.C., Shaffer, H.B. and Moyle, P.B. (1998) Biochemical identification and assessment of population structure of morphologically similar native and invading smelt species (*Hypomesus*) in the Sacramento–San Joaquin Estuary, California. *Transactions of the American Fisheries Society,* **127,** 417–424.

Vermeij, G.J. (1991) When biotas meet: understanding biotic interchange. *Science,* **253**, 1099–1104.

Vincent, E.R. (1996) Whirling disease and wild trout: The Montana experience. *Fisheries (Bethesda),* **21**, 32–33

Wikramanayake, E.D. (1990) Conservation of endemic rain forest fishes of Sri Lanka: results of a translocation experiment. *Conservation Biology,* **4**, 32–37.

Williams, J.D., Warren, M.L., Cummings, K.S., Harris, J.L. and Neves, R.J. (1992) Conservation status of freshwater mussels of the United States and Canada. *Fisheries (Bethesda),* **18**, 6–22.

Part 3

International pathways

13 The scale and ecological consequences of biological invasions in the World's oceans

JAMES T. CARLTON
Williams College – Mystic Seaport, Mystic, Connecticut, USA

Abstract

Marine organisms have been moved around the world by humans for centuries if not millennia. These introductions have led to a profound alteration of the diversity and structure of many coastal communities, including exposed rocky shores, sublittoral soft bottom habitats, sandy beaches, marshes, and estuaries. More than 1 000 species of nearshore marine plants and animals that are now regarded as naturally cosmopolitan may represent overlooked pre-1800 invasions. At least some of these early invasions are likely to now be predominant species where they were introduced, and thus perhaps some of the most important organisms regulating community structure. Testing the hypothesis that such species are part of a pre-modern invasive biota is possible through a variety of techniques. In contrast to ancient vessels, modern vessels may be playing an even larger role, carrying between 3 000 and 10 000 species globally in any given 24 hour period in their ballast water. Recognizing the scale of these modern invasions may be as difficult as recognizing the pre-1800 invasive biota, in the former case because of the decline of scientists who are able to assess alterations in coastal biodiversity.

Introduction: early human-mediated marine invasions

Marine animals, plants, and other organisms have been accidentally or intentionally moved among the world's seas as long as humans have crossed the oceans for exploration, colonization and commerce. This process has led to a profound (but in many, if not most, regions of the world, largely still unrecognized) alteration of the diversity and structure of many shallow coastal marine and estuarine communities (Carlton, 1987, 1989, 1996; Ruiz *et al.*, 1997; Cohen and Carlton, 1998).

Since the 1400s on an interoceanic scale – and much longer on an *intra*oceanic basis – ships have been an effective transport vector not only for humans but also for the movement of other mammals, birds, and plants for continent- or island-seeding. These early intentional translocations were almost always accompanied, however, by a far greater number of secondary species carried inside the ship – a terrestrial complement consisting of

O. T. Sandlund et al. (eds.), Invasive Species and Biodiversity Management, 195–212.

insects, spiders, mites, centipedes, other arthropods, and small mammals (particularly mice and rats), and a semi-terrestrial, or maritime, component, consisting of supralittoral or littoral organisms that accompanied the sand, rock, and other seaside "dry" ballast placed into the holds of ships. This latter element included shoreside plants, arthropods, mollusks, and other organisms.

While this view of the translocation of terrestrial and semiterrestrial animals and plants during the course of the ebb and flow of human populations is generally well known and understood, far less is known about the diversity of the fouling organisms that were attached to the hull of – or the organisms (such as shipworms and limnoriid isopods, or gribbles) that bored into – these same ships.

When a Spanish galleon departed its southern European theater of operations in the 17th century, presumably with at least a modest development of an attached fouling and perhaps boring community, what proportion of the biota – in terms of both numbers of species and numbers of individuals – survived the first major voyage leg through the tropics to southern hemisphere temperate waters? What proportion of the barnacles, bryozoans, sea squirts, mollusks, algae, and other organisms survived the next voyage from the Atlantic Ocean to the Indian or Pacific Oceans? What fraction of the original eurobiota on the hull of this vessel was not only alive but able to reproduce when the vessel put into the harbor at Guayaquil, Ecuador, in the eastern Pacific Ocean?

With uncounted vessels beginning to roam the coasts and oceans of the world in increasing numbers at the dawn of the 17th century (Haws, 1975; Keay, 1991), it is clear that humans had begun to set in motion what was to be a revolution in the structure of marine ecosystems: the movement and introduction around the world in **ecological** time of organisms that had been isolated for millions of years over **evolutionary** time. Most of the organisms being so moved on ships' hulls or in ships' dry ballast appear to be excluded from natural global transport. Thus, the coast-dwelling species that characterize ship fouling communities have never been found on wood drifting on the high seas. The high seas, lacking extensive food resources in surface waters and with a unique ensemble of physical-chemical characteristics, do not offer a viable environment for neritic (shelf-dwelling) taxa that have evolved under more eutrophic and distinctly different physical-chemical conditions. There are certainly neritic taxa with long-lived planktotrophic larvae found in the open ocean (so-called teleplanic larvae), but these are not species which characterize ship fouling or boring communities.

Ships transport neritic organisms through hostile oceanic conditions. These organisms may undergo transitory (if not lethal) physiological stress and/or undergo reduced metabolic functioning. However, ships periodically return to coastal waters, especially shallow embayments and estuaries. This return to inshore waters may provide sufficient renewal to the fouling assemblage, all the more so in earlier centuries when such port visits were prolonged stopovers of weeks or months. This renewal may take the form of the resumption of more active feeding by suspension-feeding animals in the fouling community, triggered by the denser phytoplankton concentrations that characterize shallow waters. In this manner, these organisms would potentially build energy reserves for the next open-ocean leg if such should come to pass. Similarly, attached algae on ships' hulls would also now be renewed in such coastal regions of higher nutrient concentrations.

For many ancient vessels that put into shallow coastal waters, it was the end of the voyage (for many reasons), and thus the entire ship fouling assemblage became a large and instantaneous inoculum. These unidirectional voyages may have been particularly effective in introducing **communities** of species, as the vessel would remain (floating or sunk) in a novel region long enough for the non-native organisms to mature and for the appropriate season to come to facilitate reproduction.

Which taxa may have been involved in these early invasions? Examples are shown in Table 13.1. It is important to note the potential diversity of methods by which species could be transported by early sailing vessels. Wooden vessels were thus floating marine biological islands. An imaginary example of the biological complement of such a wooden sailing vessel of the mid-18th century, assuming a maximum number of habitats "in use" – but also assuming a conservative number of species transported – is shown in Table 13.2. The hull fouling community, if well-developed, and composed of a diverse morphological array of sessile organisms, could harbor errant (vagile) species deep within its matrix. Vessels that supported rich communities of boring invertebrates could further host additional species within the borer galleries. Furthermore, as Cheng (1989) has noted, neustonic species could potentially lay their eggs directly on ship hulls (see Table 13.1). The sand and rock ballast inside the ship also may have supported a host of species. Anchors and anchor chains, especially in older vessels, may have transported a unique subassemblage of species.

In short, at one time, an older, well-fouled and well-bored vessel with muddy or colonized anchor systems and with mixed sand and rock ballast could have easily transported 150 or more species of marine protists, invertebrates and plants on a single voyage. Two hundred and fifty years later – at the turn of the 21st century – that number of species transported by one ship on one voyage may be easily rivalled or exceeded, despite the fact that ship fouling communities are doubtless only a shadow of what they used to be, as discussed below.

Carlton and Hodder (1995) have shown through experimental studies on a fouling community on a replica of Sir Francis Drake's ship "Golden Hinde" – whose speed and port-residency mimicked 16th century conditions – the nature of the losses and gains in species in the fouling community as the ship transited open coastal waters and visited several ports. However, no experimental studies have been conducted on replicas of ancient vessels that have crossed oceans or sailed between oceans, nor have any studies been undertaken to specifically address the above hypothesis of "port renewal". Finally, it may be noted that the ability of ships to introduce species is on occasion viewed as if the phenomenon were limited to only shipping harbors and ports, and indeed even to regions within the ports where ships concentrate (or, in modern times, where ships release most of their ballast water). There is no evidence that this limitation exists. Two processes may be considered here: (1) the ability of species to be released from ships as the vessel proceeds through open water, such as along coastlines, and (2) the ability of species to spread after inoculation.

Relative to the first, Carlton and Hodder (1995) considered the phenomenon where the common fouling hydroid *Ectopleura crocea* (= *Tubularia crocea*) shed many of its hydranths as a vessel sailed through open waters. These hydranths, potentially capable of bearing viable larvae, may thus be scattered broadly along a coastline, entering small river mouths and lagoons where no shipping exists. Similarly, the attached scyphistomae of scyphozoan jellyfish may release larvae (ephyrae) as a ship sails across an ocean or along a coastline, thus widely seeding a region.

Relative to the second, natural dispersal (or "secondary" dispersal) of an introduced species often proceeds after establishment. Thus, planktotrophic larvae or water-borne juveniles (of brooding taxa such as peracarid crustaceans) may be carried broadly within estuaries and bays, and equally widely *along* coastlines, entering many other estuaries and bays.

Table 13.1 Higher-level taxonomic groups with now-"cosmopolitan" species likely to have been dispersed prior to the year 1800 by ships.

Group	Examples/Remarks
"Protozoa" (protozoans)	Ciliates such as folliculinids, *Vorticella, Zoothamnium* and *Cothurnia*; meiofaunal protozoa in sand ballast; foraminiferans in fouling and in sand ballast.
Nematoda (roundworms)	In ship fouling and in sand ballast.
Porifera (sponges)	Encrusting species, and boring *Cliona* in barnacles and mollusks on hull fouling.

Table 13.1 continued

Group	Examples/Remarks
Cnidaria (coelenterata)	
Hydrozoa (hydroids)	Potentially a great many species transported and introduced by ship fouling.
Scyphozoa (jellyfish)	Transported as fouling scyphistomae (with subsequent release of ephyrae).
Anthozoa (sea anemones)	Smaller crevicolous sea anemones, such as diadumenids and sagartiids, and larger anemones such as *Metridium*.
Platyhelminthes (flatworms)	Turbellarians in ship fouling and sand ballast; parasitic trematodes in invertebrate hosts.
Rhynchocoela (nemerteans)	Semiterrestrial, supralittoral species carried in ships' ballast; *Lineus* and other genera in fouling.
Annelida (true worms)	
Oligochaeta (oligochaetes)	Fouling species and shore-dwelling species (such as enchytraeids) transported in ballast.
Polychaeta (polychaetes)	Many species of phyllodocids, nereids, spionids, syllids, cirratulids, polynoids, capitellids, terebellids, sabellids, spirorbids, serpulids, and other errant and sedentary families in fouling communities; supralittoral species (such as nereids) transported with rock ballast; archiannelids transported in sand ballast.
Crustacea	
Cirripedia (barnacles)	Many now widespread barnacle species, especially throughout the Indo-Pacific, likely owe many outlier populations to dispersal via ships.
Amphipoda (amphipods)	Many species of caprellid amphipods, associated with hydroid fouling, and even more species of gammarid amphipods; shore-dwelling talitrids in ballast.
Isopoda (isopods)	Many species of burrowing, boring, and nestling isopods, especially sphaeromatids, cirolanids, and limnoriids, as well as isopods associated with hydroid and other fouling; shore-dwelling ligiids carried in ballast, or in cracks in wooden ship hulls above the waterline.
Tanaidacea (tanaids)	As with fouling isopods.
Copepoda (copepods)	For example, benthic harpacticoid copepods, including those associated particularly with wood and those in sand ballast.
Ostracoda (ostracods)	Potentially many species of fouling ostracods.
Decapoda (crabs and shrimps)	Grapsid, xanthid, majiid and other nestling crabs; shrimp transported deep within the matrix of fouling communities, and perhaps in the galleries of ship-wormed vessels.

Table 13.1 continued	
Group	**Examples/Remarks**
Insecta (insects)	In particular marine Diptera (flies) as larvae in fouling communities; many littoral and supralittoral species transported with sand and rock ballast; neustonic seastrider *Halobates* laying eggs on ships' hull (Cheng, 1989) or entrained in fouling communities (Carlton et al. 1995).
Pycnogonida (sea spiders)	In hydroid fouling.
Mollusca	
Gastropoda (snails and sea slugs)	Smaller gastropods that may be carried in rock ballast (such as littorinids or "periwinkles"); nudibranch opisthobranchs associated with hydroid and bryozoan fouling; limpets on ships' hulls.
Bivalvia (clams, mussels, and oysters)	Nestling bivalves, such as hiatellids, venerids, petricolids, and pholads, and fouling bivalves such as chamids, mussels and oysters; the "naturalness" of the biogeography all of these groups should be reconsidered, given their ubiquitousness in hull fouling.
Polyplacophora (chitons)	While few chitons are noted as introductions, the presence of chitons in ship fouling communities and on ships' anchors (which may remain wet through wave splash over considerable distances) suggests the potential for former but now overlooked introductions.
Ectoprocta (bryozoans)	As with the hydroids, a potentially very large number of fouling species transported and introduced globally.
Kamptozoa (entoprocts)	*Barentsia* (and other genera?) in fouling.
Chordata	
Ascidiacea (sea squirts)	A great many fouling species transported globally, a phenomenon well underestimated.
Algae	It seems probable that many of the common fouling algae, such as *Ulva, Enteromorpha, Polysiphonia,* and *Bryopsis*, and many of the smaller fouling microalgae, owe a large portion of their modern distributions to ship transport. Larger algae, such as the kelps (*Macrocystis, Undaria, Laminaria* and others) which can colonize boat bottoms or become entangled in anchors, may be transportable long distances by boats.
Sea grasses	Possibly entangled and transported on anchors and anchor chains.

Table 13.2 An imaginary assemblage of marine organisms on a wooden sailing vessel of 1750. Note that the number of species in this list would increase considerably if parasites, commensals, and other symbionts are included.

Position on/in vessel	Examples of taxa and conservative diversity estimate (numbers of species)	Total species richness
Hull fouling	Protozoa (10), sponges (2), scyphozoans as scyphistomae (1), hydroids (5), sea anemones (1), flatworms (5), nemerteans (2), polychaetes (20), barnacles (3), isopods (5), tanaids (1), amphipods (10), ostracods (3), harpacticoid copepods (3), insects (1), crabs (2), pycnogonids (2), bivalves (5), nudibranchs (2), entoprocts (1), bryozoans (8), seasquirts (5), algae (10), lignicolous fungi (1)	108
Hull boring	Shipworms (2), boring isopod *Limnoria* (2), associated borer-gallery invertebrate biota (protists, amphipod *Chelura*, flatworms, polychaetes, ostracods) (6)	10
Hull nestlers (cracks in wood above waterline)	Ligiid isopods (1), snails (e.g., *Assiminea* or *Myosotella*) (1)	2
Anchor, anchor chain, anchor locker	Protozoa (5), capitellid polychaetes (1), oligochaetes (1), chitons (on anchor) (1), microalgae (3)	11
Sand ballast	Meiofaunal invertebrates (5)	5
Supralittoral sand/rock ballast	Higher plants (5, as seeds), talitrid amphipods (1), insects (10), nemerteans (1), enchytraeid oligochaetes (1), nereid polychaetes (1), snails (1)	20
	Total:	156

The potential scale of cryptic invasions

How many species were successfully introduced around the world in these early ship voyages over a period of at least 500 or 600 years? Dozens? Hundreds? Thousands?

Ships were of course moving around the world for centuries prior to the advent of any biological expeditions and collections; as a result, we have virtually no picture of the number or diversity of invasions prior to the mid-19th century. We can, however, assume that ship fouling communities were in motion prior to the 19th century, as discussed above – and extensive evidence since the 19th century clearly indicates that vessels acted as successful "conveyor belts" that led to exotic species invasions.

Thus, if we were to hypothesize that in the 300 year period between, for example, 1500 and 1800, a minimum of 3 to 5 species per year were transported and introduced successfully by ships to a new location in the world, then between 900 and 1 500 coastal species of marine plants and animals that are now regarded as naturally widespread – if not cosmopolitan – in fact represent overlooked invasions. It follows that at least some of these pre-1800 invasions are likely to be common if not abundant species where they were introduced, and thus perhaps some of the most important organisms regulating community structure.

In classical biogeography, these now widespread species are viewed as "native" or "natural" – that is, in the absence of evidence that they **were** introduced by human activity, their global distribution is seen to be the result of natural processes (natural dispersal of larvae or rafted adults, the former presence of now-closed seaways across the Isthmus of Panama, now-extinct circumglobal seas such as the Tethys Sea, and, indeed, plate tectonics – despite the fact that some of these processes would have created disjunct populations that would now be millions or tens of millions of years old that failed to undergo allopatric speciation).

Carlton (1996b) has argued, however, that the absence of evidence that now-widespread coastal species were not introduced does not logically default the organisms in question to a "naturally distributed" status, but, rather, the lack of evidence defaults to an unknown status. Thus, species that could have been transportable by human activity – or could have been transported by natural processes – are **cryptogenic** species, taxa whose status as native or exotic remains undetermined. Bertelsen and Ussing (1936) heralded this concept many years ago with their prophetic statement that "If an animal form with a very wide and scattered distribution has only the slightest possibility of [having been transported by wooden ships], this possibility should also be taken into consideration."

Testing this hypothesis that 100s or 1 000s of cryptogenic species are part of a pre-modern cryptic invasive biota is possible through a variety of techniques, as discussed in part by Chapman and Carlton (1991, 1994). For many now-widespread taxa, recognizing early invasions will be difficult at best, and will require painstaking reviews of museum collections accompanied by genetic analyses of widespread populations. For other species, while recognizing the probability that their modern distribution has been ship-created in large part, establishing an original provenance and thus a picture of invasions may be nearly impossible.

Potential examples of cryptic invasions abound. As noted in Table 13.1, for example, even very large brown kelps (Phaeophyta), such as the

Northeastern Pacific Ocean *Macrocystis pyrifera*, can attach to boat bottoms (J. Carlton, personal observation). This species appears to be native to the Californian and Mexican coasts, where both marine vertebrates (sea otters) and invertebrates (such as certain mollusks) appear to be closely linked evolutionarily with this algae. During the 17th and 18th century Spanish and other vessels with lengthy anchoring periods on the central or southern California coasts would have become fouled with this kelp. In addition, ship anchors pulled up from the bottom may have entangled *Macrocystis*, although seaweeds could get entrained in other external surfaces of a vessel as well and be carried long distances. Carlton *et al.* (1995) has thus noted that a staysail schooner (35 meters in length) was observed to transport in the North Atlantic Ocean the same individuals of living brown algae (*Ascophyllum nodosum*) for two weeks at the base of a bobstay chain, below the ship's bow, over a distance of 600 kilometers, through sea states that reached conditions of Beaufort 7 (wave heights ranging up to four meters).

What is now identified as *Macrocystis pyrifera* is now common in certain coastal areas in the Southern Hemisphere, such as Australia (Tasmania only), New Zealand, and the east and west coasts of South America (Womersely, 1987) – all regions where Spanish and other vessels, having earlier visited Californian waters, would have later found themselves at extended anchorages. Indeed, it is compelling to note that the first specimens of *Macrocystis pyrifera* were found in the southern Atlantic Ocean (Womersely, 1987) in the 18th century. Such "early records" have been argued to suggest that such species must be naturally distributed – but, clearly, ships had the potential to move and relocate this kelp around the world for 200 or more years prior to the 1700s! The distribution of *Macrocystis* is now held to be natural. It is, at the least, a cryptogenic species in the southern hemisphere.

In a further example, a guild of wood-boring sphaeromatid isopods with sculptured pleotelsons, some of which species may have arisen in the Indian Ocean, were widely dispersed by ships, but the extent of this activity is only now becoming clear. The small (0.5 cm length) isopod *Sphaeroma terebrans* bores into the growing root tips of living red mangroves (*Rhizophora mangle*) in the tropical and subtropical Western Atlantic Ocean from South America to Mexico and into Florida. In many regions it appears to control the seaward extent of the mangrove forests by destroying the prop roots. Only recently has it been realized that this isopod is in all likelihood native to the Indian Ocean (Carlton and Ruckelshaus, 1997), having arrived, perhaps in the 1870s, as a fouling or boring organisms in ships.

This one non-native isopod species has thus virtually "reset" the seaward history of the mangrove ecosystems of a large part of the tropical Atlantic coast – and yet there were so few biologists in these regions at the time that this remarkable introduction, and the subsequent ecosystem-level changes it caused, went unnoticed. An example such as this further suggests that other important community- or ecosystem-level regulatory processes – such as this one, the control of the seaward progression of mangrove communities by a single species – may in fact have been non-existent only 150 years ago.

The modern manifestation of shipping: ballast water

While organisms continue to attach to ships' hulls – albeit in lesser biomass than many centuries ago, because of antifouling paints, increased ship speeds, and decreased harbor residencies, modern cargo vessels still play a significant role in the dispersal of nonindigenous species. In contrast to ancient vessels which carried organisms on the **outside** of the vessel, modern vessels carry organisms in far greater quantities on the **inside** of the vessel in ballast water.

Ballast water is the water that a ship intentionally takes aboard for stability, trim, and other purposes; it is carried both in empty cargo holds and in dedicated ballast tanks. It is not a simple world from the biologists' point of view, and as a transport mechanism it appears to have few if any parallels on land or at sea for its biological breadth and efficiency (Carlton, 1985: Carlton and Geller, 1993; Carlton *et al.*, 1995). A ship may have aboard ballast water from one source, or from several different sources in several different tanks, or from different sources mixed together in one tank. A ship may carry a few tonnes (cubic meters) of ballast water to many tens of thousands of tonnes of ballast water. At the bottom of ballast tanks there may be extensive sediment accumulations that facilitate the development of benthic invertebrate communities.

When crossing the sea or running up a coast to pick up cargo, a ship carries enough water to compensate proportionally for the lack of cargo; when it arrives at its destination it discharges this water to load the cargo. Thus, while countries, or specific regions, or individual ports, that are net exporters receive the most water, countries or regions or ports that are net importers may also receive a good deal of ballast water, not only from the ships that actually do arrive to pick up some cargo, but because most ships carry some water most of the time. One thousand metric tonnes of water in a forepeak ballast tank means virtually "no ballast on board" to the ship's crew: one thousand metric tonnes of water means a great many marine organisms to a marine biologist.

There are many dedicated trade routes upon which ships and their ballast water and its plankton assemblages are moved. There are also many other routes that capture the cargo-of-the-day, and send ships in complex routes around the world. Thus a vessel may depart Morocco in ballast for Argentina to load a cargo of black beans for Mexico; having delivered its cargo to Mexico, ballast water from the subtropical Vera Cruz River is then carried up to Long Island Sound near New York, where the water is discharged as the vessel loads scrap metal, bound for India. In the meantime, another ship departs Odessa in the Black Sea, in ballast, for West Africa, to load a cargo of bauxite for Texas in the Gulf of Mexico. But this time Texas provides another cargo – wheat – and thus perhaps only the very smallest amounts of water, used only for trim, now finds its way down to Chile. This water may have come from Texas, or from Africa, or from the Black Sea. Its cargo offloaded in Chile, the vessel is then off – with a load of Chilean plankton – to the Columbia River, between Oregon and Washington states with a potentially excellent match between the temperate waters at either end of the globe, and thus perhaps a good match for the life on board with its new target environment in the Northwest Pacific Ocean.

As a result of the extensive modern-day movement of ballast water around the world, it is not surprising that invasions have occurred, although the recognition of these newer invasions tends to be restricted to the larger and more abundant and thus more obvious animals. Some of the better known examples of the late 1980s and 1990s include: (1) the establishment of the Chinese clam *Potamocorbula* in San Francisco Bay, where by the 1990s it became so abundant – tens of thousands of clams per square meter of bay floor – that the spring phytoplankton bloom (measured as water column chlorophyll), one of the core elements of the bay's food web, has virtually disappeared (Alpine and Cloern, 1992, and J. Thompson, personal communication, 1998); (2) the invasion of the Japanese starfish (seastar) *Asterias amurensis* in southern Australia (Buttermore *et al.*, 1994), whose ecological consequences are not yet known, and (3) the establishment of the comb jelly (ctenophore) *Mnemiopsis leidyi* in the Black Sea, and the subsequent demise of the already weakened Azov and Black Sea anchovy fisheries (Shushkina *et al.*, 1990; Malyshev and Arkhipov, 1992).

There have been a great many suggestions of "what to do about ballast water" (National Research Council, 1996), many of which inadvertently oversimplify the global complexities of having tens of thousands of ships at sea of many different types, ages, voyage lengths, and operating under many sea conditions. The primary and immediate "stop-gap" measure for ballast management is what is variously known as midocean, open ocean, deep ocean, or high seas exchange. Thus, a vessel sailing from, for example, the Weser River in Germany to Chesapeake Bay on the American Atlantic coast would attempt to release (**deballast**) the original European ballast water in the open ocean (where coastal species would not survive) and then **reballast** with oceanic water, releasing the latter into Chesapeake Bay (where the oceanic species would not survive).

There are, however, a number of potential challenges with high seas exchange. For example, when sea conditions are excessively rough, deballasting could lead to ship destabilization, a self-defeating process. In addition, ballast exchange in and of itself may not sufficiently remove bottom sediments (noted earlier to be able to support benthic communities) nor kill fouling organisms. Ballast exchange may also be of limited value for ships moving along coastlines between distant bays. Such vessels could translocate a new invasion in one estuary many 100s or 1 000s of kilometers to another bay in a matter of hours or days. However, such vessels do not normally divert to the open ocean, and ballast exchange (releasing water on the move) as the ship transits in a parallel fashion a coastline could have an inadvertent "seeding" effect, whereby larvae of a species are spread along a long length of coast. On the one hand, the dilution of such released larvae, even if only a few miles off a coastline, should be huge; arguing against this would be the release of a "cloud" of millions of larvae that would be transported and survive as a cohort.

These and other challenges associated with ballast exchange are some of the major driving factors that are stimulating research in ballast management technologies, such as equipping individual vessels with the ability to remove or kill the biota in ballast water. These technologies include filtration, heating and ultraviolet light (National Research Council, 1996). Individual nations are, in the meantime, requesting vessels to undertake open ocean exchange, while the United Nations International Maritime Organization in 1998–99 considers a possible instrument (such as a MARPOL Annex) on ballast water management to reduce the spread of nonindigenous organisms.

The potential scale of modern ballast water transport

Ballast water has been found to carry hundreds of species, in virtually every kingdom and phylum – from cholera and botulism bacteria to protists, diatoms, dinoflagellates, algae, invertebrates, and fish. Carlton and Geller (1995, footnote 12) estimated that "on any one day, several thousand species" may be carried around the world in ballast water. This estimate was later converted to "more than 3 000 species... at any given moment" (National Research Council, 1995). In retrospect, this number would appear to be too low of an estimate. There are more than 35 000 vessels at sea at any given time; if only 10 percent of these vessels (3 500) are assumed to be carrying a full ballast load on any one day, and if only two unique species are assumed to be in each vessel, then more than 7 000 species may in fact be "in motion" on any given 24 hour-period.

This revised estimate, of course, assumes that 90 percent of the world's fleet is without ballast at any given time, certainly a grossly conservative approach, especially if one considers the container ship trade, whose vessels *always* have some water all of the time. If one adds the ballast water carried by military vessels (such as submarines), ocean-going tugs and barges, self-propelled exploratory platforms, ballastable yachts, and other vessels, it may be that the number of taxa transported daily on a global basis exceeds 10 000 species. Of course, the next day, while the diversity may be comparable, different species will be involved; if we estimate that between 5 % and 10% of the species (500 to 1 000 species) are changed daily, total movement on a 7-day (weekly) basis could then approach or exceed 15 000 species.

While these are merely estimates based upon assumptions and scenarios, it is not so much the exact number of species that is of concern, but rather the sheer scale of the process, and its implications relative to the scale of daily inoculations that could lead to introductions. As human populations grow, the world trade of commercial goods increases, and so does the world movement of ballast water. The number of daily inoculations increases, inevitably leading to more successful introductions. It is thus not surprising that there have been an increased number of ballast-mediated invasions around the world with each succeeding decade (Carlton *et al.*, 1990, 1995; Mills *et al.*, 1993; Carlton, 1996). Indeed, Cohen and Carlton (1998) show that invasions in San Francisco Bay, California, now average one newly established species every 14 weeks – a result of both ballast water and other vectors, noted below.

Other human-mediated dispersal vectors in the ocean

There are many other human-mediated transport mechanisms other than ship fouling or ship's dry ballast; these are reviewed by Carlton (1985, 1992a, 1993, 1994). As humans move so does their food or their potential food. The archeological record indicates, for examples, that the Vikings returned to the northlands of ancient Europe with the American soft shell clam, *Mya arenaria*, now a common European species (Carlton, 1996a). Centuries later, in the early 1800s, other Europeans apparently returned the favor, transporting the popular and very edible European periwinkle, *Littorina littorea,* to North American shores – where, while it is not much eaten in America, it has by its sheer abundance revolutionized the ecological structure of American shores from Canada to New Jersey, ranging from profoundly influencing the abundance of seaweeds to altering the development of marshes (Carlton, 1992b).

In the past five to six centuries, also, humans began to move oysters around the world. The evidence now suggests, for example, that the well-known Portuguese oyster *Crassostrea angulata* is in fact nothing more than the Japanese oyster *Crassostrea gigas*, which appeared in Europe for the first time soon after Portuguese explorers returned from Asia in the 16th century (Carlton, 1985). Three hundred years later, in the 19th century, American Atlantic oysters, *Crassostrea virginica*, crossed the North American continent in the new Transcontinental Railroad in 1869, simultaneously and unintentionally transporting much of the Western Atlantic Ocean estuarine biota to Eastern Pacific Ocean estuaries (Cohen and Carlton, 1995). Oyster movements continued well into the twentieth century. The Pacific (Japanese) oyster *Crassostrea gigas* was moved virtually all over the world to scores if not hundreds of new locations. All of these movements of adult oysters were inevitably accompanied by the translocation of a rich associated biota, especially if one considers the diversity of smaller organisms in the silt and mud on and amongst the oyster shells. These movements caused Elton (1958) to consider that oyster movements were one of the most profound mechanisms leading to the reorganization of marine life in modern times!

In one of the lasts major events of the movement of massive numbers of adult oysters, *C. gigas* was brought (again!) to Europe, from Japan and from British Columbia, in the early 1970s to France. These oyster transplantations led to the establishment in Europe of a number of Asian invertebrates and algae, including the well-known Asian edible seaweed *Undaria*, a number of other smaller Western Pacific algae, and the not-so-edible Asian fouling seaweed *Sargassum muticum* (Verlaque, 1996). Aquaculture and mariculture activities are blossoming throughout the world, and along with that expansion a concomitant ability to move virtually any organism anywhere in the world in 24 hours – including the viruses and other pathogenic organisms that often accompany such intensively-cultured stocks (and can in turn potentially destroy the very industry that is trying to grow!).

While we strive to control ballast invasions, it is important to note that the increase in aquaculture and mariculture activities around the world has necessitated equal levels of concern relative to the potential for accidental introductions and intentional releases. An important step in this regard has been the evolution since the 1970s of the "Code of Practice on the Release of Marine and Aquatic Organisms", of the International Council for the Exploration of the Sea (ICES, 1995). This Code argues clearly for reasoned caution, extensive scientific work, and careful and exhaustive consideration prior to any intentional releases.

Finally, translocation of "products from the sea" are not limited to direct human consumption. A remarkable example is the movement of coastal algae as "packing" material to ship living marine worms (polychaetes) to be used as bait. One such industry, located in the state of Maine, USA, ships worms-and-algae around the world (Cohen and Carlton, 1995; Cohen *et al.*, 1995). These algae (and no doubt many of the worms) are released directly into novel coastal environments, and have doubtless led to a number of unrecognized invasions of crustaceans, mollusks, and others invertebrates. One recognized example of an invasion resulting from this transport vector is the establishment of the Atlantic snail *Littorina saxatilis* in San Francisco Bay, California (Carlton and Cohen, 1998).

Epilogue

A few concluding remarks concerning the overlooked scale of the ecological consequences of marine biological invasions are appropriate here, especially given the proposed scale of earlier translocations by wooden ships (1 500 (or more?) species of cryptic invasions of earlier centuries) on the one hand, and the estimated scale of modern daily inoculations by modern ships (7 000 (or more?) taxa per day) on the other hand.

Relative to earlier invasions of the 18th and 19th centuries, it is clear that the insertion of single species early on into natural ecosystems resulted in profound community- and ecosystem-level modifications, by the sudden arrival of a neokeystone species. The role of the Indian Ocean isopod *Sphaeroma terebrans* inserted into and altering the mangrove systems of the tropical Western Atlantic Ocean was noted earlier, as was the arrival of the European snail *Littorina littorea* and its regulatory nature in coastal ecosystems of the temperate Western Atlantic Ocean. That these are likely the "tip of the invasion iceberg" seems eminently probable – it would be presumptive at best to assume that we have recognized all of the important earlier invasions! What of, for example, the probable introductions of the European mussel *Mytilus edulis* to the southern hemisphere by shipping in the 18th and 19th centuries? These mussels, some still going under regional names in southern waters, are now aspect dominant members of some intertidal and shallow subtidal communities of South America and Australia. While we have yet to perhaps "catch up" with these earlier invasions, the 20th century invasion of the Mediterranean mussel *Mytilus galloprovincialis* is recognized in South Africa (Griffiths *et al.*, 1992) – the difference being that the arrival of this northern hemisphere species could be witnessed by a resident scientific community, whereas earlier invasions of mussels of course passed in silence.

Relative to later invasions of the 20th century, larger and more conspicuous invaders are more easily recognized; the examples of *Potamocorbula*, *Asterias*, and *Mnemiopsis* were noted earlier. If our estimates of the total number of taxa now being moved globally are even fractionally correct, and even with the recognition that most inoculations of most species likely lead to 100% mortality most of the time, there is a curious silence from many corners of the world relative to new invasions, and even more of a silence from all corners relative to new invasions of smaller organisms such as polychaetes, amphipods, flatworms, nemerteans, hydroids, nematodes, rotifers, gastrotrichs, protozoans, diatoms, dinoflagellates, and so forth. Rather than being reflective of an ocean generally resistant to invasions, a reasonable assumption is that this lack of reports is due the concomitant extraordinary decline of marine ecologists with expertise in regional biodiversity, with expertise in systematics (which knowledge is not mutually exclusive with expertise in, for example, experimental manipulative ecology), and with the interest and time to document alterations in the coastal flora and fauna.

The world, however, with increasing human populations and thus with increasing burdens and stresses on marine life everywhere, now requires greater – not lesser – attention by scientists who can document the scale of change in time and space in marine ecosystems in the 21st century.

References

Alpine, A.E. and Cloern, J.E. (1992) Trophic interactions and direct physical effects control biomass and production in an estuary. *Limnology and Oceanography,* **37**, 946–955.

Bertelsen, E. and Ussing, H. (1936) Marine tropical animals carried to the Copenhagen Sydhavn on a ship from the Bermudas. *Videnskabelige Meddelelser fra Dansk Naturhistorisk Forening i København,* **100**, 237–245.

Buttermore, R.E., Turner, E. and Morrice, M.G. (1994) The introduced northern Pacific seastar *Asterias amurensis* in Tasmania. *Memoirs of the Queensland Museum,* **36**, 21–25.

Carlton, J.T. (1985) Transoceanic and interoceanic dispersal of coastal marine organisms: the biology of ballast water. *Oceanography and Marine Biology, an Annual Review,* **23**, 313–371.

Carlton, J.T. (1987) Patterns of transoceanic marine biological invasions in the Pacific Ocean. *Bulletin of Marine Science,* **41**, 452–465.

Carlton, J.T. (1989) Man's role in changing the face of the ocean: biological invasions and implications for conservation of near-shore environments. *Conservation Biology,* **3**, 265–273.

Carlton, J.T. (1992a) Dispersal of living organisms into aquatic ecosystems as mediated by aquaculture and fisheries activities, in *Dispersal of Living Organisms into Aquatic Ecosystems,* (eds A. Rosenfield and R. Mann), Maryland Sea Grant Publication, College Park, Maryland, pp. 13–45.

Carlton, J.T. (1992b) Introduced marine and estuarine mollusks of North America: an end-of-the-20th-century perspective. *Journal of Shellfish Research*, **11**, 489–505.

Carlton, J.T. (1992c) Blue immigrants: the marine biology of maritime history. *The Log* (*Mystic Seaport Museum, Mystic CT*), **44**, 31–36.

Carlton, J.T. (1993) Dispersal mechanisms of the zebra mussel (*Dreissena polymorpha*), in *Zebra Mussels: Biology, Impacts, and Control*, (eds T.F. Nalepa and D.W. Schloesser), CRC Press, Inc., Boca Raton, Florida, Chapter 40, pp. 677–697.

Carlton, J.T. (1994) Biological invasions and biodiversity in the sea: the ecological and human impacts of nonindigenous marine and estuarine organisms. Keynote Address, in *Nonindigenous Estuarine and Marine Organisms (NEMO), Proceedings of the Conference and Workshop, Seattle, Washington, April 1993,* U.S. Department of Commerce, National Oceanic and Atmospheric Administration, pp. 5–11.

Carlton, J.T. (1996a) Marine bioinvasions: the alteration of marine ecosystems by nonindigenous species. *Oceanography,* **9**, 36–43.

Carlton, J.T. 1996b Biological invasions and cryptogenic species. *Ecology,* **77**, 1653–1655.

Carlton, J.T. and Cohen, A.N. (1998) Periwinkle's progress: the Atlantic snail *Littorina saxatilis* (Mollusca: Gastropoda) establishes a colony on a Pacific shore. *The Veliger,* **41**.

Carlton, J.T. and Geller, J.B (1993) Ecological roulette: The global transport of nonindigenous marine organisms. *Science,* **261**, 78–82.

Carlton, J.T. and Hodder, J. (1995) Biogeography and dispersal of coastal marine organisms: experimental studies on a replica of a 16th-century sailing vessel. *Marine Biology,* **121**, 721–730.

Carlton, J.T., Reid, D.M. and van Leeuwen, H. (1995) *Shipping Study. The Role of Shipping in the Introduction of Non-Indigenous Aquatic Organisms to the Coastal Waters of the United States (other than the Great Lakes) and an Analysis of Control Options.* The National Sea Grant College Program/Connecticut Sea Grant Project R/ES-6. Department of Transportation, United States Coast Guard, Washington, DC. and Groton, Connecticut. Report Number CG-D-11-95. Government Accession Number AD-A294809, 213 pp. and Appendices A-I (122 pages).

Carlton, J.T. and Ruckelshaus, M.H. (1997) Nonindigenous marine invertebrates and algae, in *Strangers in Paradise. Impact and Management of Non-Indigenous Species in Florida,* (eds D. Simberloff, D.C. Schmitz, and T. C. Brown), Island Press, Washington, DC. and Covelo CA, pp. 187–201.

Carlton, J.T., Thompson, J.K., Schemel, L.E. and Nichols, F.H. (1990) Remarkable invasion of San Francisco Bay (California, USA) by the Asian clam *Potamocorbula amurensis*. I. Introduction and dispersal. *Marine Ecology Progress Series,* **66**, 81–94.

Chapman, J.W. and Carlton, J.T. (1991) A test of criteria for introduced species: the global invasion by the isopod *Synidotea laevidorsalis* (Miers, 1881). *Journal of Crustacean Biology,* **11**, 386–400.

Chapman, J.W. and Carlton, J.T. (1994) Predicted discoveries of the introduced isopod *Synidotea laevidorsalis* (Miers, 1881). *Journal of Crustacean Biology,* **14**, 700–714.

Cheng, L. (1989) Factors limiting the distribution of *Halobates* species, in *Reproduction, Genetics and Distribution of Marine Organisms*, (eds J.S. Ryland and P.A. Tyler), 23rd European Marine Biology Symposium, Olsen and Olsen, Fredensborg, Denmark, pp. 357–362.

Cohen, A.N. and Carlton, J.T. (1998) Accelerating invasion rate in a highly invaded estuary. *Science,* **279**, 555–558.

Cohen, A.N., Carlton, J.T. and Fountain, M.C. (1995) Introduction, dispersal, and potential impacts of the green crab *Carcinus maenas* in San Francisco Bay, California. *Marine Biology,* **122**, 225–237.

Chu, K.H., Tam, P.F., Fung, C.H. and Chen, Q.C. (1997) A biological survey of ballast water in container ships entering Hong Kong. *Hydrobiologia,* **352**, 201–206.

Cohen, A.N., Carlton, J.T. and Fountain, M.C. (1995) Introduction, dispersal and potential impacts of the green crab *Carcinus maenas* in San Francisco Bay, California. *Marine Biology,* **122**, 225–237.

Elton, C.S. (1958) *The Ecology of Invasions by Animals and Plants.* Methuen, London.

Galil, B.S. and Huelsmann, N. (1997) Protist transport via ballast water – biological classification of ballast tanks by food web interactions. *European Journal of Protistology,* **33**, 244–253.

Griffiths, C.L., Hockey, P.A.R., Van Erkom Schurink, C. and Le Roux, P.J. (1992) Marine invasive aliens on South African shores: implications for community structure and trophic functioning. *South African Journal of Marine Science,* **12**, 713–722.

Haws, D. (1975) *Ships and the Sea. A Chronological Review.* Thomas Y. Crowell Co., Inc., New York.

ICES (1995) *Code of Practice on the Introductions and Transfers of Marine Organisms, 1994. Preamble and a Brief Outline of the ICES Code of Practice, 1994.* International Council for the Exploration of the Sea, Copenhagen, Denmark, 5 pp.

Keay, J. ed. (1991) *History of World Exploration.* The Royal Geographical Society. Mallard Press, BDD Promotional Book Co., Inc., New York NY.

Malyshev, V.I. and Arkhipov, A.G. (1992) The ctenophore *Mnemiopsis leidyi* in the western Black Sea. *Hydrobiological Journal,* **28**, 33–39.

Mills, E.L., J.H. Leach, J.T. Carlton, and Secor, C. (1993) Exotic species in the Great Lakes: a history of biotic crises and anthropogenic introductions. *Journal of Great Lakes Research,* **19**, 1–54.

National Research Council (1995) *Understanding Marine Biodiversity: A Research Agenda for the Nation.* National Academy Press, Washington, DC.

National Research Council. (1996) *Stemming the Tide. Controlling Introductions of Nonindigenous Species by Ships' Ballast Water.* National Academy Press, Washington, DC.

Ruiz, G.M., Carlton, J.T., Grosholz, E.D. and Hines, A.H. (1997) Global invasions of marine and estuarine habitats by non-indigenous species: mechanisms, extent, and consequences. *American Zoologist,* **37**, 621–632.

Shushkina, E.A., Nikolaeva, G.G. and Lukasheva T.A. (1990) Changes in the structure of the Black sea planktonic community at mass reproduction of sea gooseberries *Mnemiopsis leidyi* (Agassiz). *Oceanology,* **51**, 54–60.

Verlaque, M. (1996) L'etang de Thau (France), un site majeur d'introduction d'especes en Mediterrannee – relations avec l'ostreiculture, in *Second International Workshop on* Caulerpa taxifolia, (eds M. Ribera *et al.*), Publicacions Universitat Barcelona, Spain, pp. 423–430.

Womersely, H.B.S. (1987) *The Marine Benthic Flora of Southern Australia.* Part II. South Australia Government Printing Office, Adelaide.

14 The Red Sea – Mediterranean link: unwanted effects of canals

CHARLES F. BOUDOURESQUE
Université de la Méditerranée, Marseilles, France

Abstract

The opening of the Suez Canal in 1869 was the cause of the most important biogeographic phenomenon witnessed in the contemporary oceans. Nearly 300 species of Red Sea and Indo-West Pacific origin invaded and settled in the Mediterranean (the "Lessepsian migrants"). Most settlers were benthic and demersal species, including fishes. The migration was almost exclusively from the Red Sea to the Mediterranean.

Migration is easier since the deepening of the Suez Canal which has removed the hyperhalinity of the Bitter Lakes as a barrier. The reduction of the Nile flow has created more stenohaline conditions at the western Canal outlet. As a result there are no signs that the influx of Lessepsian migrant is nearing a plateau, and they now represent about 4% of the Mediterranean species diversity and 10% of the Levantine basin diversity. Several hypotheses to explain the success of Lessepsian migrants are discussed.

Little attention has been paid to the ecological impact of Lessepsian migrants, both at the species and at the ecosystem level. Some evidence suggests drastic changes in species abundance and niche displacement that can be attributed to the competition with Red Sea species. The presence of large herbivore fishes among the migrants has probably had a strong impact on the functional processes of the Mediterranean ecosystems, where herbivory was low. Some species (e.g. the jellyfish *Rhopilema nomadica*) have exerted a significant negative impact on fisheries and tourism. Some migrant species are now of economic importance and are being exploited by local fisheries (e.g. off the Israeli coast, where migrant fishes constitute a third of the trawl catches).

Migration across a canal is probably difficult to control. Possible mechanisms capable of slowing down the invasion of Red Sea species (e.g. the setting up of a lock gate, and/or reactivation of haline barriers) should be studied.

Introduction

The access routes for exotic marine species to a new area are diverse (Zibrowius, 1991; Boudouresque and Ribera, 1994; Farnham, 1994; Verlaque, 1994; Ribera and Boudouresque, 1995; see also Carlton, 1998). (i) They may occur by transportation of sessile (fouling) and vagile

O. T. Sandlund et al. (eds.), Invasive Species and Biodiversity Management, 213–228.

(clinging) species on ship hulls or drilling platforms. (ii) Transportation of species in ballast water and formerly, transport with solid ballast such as sand and stones. Ballast water has been responsible of the introduction species such as the zebra mussel *Dreissena polymorpha* to North America (Carlton, 1993) and the medusa *Mnemiopsis leidyi* to the Black Sea (Travis, 1993). (iii) Intentional introductions of species for commercial use, for example the Japanese oyster *Crassostrea gigas* and the venerid *Ruditapes philippinarum* (Bodoy *et al.,* 1981). (iv) Accidental introductions of species accompanying intentionally introduced species, for example the importation of spat of *Crassostrea gigas* that resulted in the introduction of nearly 15 species of Japanese algae to the Mediterranean (e.g., Riouall, 1985; Riouall *et al.,* 1985; Rueness, 1989; Verlaque and Riouall, 1989; see reviews in Verlaque, 1994; Ribera and Boudouresque, 1995). (v) Discarding of intentionally imported live specimens from the market to the sea have included species used as bait and algae used as packing material (e.g., *Fucus spiralis* introduced from Brittany to a French Mediterranean brackish lagoon; Sancholle, 1988). (vi) Escape from aquaria, especially through open systems that do not have treatment. In the North-Western Mediterranean, the green tropical alga *Caulerpa taxifolia* was introduced this way (Boudouresque *et al.,* 1992, 1995; Meinesz *et al.,* 1993; Meinesz and Hesse, 1991). (vii) Species introduced in association with scientific research. Many scientists who use non-indigenous strains or species are often unaware of, or underestimate the risks of introduction. They fail to take the elementary precautions that are required to prevent these species from escaping from their cultures or breeding sites. An example of this is the red alga *Mastocarpus stellatus* introduced into Germany (Helgoland; North Sea) by a visiting scientist (Ribera and Boudouresque, 1995).

In the Mediterranean, the Suez Canal is an additional and unique route for the introduction of species. The canal, 163 km long running from Port Said to the Gulf of Suez (Egypt), was opened in 1869. It linked two biogeographical provinces, the Mediterranean and the Red Sea that had been partially separated since the early Miocene (ca 20 Ma) and completely separated since the late Miocene (Messinian), ca 5 Ma ago (Robba, 1987). The Suez Canal was deepened and widened several times and now reaches a navigational depth of 14.5 m and is 365 m wide (Halim, 1990). Mass migration of species from the Red Sea to the Mediterranean ("Lessepsian migrants"; Por, 1969) and on a much smaller scale from the Mediterranean to the Red Sea (referred to as "anti-Lessepsian migrants") are considered the most spectacular biological invasions witnessed in the contemporary oceans (Por, 1978, 1989, 1990; Spanier and Galil, 1991; Zibrowius, 1991; Galil, 1994).

Lessepsian migrants

An exhaustive inventory of Lessepsian migrants cannot be presented, because it would require the teamwork of many specialists. Furthermore, reliable inventories of the Mediterranean flora and fauna are not available for all taxonomic groups. Titles and abstracts of papers do not always provide an indication that aliens are mentioned in the text. The questions exist as to whether the identifications are reliable and previously known geographical ranges are representative (see discussion in Por, 1978; Zibrowius, 1991). As a result, our inventory of Lessepsian migrants (Table 14.1) does not intend to be exhaustive, but emphasises the scale of this unique inflow of migrants to the Eastern Mediterranean. Nearly 300 Red Sea species are known to have entered the Mediterranean through the Suez Canal.

Table 14.1 Lessepsian migrants from the Red Sea to the Mediterranean, as number of species per taxonomic unit. The caution taken by Por (1978) in separating "High probability" and "Low probability" Lessepsian migrants has no meaning in the present context, according to Por (1989).
* Including 10 species of serpulid Polychaeta. Zibrowius (1983, 1991) stated that considerably more serpulid species of Indo-Pacific origin had settled on the Levantine coast.
** According to Zibrowius (1991), Por's list of Lessepsian ascidians, mainly based on Pérès (1958), could provide an over-estimation of alien species.

Taxon	No. of spp.	References
Algae	21-25	Por (1978), Ribera and Boudouresque (1995), Verlaque (1994)
Seagrasses	1	Hartog (1972), Hartog and Van der Velde (1993)
Porifera	7	Por (1978)
Hydroidea	3	Por (1978)
Scyphozoa	3	Zibrowius (1991)
Polychaeta	31*	Ben-Eliahu (1986), Por (1978), Zibrowius (1991)
Pycnogonida	1	Por (1978)
Crustacea Decapoda	35	Galil et al. (1989), Galil and Golani (1990), Por (1978, 1989)
Crustacea (others)	23	Lakkis (1976), Morri et al. (1982), Por (1978, 1989)
Mollusks	105	Barash and Danin (1986), Aartsen et al. (1989, 1990)
Bryozoa	8	Por (1978), Zibrowius (1991)
Chaetognata	1	Guergues and Halim (1973), Por (1989)
Echinodermata	5	Cherbonnier (1986), Por (1978)
Tunicata (ascidians)	7**	Por (1978)
Enteropneusta	1	Por (1978)
Pisces	45	Ben-Tuvia (1985), Fredj and Meinardi (1989), Golani (1987), Golani and Ben-Tuvia (1986), Spanier and Goven (1988)

Decapod crustaceans, mollusks and fishes are the principal taxonomic groups represented among the Lessepsian migrants (Por, 1989). It should be emphasised that this is unusual because fishes are seldom a major component of species introductions in the marine environment. Planktonic species, which are generally stenohaline, are more poorly represented among the Lessepsian migrants. The fluctuant and rather high salinity of the canal, especially the Bitter Lakes, remain barriers to many migrants (Godeaux, 1974; Kimor, 1990; Por, 1990).

Por (1978) expressed the view that Lessepsian migration will eventually approach a plateau. This is indeed to be expected, but has not yet happened. The number of introduced species continues to increase at an exponential rate (Figure 14.1). Of course, these figures are to be taken with some caution, because the first observation of an introduced species always occurs some time after its real introduction. In addition, these figures may be indicative of an increase in the intensity of observation and our knowledge of biogeography. Nevertheless, a similar kinetics of increase in the rate of marine species' introduction has been evident among other vectors of introduction than the Suez Canal (Boudouresque and Ribera, 1994; Ribera and Boudouresque, 1995).

Most of Lessepsian migrants are still confined to the Levantine coast (from Egypt to Syria). Some are also present in other parts of the Eastern Mediterranean basin. For example, the opistobranch *Bursatella leachi* is now common not only in the Levantine basin, but also has progressed to Greece, the Aegean Sea, the Ionian Islands, Malta, Southern Italy and the Adriatic Sea (Zibrowius, 1991). Very few species extend to the Western Mediterranean, for example the mollusks, *Brachydontes variabilis* and *Pinctada radiata,* observed in Corsica (J. Godeaux, pers. comm.) and the green alga *Caulerpa racemosa* in Leghorn, Northern Italy (Piazzi *et al.*, 1994). Por (1983) proposed the delimitation of a biogeographical "Lessepsian province" (Figure 14.2). The limits of this province presently are the entrance to the Aegean Sea and the Eastern coast of Sicily.

Lessepsian migrants are not the only alien species in the Mediterranean. Other routes of access to the Mediterranean have occurred. For example, 60 macroalgae are considered as probably introduced to the Mediterranean (Ribera and Boudouresque, 1995), of which 21–25 species are Lessepsian migrants (Table 14.1).

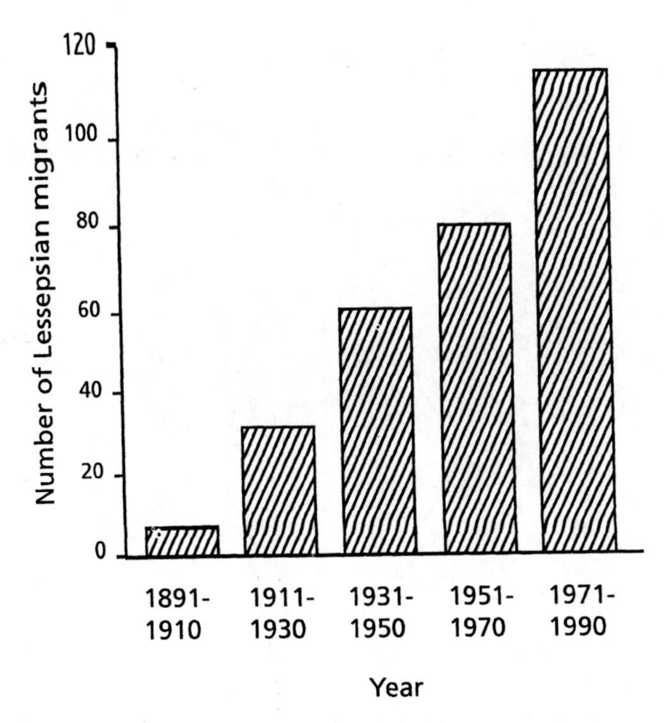

Figure 14.1 Number of Lessepsian migrants to the Mediterrenean, by periods of 20 years (non-cumulative data), according to the date of first observation (or when not specified, the date of publication).

Anti-Lessepsian migrants

The flow of migration through the Suez Canal is almost exclusively unidirectional from the Red Sea to the Mediterranean (Por, 1978). Very few Mediterranean species have moved in the reverse direction ("anti-Lessepsian migrants"), for example, the sea star *Sphaeriodiscus placenta* and the fishes *Liza aurata* and *Dicentrarchus punctatus* (Ben-Tuvia, 1975; Fouda and Hellal, 1987; Por, 1978, 1990). A total of 53 species at different times have been considered as Mediterranean immigrants to the Red Sea. A critical evaluation of the data leads to a reduction in the number of species to 10–20; most of them are still confined to the vicinity of Suez harbour and lagoons (Por, 1978).

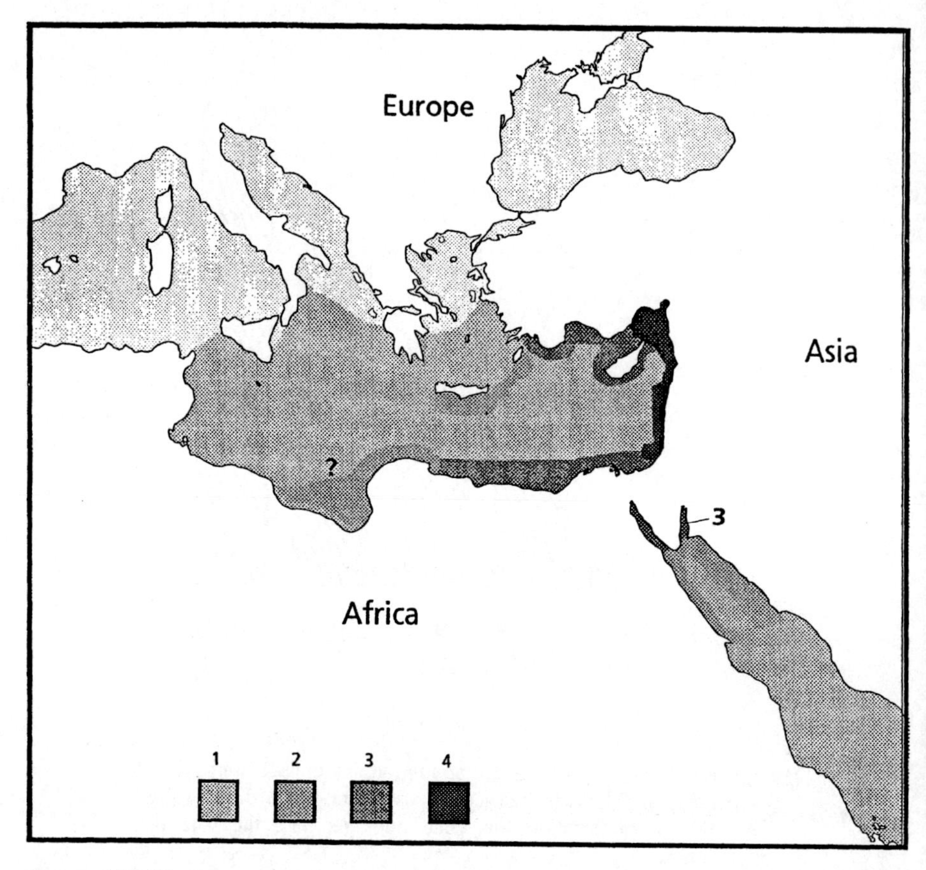

Figure 14.2 The Lessepsian province of the Mediterranean in relation to the Northern Red Sea. Four scales of shading depict the density of the settlement by migrants (from Por, 1990).

Why such a success for Lessepsian migrants?

The most important factors liable to hinder migrations through the Suez Canal are (Halim, 1990; Morcos, 1980; Spanier and Galil, 1991): (i) temperature fluctuations and turbidity of the canal water, which is higher than in the adjacent sea (this barrier has gradually disappeared); (ii) the degree of salinity of the hypersaline Bitter Lakes, which for a long time hindered migrations, but has gradually fallen from 70 to 43–48 *per mille* (similar to that of the Northern Gulf of Suez); and (iii) the Nile fresh-water dilution plume. Since the damming of the Nile at Aswan in 1965–1967, which drastically reduced the river's outflow, this barrier has almost disappeared.

Several causes can be put forward to explain the success of Lessepsian migrants. (i) A relatively low species diversity exists in the Levantine basin (when compared with the western basin), which probably results from its geological history. During the Messinian, the Mediterranean was periodically cut off from the Atlantic, resulting in hypersaline conditions until Atlantic waters refilled it 5 Ma ago (Hsu *et al.*, 1978). The effects on species diversity were severe, even if not as catastrophic as it appeared when it was first discovered (Por, 1990). Recolonization of the Mediterranean involved species of Atlantic origin, which still dominate the Mediterranean biota (Fredj *et al.*, 1992). (ii) The Pleistocene Glacial periods probably contributed to the shortage of thermophilic species in the Levantine basin. In some areas of the Levantine basin, summer surface temperatures can reach tropical conditions of 28–29 °C, and winter temperatures, always below 20 °C, are characteristic of warm temperate regions. The number of successful Lessepsian colonists demonstrates that habitats suitable for tropical species were available (Stephenson,1948; Lüning, 1990; Spanier and Galil, 1991). (iii) The pre-adaptation of the Red Sea species to the Suez Canal salinity is to be considered. The Gulf of Suez, a shallow body of water with salinity values as high as 45 *per mille* and wide temperature fluctuations could act as a virtual lock chamber (Spanier and Galil, 1991). (iv) Finally, the water of the canal tends to flow towards the Mediterranean for ten months (the mean sea level of the Red Sea in Suez Gulf is 30–40 cm higher than at the Mediterranean canal outlet from January to June), reversing only in August-September (Morcos, 1967, 1980).

Could the possible recent warming of Mediterranean waters resulting from global climate change contribute to the explanations of such a success? Temperatures in deep waters of the Western Mediterranean basin have risen by 0.12°C since 1960 (Béthoux *et al.*, 1990). Some species of fishes and algae with established thermal preferences have recently extended their distribution area toward the north-western Mediterranean, where a warming trend for surface water temperatures exists (Bianchi and Morri, 1993, 1994; Francour *et al.*, 1994). Nevertheless, these changes could prove to be simple oscillations, which may have already existed since the opening of the Suez Canal. The kinetics of invasion (Figure 14.2) do not show any conspicuous changes with time, and the rise of surface temperatures in the Eastern Mediterranean basin has not been clearly established.

According to Por (1990), the conclusions reached by Safriel and Ritte (1985) that Lessepsian migrants are usually "r-strategists" should be re-examined.

As far as the anti-Lessepsian migrants are concerned, the reasons why they were so unsuccessful could be the same as above, but taken on the reverse way. According to Por (1978), no favourable preconditions exist for the Levantine Mediterranean fauna (which is already impoverished) to migrate even further along an increasing salinity and temperature gradient. Moreover, the Red Sea fauna is particularly rich.

Unlike the Suez Canal, the opening of the Panama Canal, which links the Pacific to the Atlantic across Central America, has only resulted in the passage of a limited number of euryhaline species. According to Hay and Gaines (1984), it is not the lock system that segments the canal, or the freshwater lakes along its course that explain the low exchange of flora and fauna between the two oceans. Many Caribbean species can attach themselves to the hulls of ships and withstand a 6 to 12 hour passage through fresh water. As far as algae are concerned, the primary barrier to successful transport and establishment of these Caribbean species appears to be herbivore activity and lack of reef-generated refuge areas on the Pacific coast.

Ecological and economic consequences

Most attention has focused on environmentally damaging consequences and short-term economic losses resulting from alien invasive species in terrestrial and freshwater environment (see Mooney and Drake, 1986; Drake *et al.*, 1989; Pieterse and Murphy, 1990; Groves and di Castri, 1991). On the other hand, the marine environment, especially the Mediterranean one, has not been examined extensively.

Little attention has been paid to the ecological impact of Lessepsian migrants, both at the species and at the ecosystem levels. There is a common empirical opinion that the Red Sea species invasion has not resulted in Mediterranean species depletion, but instead in species enrichment. This simplistic interpretation (likely to appeal to politicians) whose logical result would be the world-wide standardisation (referred to as "mcdonaldisation" by Boudouresque, 1996) is diametrically opposed both to the concept of biodiversity and to the ethics of conservation of the environment. If this were in fact the case, zoos and botanical gardens would be the paradigm of biodiversity.

The hypothesis that Lessepsian migrants enrich the Mediterranean biodiversity is poorly founded on scientific data. The fate of native species has been poorly studied. On the contrary, evidence exists of drastic changes in abundance and niche displacement of native species that can be attributed to competition with Red Sea species. It is highly probable that a strong impact on the functional processes of the ecosystems of the Mediterranean, a sea characterised by a low herbivory, exists because of the presence of large herbivorous fishes (the rabbit fishes, *Siganus rivulatus* and *S. luridus*) among the migrants.

The arrival of many Red Sea species has had a great impact on the composition of the south-eastern Mediterranean fish fauna. According to Fredj and Meinardi (1989) and Ben-Tuvia (1985), they represent 7% of the 648 fish species known from the Mediterranean, about 10% of the 469 species reported from the Eastern basin and 12% of the species of the Levantine basin. In benthic hard bottom ecosystems of the south-eastern shelf, they contribute nearly half of the total fish abundance and biomass (Spanier *et al.,* 1989), with the dominant species now being the rabbit fishes, and the red soldier fish, *Sargocentron rubrum*. Lessepsian species represent 9% of polychaetes, 9% of mollusks and 20% of decapod crustaceans along the Levantine coast (Barash and Danin, 1986; Ben-Eliahu, 1989; Galil, 1986). The microplankton communities of the northern Red Sea and of the Levantine basin now share a number of basic characteristics (Kimor, 1990). Currently, Lessepsian migrants represent about 4% of the Mediterranean species diversity and 10% of the Levantine basin diversity (Por, 1978, 1990).

Most settlers are benthic or demersal species and found at intermediary depths, between 20–40 m, on mixed sandy-muddy bottoms. This is a result of the cooling of the shallow waters in the winter and the exposure to cold at higher intertidal levels. Usually, the depth limit of the Lessepsian migrants is 70–80 m. A year-round temperature of 18 °C exists at this depth (Por, 1978).

The large aplisiid opistobranch *Bursatella leachi* does not seem to have a native competitor (Zibrowius, 1991). The introduced asteroid *Asterina wega* appears to have locally replaced the native, ecologically similar *Asterina gibbosa*. The native prawn *Penaeus kerathurus,* which supported a commercial fishery throughout the 1950s, has nearly disappeared; it has been replaced by *P. japonicus* (Geldiay and Kocatas, 1972; Spanier and Galil, 1991). The recent decrease of the previously prevalent indigenous jellyfish *Rhizostoma pulmo* may also be a case of competitive displacement by the Lessepsian *Rhopilema nomadica,* which belongs to the same family (Spanier and Galil, 1991). Several native species have been competitively displaced to greater depths by introduced competitors. For example, the red snapping shrimp *Alphaeus glaber,* the red mullet *Mullus barbatus* and the hake *Merluccius merluccius* have been displaced respectively by the shrimp *Alphaeus rapacida,* the goldband goatfish *Upeneus moluccensis* and the brushtooth lizardfish *Saurida undosquamis* (Por, 1978). According to Por (1990), the conclusions reached by Safriel and Ritte (1985) that there is a smooth gene-flow between conspecific populations on both sides of the Suez Canal should be rechecked. Golani (1990) showed that some species of

Lessepsian fishes show significant morphologic differences compared with their Red Sea source populations. The causes of these differences could stem from species' spawning season, differences in the annual temperature regime at the source and target sites, and founder effects.

It is not unusual for an introduced species to present an early exuberant development and then decline. This is the case for the gastropod *Rhinoclavis kochi,* which was first reported in the mid-1960s and became one of the dominant species in the late 1970s, but has since become relatively rare (Galil and Lewinsohn, 1981; Spanier and Galil, 1991). Nevertheless, there are no known examples of disappearance of a Lessepsian species in the Mediterranean (Por, 1978). As for other species, changes in abundance seem related to climatic fluctuations. For example, the fishes *Saurida undosquamis* and *Upeneus moluccensis* experienced peaks of abundance following an exceptionally warm winter (Oren, 1957; Ben-Tuvia, 1973; Ben-Yami and Glaser, 1974). Finally, the exponential expansion phase of several species reached a plateau, but a decline has not yet occurred. This is exemplified by the prawn *Penaeus japonicus* and by the swimming crab *Charybdis longicollis*, which dominate the macrobenthic fauna of silty sand bottoms, forming up to 70% of the biomass in certain places (Galil, 1986; Spanier and Galil, 1991).

Some migrant species are now of economic importance, being exploited by local fisheries (Oren 1957; Galil 1986; Spanier and Galil 1991; Zibrowius 1991). The crab *Portunus pelagicus* has become the dominant crab in commercial catches throughout the eastern Mediterranean, especially in Egypt. The prawns *Penaeus japonicus* and *P. monoceros* are commercially fished, making up most of the shrimp catches in Israel and Egypt. Off the Israeli coast, migrant fishes constitute a third of the trawl catches (e.g. the brushtooth lizardfish *Saurida undosquamis*). The migrant goldband goatfish, *Upeneus moluccensis,* represents a third of the mullid catches (up to 83% in 1954–1955). It is unclear, however, whether the total prawn and fish stock has increased or decreased. The economic benefits of an introduction should not be assessed simply on the basis of strict sale price, but should also take into account losses from other business activities and the costs of any damage that may result. Furthermore, some species (e.g. the jellyfish *Rhopilema nomadica*) have exerted a significant negative impact on fisheries and tourism. *Rhopilema nomadica,* whose umbrella measures 20–60 (100) cm, can assemble in "jellyfish belts" at a distance of 1–5 km offshore from the Israeli coast, especially in summer. Maximal density was estimated at 25 individuals per m^3. For several weeks in the summer of 1990, the coastal fishing was disrupted due to damage to nets and inability to sort the yield. Local municipalities reported a decrease in beach frequentation, due to the painful stings inflicted by the jellyfish. Nets strung along bathing beaches proved ineffective, since jellyfish fragments passed through (Spanier and Galil, 1991). In Cyprus, an introduced green alga of the genus *Cladophora* (perhaps *C. patentiramosa*) proliferated off the beaches, and its flotsam accumulated on the beaches themselves, hindering bathing (Demetropoulos and Hadjichristophorou, in Ribera and Boudouresque, 1995).

Possible mechanisms of control

The complete or partial eradication of an introduced marine species is certainly very difficult, if not impossible given the present state of our knowledge and techniques. In addition, there has been little research in this field. So, the introduction of a species is to be considered as a probably definitive act that we impose on future generations.

Migration across a canal is probably the most difficult to control, especially the Suez Canal which handles some 20% of the world maritime traffic. Nothing has been done to study possible mechanisms capable of slowing down the invasion of Red Sea species. The setting up of a lock gate in the course of the canal should be considered. The possibility of technically treating the lock water (chemical or other treatment) in order to reduce its biological load should be examined. The reactivation of the saline barriers (Bitter Lakes) could also be studied, for example through the coupling with a desalinisation plant.

Conclusions

The Suez Canal has been compared to a thin capillary tube connecting two huge marine basins (Halim, 1990). Although its role in the exchange of water has been insignificant, it became the site of an immense biological invasion from the Red Sea to the Mediterranean.

Migration through the Suez Canal is an ongoing process, and there are no signs that the influx is nearing a plateau. Most of Lessepsian migrants are concentrated in the Eastern Mediterranean basin (especially along the Levantine coast). The area, therefore, could be considered as a "Lessepsian biogeographical province". The fauna and flora, as well as the functional processes at the ecosystem level have already dramatically changed in this part of the Mediterranean, even if the magnitude and the consequences of these changes have not been measured. No concentrated effort to follow the deployment of this unique biogeographical phenomenon exists. New records of Lessepsian migrants are generally the result of occasional observations (Por, 1989).

The rate of marine species introduction in most parts of the world's oceans is rapidly increasing. What will happen if the introduced species become the majority, overwhelm the indigenous species, and build new communities whose composition, structure and functioning we are incapable of predicting? Carlton and Geller (1993) refer to this as "ecological roulette". Lessepsian migration and the south-eastern Mediterranean could prefigure the fate of other seas. It should be of interest to carefully study such a model and to try to slow or stop the invasion.

Acknowledgments

The author wishes to thank Michèle Perret-Boudouresque who provided documentation and Simone Fournier and Lorraine Fleming for checking the English text.

References

Aartsen, J.J., van Barash, A. and Carozza, A. (1989) Addition to the knowledge of the Mediterranean Mollusca of Israel and Sinai. *Bolletin Malacologia*, **25**, 63–76.

Aarsten, J.J. van Carozza, F. and Lindner, G. (1990) *Acteocina mucronata* (Philippi, 1849), a recent Red Sea immigrant species in the eastern Mediterranean. *Bolletin Malacologia*, **25**, 285–288.

Barash, A. and Danin, Z. (1986) Further additions to the knowledge of Indo-Pacific Mollusca in the Mediterranean Sea (Lessepsian migrants). *Spixiana,* **9**, 117–141.

Ben-Eliahu, M.N. (1986) Red Sea serpulids in the Eastern Mediterranean, in *Proceedings of Second International Polychaete Conference*, Copenhagen.

Ben-Eliahu, M.N. (1989) Lessepsian migration in Nereidae (Annelida: Polychaeta): some case histories. in *Environmental Quality and Ecosystem Stability*, (eds E. Spanier, E. Steinberger and M. Luria), ISEEQS Publications, Jerusalem, Israel, volume IV-B, pp. 124–134.

Ben-Tuvia, A. (1973) Man-made changes in the Eastern Mediterranean sea and their effect on the fishery resources. *Marine Biology,* **19**, 197–203.

Ben-Tuvia, A. (1975) Comparison of the fish-fauna in the Bardawil lagoon and in the Bitter Lakes. *Rapports P.V. Réunions Commission internationale Exploration scientifique Méditerranée*, **23**, 125–126.

Ben-Tuvia, A (1985) The impact of the Lessepsian (Suez Canal) fish migration on the eastern Mediterranean ecosystem, in *Mediterranean Marine Ecosystems*, (eds M. Moraitou-Apostolopoulou and V. Kiortsis), Plenum Press, pp. 367–375.

Ben-Yami, M. and T. Glaser (1974) The invasion of *Saurida undosquamis* (Richardson) into the Levant Basin – an example of biological effect of interoceanic canals. *Fishery Bulletin,* **72,** 359–373.

Béthoux, J.P., Gentili, B., Raunet, J. and Taillez, D. (1990) Warming trend in the Western Mediterranean deep water. *Nature,* **347,** 660–662.

Bianchi, C.N. and Morri, C. (1993) Range extensions of warm-water species in the northern Mediterranean: evidence for climatic fluctuations? *Porcupine Newsletter,* **5**, 156–159.

Bianchi, C.N. and Morri, C. (1994) Southern species in the Ligurian Sea (Northern Mediterranean): new records and a review. *Bolletino Museo Istituto Biologia Università Genova,* **58–59,** 181–197.

Bodoy, A., Maître-Allain, T. and Riva, A. (1981) Croissance comparée de la palourde européenne *Ruditapes decussatus* et de la palourde japonaise *Ruditapes philippinarum* dans un écosystème artificiel méditerranéen. *Vie marine,* **2**, 39–51.

Boudouresque, C.F. (1996) *Impact de l'Homme et Conservation du Milieu Marin en Méditerranée.* 2° édition. GIS Posidonie Publishers, Marseille.

Boudouresque, C.F., Meinesz, A., Ribera, M.A. and Ballesteros, E. (1995) Spread of the green alga *Caulerpa taxifolia* (Caulerpales, Chlorophyta) in the Mediterranean: possible consequences of a major ecological event. *Scientia Marina,* **59 (suppl. 1**), 21–29.

Boudouresque, C.F., Meinesz, A., Verlaque, M., and Knoepffler-Peguy, M. (1992) The expansion of the tropical alga *Caulerpa taxifolia* (Chlorophyta) in the Mediterranean. *Cryptogamie-Algologie,* **13**, 144–145.

Boudouresque, C.F. and Ribera, M.A. (1994) Les introductions d'espèces végétales et animales en milieu marin. Conséquences écologiques et économiques et problèmes législatifs, in *First International Workshop on* Caulerpa taxifolia, (eds C.F. Boudouresque, A. Meinesz and V. Gravez), GIS Posidonie Publishers, Marseilles, pp. 29–102

Carlton, J.T. (1993) Dispersal mechanisms of the zebra mussel (*Dreissena polymorpha*), in *Zebra Mussel: Biology, Impact and Control,* (eds T.F. Nalepa and D.W. Schloesser), Lewis publ., Boca Raton, Florida, USA.

Carlton, J.T. (1998) The scale and ecological consequences of biological invasions in the World's oceans, in *Invasive Species and Biodiversity Management*, (eds O. T. Sandlund, P. J. Schei and Å. Viken), Kluwer Academic Publishers, Dordrecht, The Netherlands.

Carlton, J.T. and Geller, J.B. (1993) Ecological roulette: the global transport of non-indigenous marine organisms. *Science,* **261**, 78–82.

Cherbonnier, G. (1986). Holothuries de Méditerranée et du Nord de la Mer Rouge. *Bulletin du Muséum national d'Histoire naturelle, Sér. 4*, **8**, 43–46.

Drake, J.A., Mooney, H.A., di Castri, F., Kruger, F., Groves, R., M. Rejmanek and Williamson, W., eds (1989) *Biological Invasions: a Global Perspective.* SCOPE **37,** John Wiley and Sons, Chichester, UK.

Farnham, W.F. (1994) Introduction of marine benthic algae into Atlantic European waters, in *Introduced Species in European Coastal Waters*, (eds C.F. Boudouresque, F. Briand and C. Nolan), European Commission publications, Luxemburg, pp. 32–36

Fouda, M.M. and Hellal, A.M. (1987) The Echinoderms of the north-western Red Sea, Asteroidea. *Fauna and Flora of Egypt,* **2**, 1–71.

Francour, P., Boudouresque, C.F., Harmelin, J.G., Harmelin-Vivien. M.L and Quignard, J.P. (1994) Are the Mediterranean waters becoming warmer? Information from biological indicators. *Marine Pollution Bulletin,* **28**, 523–526.

Fredj, G., Bellan-Santini, D. and Meinardi, M. (1992) Etat des connaissances sur la faune marine méditerranéenne. *Bulletin de l'Institut océanographique,* **num. spéc. 9**, 133–145.

Fredj, G. and Meinardi, M. (1989) Inventaire faunistique des ressources vivantes en Méditerranée: intérêt de la banque de données MEDIFAUNE. *Bulletin de Societé Zoologique de France,* **114**, 75–87.

Galil, B. (1986) Red Sea Decapods along the Mediterranean coast of Israel: ecology and distribution, in *Environmental Quality and Ecosystem Stability, vol. III A/B*, (eds Z. Dubinsky and Y. Steinberg), Bar Ilan University Press, Ramat Gan, Israel, pp. 179–183.

Galil, B.S. (1994) Lessepsian migration. Biological invasion of the Mediterranean, in *Introduced Species in European Coastal Waters* (eds C.F. Boudouresque, F. Briand and C. Nolan), European Commission publications, Luxemburg, pp. 63–66.

Galil, B.S. and Golani, D. (1990) Two migrant Decapods from the eastern Mediterranean. *Crustaceana*, **58**, 229–236.

Galil, B. and Lewinsohn, C. (1981) Macrobenthic communities of the Eastern Mediterranean continental shelf. *Marine Ecology, PSZN*, **2**, 343–352.

Galil, B.S., Pisanty, S., Spanier, E. and Tom, M. (1989) The Indo-Pacific lobster *Panulirus ornatus* (Crustacea : Decapoda): a new lessepsian migrant to the eastern Mediterranean. *Israel Journal of Zoology*, **35**, 241–243.

Geldiay, R. and Kocatas, A. (1972) *A Report on the Occurrence of Penaeidae (Decapoa Crustacea) along the Coast of Turkey from Eastern Mediterranean to the Vicinity of Izmir, as a Result of Migration and its Factors.* 17° Congrès international de Zoologie, Monte Carlo.

Godeaux, J. (1974) Thaliacés récoltés au large des côtes égyptiennes de la Méditerranée et de la mer Rouge (Tunicata, Thaliacea). *Beaufortia*, **22**, 83–103.

Golani, D. (1987) The Red Sea pufferfish *Torquigener flavimaculosus* a new Suez Canal migrant in the Eastern Mediterranean (Pisces : Tetraodontidae). *Senckenberg. Mar.*, **19**, 339–343.

Golani, D. (1990) Environmentally-induced meristic changes in Lessepsian fish migrants, a comparison of source and colonizing populations. *Bulletin de l'Institut océanographique de Monaco, N.S.*, **num. spéc. 7**, 143–152.

Golani, D. and Ben-Tuvia, A. (1986) New records of fishes from the Mediterranean coast of Israel, including Red Sea immigrants. *Cybium*, **10**, 285–291.

Groves R.H. and di Castri, F., eds (1991) *Biogeography of Mediterranean Invasions.* Cambridge University Press, Cambridge, UK.

Guergues, S.K. and Halim, Y. (1973) Chétognathes du plancton d'Alexandrie. II. Un spécimen mûr de *Sagitta neglecta* Aida en Méditerranée. *Rapports P.V. Réunions Commission internationale Exploration scientifique Méditerranée*, **21**, 497 pp.

Halim, Y. (1990) On the potential migration of Indo-Pacific plankton through the Suea Canal. *Bulletin de l'Institut océanographique de Monaco, N.S.*, **num. spéc. 7**, 11–27.

Hartog, C. den (1972) Range extension of *Halophila stipulacea* (Hydrocharitaceae) in the Mediterranean. *Blumea*, **20**, 154 pp.

Hartog, C. den and Van der Velde, G. (1993) Occurrence of the seagrass *Halophila stipulacea* (Hydrocharitaceae) along the Mediterranean coast of Turkey. *Posidonia Newsletter*, **4**, 5–6.

Hay, M.E. and Gaines, S.D. (1984) Geographic differences in herbivore impact: do Pacific herbivores prevent Caribbean seaweeds grom colonizing via Panama Canal? *Biotropica*, **16**, 24–30.

Hsu, K.J., Montadert, L., Bernouilli, D., Cita, M.B., Erikson, A., Garrison, R.E., Kidd, R.B., Melieres, F., Muller, C., and Wright, R. (1978) History of the Mediterranean salinity crisis. *Initial Reports of DSDP*, **42**, 1053–1078.

Kimor, B. (1990) Microplankton of the Red Sea, the Gulf of Suez and the Levantine basin of the Mediterranean. *Bulletin de l'Institut océanographique de Monaco, N.S.*, **num. spéc. 7**, 29–38.

Lakkis, S. (1976) Sur la présence dans les eaux libanaises de quelques Copépodes d'origine Indo-Pacifique. *Rapports P.V. Réunions Commission internationale Exploration scientifique Méditerranée,* **23,** 83–85.

Lüning, K. (1990) *Seaweeds? Their Environment, Biogeography and Ecophysiology.* John Wiley and Sons Inc., New York.

Meinesz, A. and Hesse, B. (1991) Introduction et invasion de l'algue tropicale *Caulerpa taxifolia* en Méditerranée nord-occidentale. *Oceanologica Acta,* **14,** 415–426.

Meinesz, A., de Vaugelas, J., Hesse, B., and Mari, X. (1993) Spread of the introduced tropical green alga *Caulerpa taxifolia* in Northern Mediterranean. *Journal of Applied Phycology,* **5,** 141–147.

Mooney H.A. and Drake, J.A., eds (1986) *Ecology of Biological Invasions of North America and Hawaii.* Ecological studies, 58, Springer Verlag, Berlin.

Morcos, S.A. (1967) The chemical composition of sea water from the Suez Canal region, Part I: the major anions. *Kieler Meeresforschung,* **23,** 80–91.

Morcos, S.A. (1980) Seasonal changes in the Suez Canal following its opening in 1869: newly discovered hydrographic records of 1870–1872, in *Oceanography: the Past,* (eds M. Sears and D. Merriman), Springer, New York, pp. 290–305.

Morri, C., Occhipinti Ambrogi A., and Sconfietti R. (1982) Specie nuove o critiche del benthos lagunare nord adriatico. *Progetto final – Prom. Qual. Amb., Roma, CNR,* **35,** 1–4.

Oren, O.H. (1957) Changes in temperature of the Eastern Mediterranean Sea in relation to the catch of the Israel trawl fisheries during the years 1955/55 and 1955/56. *Bulletin de l'Institut océanographique de Monaco,* **1102,** 1–12.

Pérès, J.M. (1958) Ascidies récoltées sur les côtes méditerranéennes d'Israel. *Bulletin of the Research Council of Israel,* **7B,** 143–150.

Piazzi, L., Balestri, E., and Cinelli, F. (1994) Presence of *Caulerpa racemosa* in the north-western Mediterranean. *Cryptogamie-Algologie,* **15,** 183–189.

Pieterse, A.H. and Murphy, K.J., eds (1990) *Aquatic Weeds. The Ecology and Management of Nuisance Aquatic Vegetation.* Oxford University Press, Oxford, UK.

Por, F.D. (1969) The Canuellidae (Copepoda, Harpacticoida) in the waters around the Sinai Peninsula and the problem of "Lessepsian" migration of this family. *Israel Journal of Zoology,* **18,** 169–178.

Por, F.D. (1978) *Lessepsian migrations. The Influx of Red Sea Biota into the Mediterranean by Way of the Suez Canal,* Springer Verlag, Berlin.

Por, F.D. (1983) The Lessepsian biogeographic province of the Eastern Mediterranean. *Rapports P.V. Réunions Commission internationale Exploration scientifique Méditerranée, Journées Etudes Systématique et Biogéographie méditerranéenne, Cagliari,* pp. 81–84.

Por, F.D. (1989) *The Legacy of Tethys. An Aquatic Biogeography of the Levant.* Kluwer Academic Publishers, Dordrecht, The Netherlands.

Por, F.D. (1990). Lessepsian migrations. An appraisal and new data. *Bulletin de l'Institut océanographique,* **7,** 1–10.

Ribera, M.A. and Boudouresque, C.F. (1995) Introduced marine plants, with special reference to macroalgae: mechanisms and impact, in *Progress in Phycological Research,* volume 11 (eds F.E. Round and D.J. Chapman), Biopress Ltd., Bristol, UK, pp. 187–268.

Riouall, R. (1985) Sur la présence dans l'étang de Thau (Hérault, France) de *Sphaerotrichia divaricata* (C. Agardh) Kylin et *Chorda filum* (L.) Stackhouse. *Botanica Marina,* **27,** 83–86.

Riouall, R., Guiry, M.D., and Codomier, L. (1985) Introduction d'une espèce foliacée de *Grateloupia* dans la flore marine de l'étang de Thau (Hérault, France). *Cryptogamie-Algologie,* **6**, 91–98.

Robba, E. (1987) The final occlusion of Tethys: its bearing on Mediterranean benthic mollusks, in *Shallow Tethys* 2, (ed. K.G. McKenzie), Balkema, Rotterdam, Netherlands, pp. 405–426.

Rueness, J. (1989) *Sargassum muticum* and other introduced Japanes macroalgae: biological pollution of European coasts. *Marine Pollution Bulletin,* **20**, 173–176

Safriel, U.N. and Ritte, U. (1986) Suez Canal migration and Mediterranean colonization. Their relative importance in Lessepsian migration. *Rapports P.V. Réunions Commission internationale Exploration Méditerranée,* **29**, 259–263.

Sancholle, M. (1988) Présence de *Fucus spiralis* (Phaeophyceae) en Méditerranée occidentale. *Cryptogamie-Algologie,* **9**, 157–161.

Spanier, E. and Galil, B.S. (1991) Lessepsian migration: a continuous biogeographical processe. *Endeavour N.S.*, **16**, 102–106.

Spanier, E. and Goven, M. (1988) An Indo-Pacific trunkfish *Tetrosomus gibbosus* (Linnaeus): first record of the family Ostracionidae in the Mediterranean. *Journal of Fish Biology*, **32**, 797–798.

Spanier,E., Pisanty, S., Tom, M. and Almog-Shtayer, G. (1989) The fish assemblage on a coralligenous shallow shelf off the Mediterranean coast of northern Israel. *Journal of Fish Biology,* **35**, 641–649.

Stephenson, T.A. (1948) The constitution of the intertidal fauna and flora of South-Africa. Part III. *Annals of the Natal Museum,* **11**, 207–324.

Travis, J. (1993) Invader threatens Black, Azov Seas. *Science,* **262**, 1366–1367.

Verlaque, M. (1994) Inventaire des plantes introduites en Méditerranée: origine et répercussions sur l'environnement et les activités humaines. *Oceanologica Acta,* **17**, 1–23.

Verlaque, M. and Riouall, R. (1989) Introduction de *Polysiphonia nigrescens* et d'*Antithamnion nipponicum* (Rhodophyta, Ceramiales) sur le littoral méditerranéen français. *Cryptogamie-Algologie,* **10**, 313–323.

Zibrowius, H. (1983) Extension de l'aire de répartition favorisée par l'homme chez les invertébrés marins. *Oceanis,* **9**, 337–353.

Zibrowius, H. (1991) Ongoing modification of the Mediterranean marine fauna and flora by the establishment of exotic species. *Mésogée,* **51**, 83–107.

15 Trade and exotic species introductions

PETER T. JENKINS
Biopolicy Consulting, Plácitas, New Mexico, USA

Abstract

This chapter addresses the trade pathway into the United States, then takes a broader look at the international context. It notes that some authors have failed to consider exotic species introductions when discussing the environmental impacts of trade liberalization. Yet, unintentional importation through international trade has been quantified as the most important pathway of harmful exotics into the United States. Trade liberalization has stimulated greater trade volume. But, international law regulating the unintentional introduction of harmful exotics through trade is weak. Existing legal instruments have the potential to protect biodiversity from harmful exotics, but they have lacked strong implementation. An international scientific advisory panel should be convened to advise the relevant convention secretariats, trade regulation bodies, and national governments regarding risky trade routes, potentially threatening species, the nations and ecosystems most vulnerable to exotic threats to biodiversity, and improvements in prevention and control efforts. Ecological risks from exotics should provide a clear legal basis to stop shipments of novel trade items or even to stop the opening of a new trade route. Finally, macroeconomic effects of global trade liberalization in relation to exotic species and biodiversity issues have not been adequately examined.

Background

Several researchers and commentators have addressed the effects of introductions of harmful exotic species (Mooney and Drake, 1989; Soulé, 1990; Temple, 1990; OTA, 1993). The only comprehensive review suggested about 4 500 exotic species occur in a free-ranging condition in the United States and, very roughly, up to one-fifth of these have caused serious economic or ecological harm (OTA, 1993). Both intentional and unintentional releases of exotics have led to serious problems. In addition to the extensive damage caused by exotic pests, pathogens, and weeds to forestry, agriculture, and other economic interests, exotics have been a major cause of native species endangerment and extinction, particularly on islands and in other isolated habitats. Exotics are recognized as one of the most pervasive and insidious threats to biodiversity (Soulé, 1990). Yet, some authors have failed to consider exotic species introductions when

O. T. Sandlund et al. (eds.), Invasive Species and Biodiversity Management, 229–235.

discussing the environmental impacts of international trade liberalization (see "Free trade is green, protectionism is not.", Yu, 1994; "Review of U.S.-Mexico environmental issues.", U.S. Interagency Task Force, 1992). This chapter initially addresses the trade pathway into the United States, then takes a broader look at the international context.

The Trade Pathway into the United States

The most important pathway of harmful exotics into the United States is neither intentional releases nor contraband smuggled in by international travelers. It is unintentional importation through international trade. Exotics frequently "stowaway" in ships, planes, trucks, shipping containers, and packing materials, or "hitchhike" on nursery stock, fruits, vegetables, seeds, and other imports (OTA, 1993). According to Table 15.1, 38 of 46, or 82.6%, of the harmful new exotics detected from 1980 through 1993, with identified pathways, were unintentional imports. The North American Free Trade Agreement (NAFTA) and the General Agreement on Tariffs and Trade (GATT) should lead to even more unintentional import of exotics. Trade liberalization stimulates greater trade volume (U.S. Interagency Task Force, 1992).

Table 15.1 Pathways of harmful introductions into the United States, 1980–1993. Note that 159 other known introductions in the United States during this time period are not included in the table because either their pathway of introduction had not been identified or they had not been characterized by experts as economically, environmentally, or otherwise harmful, as of 1992. Adapted from table 3–1, pp. 101–105, in OTA (1993).

Identified pathway of introduction	Unintentional importation through international trade	Introduced elsewhere then spread into the U.S.	Escaped ornamental plant	Escaped aquaculture contaminant	Contaminant of smuggled product	Total
Number and per cent of harmful introductions	38 (82.6%)	4 (8.7%)	1 (2.2%)	1 (2.2%)	2 (4.3%)	46 (100%)

Perhaps the assumption of those advocating trade liberalization is that risk analysis, inspection, and enforcement to prevent harmful introductions somehow will keep pace with the volume increase. However, the U.S.

government response appears inadequate to prevent increased trade from resulting in more harmful introductions. The government has been slow to prevent the introduction or spread of several severely harmful species that have arrived through international trade in recent years, such as the zebra mussel (*Dreissena polymorpha*), introduced via ship ballast water, and the Asian tiger mosquito (*Aedes albopictus*), introduced via used tire imports (OTA, 1993). This is despite warnings about these species (see Bio-environmental Services Ltd., 1981, on the zebra mussel and Moore *et al.*, 1988, on the Asian tiger mosquito).

In the past, biological considerations regarding pest risks have not necessarily outweighed economic considerations in the U.S. Department of Agriculture as its Animal and Plant Health Inspection Service (APHIS) regulated international trade: "[I]n controversial trade matters, top management outside of APHIS may "weigh" the biological position against the economic or other positions, and the short-term decision made by non-biologists may in some instances prevail regardless of the probability of long-term adverse consequences" (Kahn, 1991).

In other words, ultimate decisions about exotic species risks have a political component that can vary from administration to administration, depending on the value placed on "free trade" versus the value placed on preventing biological damage. Not only must domestic and international mechanisms be available to prevent harmful introductions, but the political will must exist to employ them.

An example of the danger faced from newly-opened trade pathways is the Siberian timber case. (A detailed case study is found in OTA, 1993.) A few U.S. timber brokers and lumber companies promoted the importation of whole logs of Siberian larch, *Larix* spp., from Far East ports to West Coast sawmills. One test shipment was allowed without prior risk analysis. Several potentially serious pests were discovered, including the Siberian gypsy moth, *Lymantria dispar*, a strain more mobile and damaging to coniferous forests than the European gypsy moth strain. Only strong criticism by academic scientists and pressure from a few members of the U.S. Congress prevented APHIS from allowing further timber shipments. After this pressure compelled a formal risk assessment, APHIS determined that no more raw log shipments from Siberia should be allowed (USDA, 1991). Yet some, such as Yu (1994), hold up the prospect of liberalized trade in timber as a benefit of trade distortion elimination under GATT.

Additional research is needed regarding the global importance of the trade pathway in introducing harmful exotics. Just one international trade pathway, the unintentional introduction of marine organisms through the dumping of ship ballast water, is well known to have caused, and to continue to cause, numerous severely harmful invasions (Carlton, 1996, 1998).

International Law

International law regulating the unintentional introduction of harmful exotic species through trade is weak. Existing legal instruments have the potential to protect biodiversity from harmful exotics, but they have lacked strong implementation in that direction. There are two major conventions with provisions on exotics. One is the International Plant Protection Convention (IPPC), which presently addresses crop pests only. The IPPC could be expanded in scope to explicitly protect native (non-agricultural) plant life from introduced pests. Efforts to do so should be undertaken.

The other major international agreement addressing exotic species, the Convention on Biological Diversity (CBD), lacks teeth. Article 8(h) addresses "alien" species by calling for the parties to: "as far as possible and as appropriate: .... Prevent the introduction of, control or eradicate those alien species which threaten ecosystems, habitats or species." Initial drafts of the CBD included a relatively strong exotics provision (IUCN, 1989). It would have established a scientific authority styled after the Convention on International Trade in Endangered Species approach, and a listing process focusing attention on high priority exotic species threats to biodiversity. However, the finally adopted, watered-down, Art. 8(h) language lacks specificity, lacks a listing process, and lacks enforceability due to its vagueness. Further, several countries have yet to join it, including the United States where it – seemingly endlessly – awaits Senate ratification.

The need for better science and policy advice

An international scientific advisory panel should be convened under the general auspices of the CBD (as was called for specifically under the earlier CBD draft) to advise the various convention secretariats, the trade regulation bodies, and national governments regarding risky trade routes, potentially threatening species, the nations and ecosystems most vulnerable to exotic threats to biodiversity, and improvements in prevention and control efforts. Two expert groups exist that could offer such advice: 1) the World Conservation Union (or IUCN), Species Survival Commission, Invasive Species Specialist Group, and 2) the International Council of Scientific Unions, Scientific Committee on Problems of the Environment (SCOPE), Global Invasives Strategy program (Mooney, 1998).

Independent assessments should be conducted for the most vulnerable areas identified by the scientific experts. Elements of the assessments should include, at least:

- national histories of exotic species impacts on native biodiversity,
- current and projected future pathways of exotic species introductions,
- laws, policies, and programs on preventing and managing exotics affecting biodiversity,
- institutional and technological capabilities for preventing and managing exotics, and
- public education and awareness strategies, under Article 13 of the CBD, which calls for efforts in these areas.

The difficult job of reducing harmful introductions carried by international trade would be served by more proactive risk analysis, stronger enforcement of existing laws, and amendments to existing laws, or entirely new laws, specifically intended to protect biodiversity. Ecological risks from exotics should provide a clear legal basis to stop shipments of novel trade items (as eventually occurred in the Siberian timber case), or even to stop the opening of a new international trade route, until the risks are reduced to a level acceptable to the importing nation. However, data gaps and uncertainty may prevent precise risk description. Thus, broad tools such as bans or restrictions of imports may be necessary to protect biodiversity in those countries that care to do so. However, such measures could be considered unfair restraints on trade under GATT, NAFTA, or other international agreements.

Under GATT and NAFTA, bans and restrictions may not discriminate needlessly against imported goods, that is, they may not be merely protectionist barriers disguised as environmental protection measures (Esty, 1994; Yu, 1994). Bans and restrictions may be challenged before the regulatory body that administers the trade agreement, e.g., the World Trade Organization (WTO) for the GATT. It will fall to these regulatory bodies to determine whether particular bans or restrictions are supportable in view of the risks posed by exotic species. Yet, it remains unclear how those bodies will make such decisions in view of the tremendous scientific uncertainty and risk assessment cost that is often involved. For example, the risk assessment for the proposed importation of Siberian larch cost the U.S. federal government approximately US$ 500 000 (USDA, 1991). Given the narrow focus of the IPPC, the WTO cannot rely on IPPC processes and standards to protect biodiversity. An international advisory panel of experts on harmful exotics, convened under the auspices of the CBD as suggested above, could inform the trade regulation organizations about the full range of biological risks and the degree of scientific uncertainty.

Macroeconomic effects

Macroeconomic effects of global trade liberalization in relation to exotic species and biodiversity issues have not been adequately examined. For example, Yu (1994) finished his "Free trade is green" paper by "speculat[ing] on possible indirect benefits" to the environment of more liberal trade, emphasizing the benefits of de-ruralization, rising standards of living, and reforestation. This assertion is unconvincing because it does not consider the following points related to exotics and biodiversity.

As developing countries pursue export markets, traditional agroecosystems are increasingly converted to large, exotic monocultures (Norgaard, 1987; McNeely, 1988). Global homogenization reduces the diversity of crops and livestock and can increase their vulnerability to both native and exotic pests, compelling increasing reliance on pesticides potentially damaging to biodiversity and to humans.

Also, increased standards of living in developing countries are associated with increased demand for imported products of all kinds, thereby increasing the likelihood of harmful unintentional introductions through the import process. Further, such imports include exotic foods, horticultural products, pets, and so on that may become invasive later. Much of Hawaii stands as an example of virtual replacement of vulnerable native plants and animals by exotics as trade volume and living standards have increased (Cuddihy and Stone, 1990). De-ruralization and the (often exotic) reforestation of abandoned farmland that are touted as benefits of increased standards of living in industrializing countries may not benefit **native** biodiversity.

Conclusion

Increased international trade has the potential to cause more harmful exotic species introductions. More proactive, more comprehensive, better-funded, international efforts are needed to ensure that widely-adapted invasive exotics do not further homogenize biological systems on a global scale.

References

Bio-environmental Services Ltd. (1981) *The Presence and Implication of Foreign Organisms in Ship Ballast Waters Discharged into the Great Lakes.* Vol. 1. Bio-environmental Services Ltd., Georgetown, Ontario.

Carlton, J.T. (1996) Marine bioinvasions: the alteration of marine ecosystems by nonindigenous species. *Oceanography*, **9**, 36–43.

Carlton, J.T. (1998) The scale and ecological consequences of biological invasions in the World's oceans, in *Invasive Species and Biodiversity Management*, (eds O. T. Sandlund, P. J. Schei and Å. Viken), Kluwer Academic Publishers, Dordrecht, The Netherlands..

Cuddihy, L.W., and C.P. Stone (1990) *Alteration of Native Hawaiian Vegetation: Effects of Humans, Their Activities and Introductions.* University of Hawaii Press, Honolulu.

Esty, D.C. (1994) *Greening the GATT: Trade, Environment, and the Future.* Institute for International Economics, Washington, DC.

Kahn, R.P. (1991) Letter to P.N. Windle, Office of Technology Assessment, Dec. 2, 1991, cited in *Harmful Non-Indigenous Species in the United States*, U.S. Congress, Office of Technology Assessment 1993, U.S. Government Printing Office, Washington, DC.

McNeely, J.A. (1988) *Economics and Biological Diversity: Developing and Using Economic Incentives to Conserve Biological Resources.* World Conservation Union, Gland, Switzerland.

Mooney, H.A. (1998) A Global Strategy for dealing with alien invasive species, in *Invasive Species and Biodiversity Management,* (eds O.T. Sandlund, P.J. Schei and Å. Viken), Kluwer Academic Publishers, Dordrecht, The Netherlands.

Mooney, H.A., and Drake, J.A. (1989) Biological invasions: a SCOPE program overview, in *Biological Invasions: a Global Perspective*, (eds J.A. Drake, H.A. Mooney, F. di Castri, R.H. Groves, F.J. Kruger, M. Rejmanek and M. Williamson), John Wiley and Sons, New York, pp. 491–506.

Moore, C.G., Francy, D.B., Eliason, D.A. and Monath, T.P. (1988) *Aedes albopictus* in the United States: rapid spread of a potential disease vector. *Journal of the American Mosquito Control Assoctation*, 4, 356–361.

Norgaard, R.B. (1987) Economics as mechanics and the demise of biological diversity. *Ecological Modelling*, **38,** 107–121.

Soulé, M.E. (1990) The onslaught of alien species, and other challenges in the coming decades. *Conservation Biology*, **4**, 233–239.

Temple, S.A. (1990) The nasty necessity: eradicating exotics. *Conservation Biology*, **4**, 113–115.

OTA (1993) *Harmful Non-Indigenous Species in the United States.* Office of Technology Assessment, U.S. Government Printing Office, Washington, DC.

USDA (1991) *Pest Risk Assessment of the Importation of Larch from Siberia and the Soviet Far East.* Miscellaneous Publication No. 1495, U.S. Department of Agriculture, U.S. Forest Service, Washington, DC.

U.S. Interagency Task Force (1992) *Review of U.S.–Mexico Environmental Issues.* Office of the U.S. Trade Representative, Washington, DC.

Vogt, D.U. (1992) *Sanitary and Phytosanitary Measures Pertaining to Food in International Trade Negotiations.* Congressional Research Service, Library of Congress, Washington, DC.

IUCN (1989) *Biological Diversity Convention Draft* (June 30, 1989). IUCN Environmental Law Centre, Bonn.

Yu, D. (1994) Free trade is green, protectionism is not. *Conservation Biology,* **8**, 989–996.

16 Commercial forestry and agroforestry as sources of invasive alien trees and shrubs

DAVID M. RICHARDSON
University of Cape Town, Cape Town, South Africa

Abstract

Alien trees and shrubs are planted over large areas in many parts of the world to provide a wide range of products and services. Although native species can, to some extent, supply these products and services in some cases, alien trees and shrubs are indispensible components of many ecosystems. Some alien tree species (currently a small sub-sample) used in forestry and agroforestry cause major problems by invading natural and semi-natural ecosystems. The magnitude of the problem has increased substantially over the past few decades, tracking the rapid increase in the scale of afforestation. Natural experiments created by the widespread planting of a fairly small number of species in different parts of the world show that, in general: (1) The species that cause the greatest problems are those that have been most widely planted and for the longest time. (2) The most impacted areas have the longest histories of intensive planting.

Pinus spp. are especially problematic as invaders in the southern hemisphere, where at least 19 species are invasive over large areas. The major invasive species have a predictable set of life-history attributes, and invasion events can be explained using models that simulate the dynamics of population growth and interactions between the invader and features of the invaded system.

Tree and shrub invasions have affected natural systems in many ways, sometimes causing major economic problems, but few thorough cost-benefit studies have been done. In South Africa where invasive pines and other species are having a major impact on watershed hydrology, studies have shown that the financial benefits of improved water yield from watersheds kept clear of alien trees outweighs the substantial costs of control.

Management of forestry trees as invaders requires actions to deal with the current problems and protocols to reduce the magnitude of the problem in the future.

Background and history of afforestation

Afforestation has a very long history in the northern hemisphere (Mather, 1993), but it is only in the 20th century that trees have been planted on a large scale over large areas in alien environments (regions far removed from their natural ranges) (Zobel *et al.*, 1987). A few alien tree species now form the foundation of commercial forestry enterprises in many parts of the world

O. T. Sandlund et al. (eds.), Invasive Species and Biodiversity Management, 237–257.

("exotic forests" *sensu* Zobel *et al*., 1987). Hundreds of other tree species have been widely planted for many purposes, including erosion and driftsand control, and to supply fuelwood and a range of other products. The bulk of alien tree plantings in the tropics and subtropics date from the second half of the 20th century. Afforestation policies and criteria for selecting alien species have been shaped by many ecological, economic, cultural and political factors in different parts of the world.

Alien trees and shrubs contribute significantly to the economies of many countries, but there are also significant costs associated with their widespread use. Costs that are increasing in importance in many parts of the world are those associated with the invasive spread of cultivated trees from planting sites into natural and semi-natural habitats where they have major impacts on a wide range of ecosystem properties and functions. Such invasions are increasingly causing conflicts of interest between foresters on the one hand, and conservationists and watershed managers on the other. This chapter explores the emergence of this problem and attempts to explain why some species have caused bigger problems than others and why some habitats are affected more than others. The magnitude of such costs are also examined with reference to several case studies. Problems associated with alien tree invasions in the southern hemisphere are highlighted, and in particular those caused by invasive pines. Options for addressing the problem are discussed.

Large-scale planting of timber- and crop-producing trees was carried out in parts of the Mediterranean Basin as long ago as 255 BC. In the Orient, commercial tree planting was recorded before the 6th century AD. Conifers were planted to stabilize sand dunes in Portugal as early as the 14th century, and in subsequent centuries large areas were afforested for this purpose on the Atlantic coast of France and along the shores of the Baltic. Afforestation for timber production and environmental management was carried out in Japan in the 17th and 18th centuries. Despite its long history, the scale of forestry remained fairly small until recently. Sustained, large-scale forestry was not widespread until the 20th century in Europe, and only expanded to other parts of the world in the last half of this century (Mather, 1993). The development of forestry with alien species in different parts of the world is well documented by Zobel *et al*. (1987).

The area afforested with pines and eucalypts in the southern hemisphere increased dramatically during the second half of the 20th century. For pines, the most dramatic growth has been in Chile where the first large-scale plantings (of *P. radiata*) were done only as recently as the early 1970s. The expansion of plantations of this species in Chile and the other Pacific-rim countries of Australia (since the early 1960s) and New Zealand (since the late 1960s) has been phenomenal; by 1996 roughly 4 million ha had been planted to this species (Lavery and Mead, 1998). Rapid expansions in pine afforestation have taken place in other South American countries (notably Brazil and Argentina) and in other tropical and subtropical regions (Le Maitre, 1998).

Besides the use of alien trees in commercial forestry (to supply roundwood and pulp), there has also been a rapid growth, mainly in developing countries, in the use of alien trees for "nonconventional forestry" uses, e.g. for fuelwood production and for restoring badly eroded or exhausted lands (Zobel *et al.*, 1987). "Agroforestry" is used in its broadest sense in this chapter, and is taken to refer to any use of trees in agriculture other than exclusively for the production of crops; such uses include windbreaks, shelter trees, and the intercropping of trees and arable crops. The 1980s saw the rapid expansion of tree planting, mainly with legumes, to meet the needs of rural communities and to support agricultural systems in agroforestry, to revegetate degraded lands, and to protect areas from soil erosion and desertification (see Hughes and Styles, 1989; Khurana and Khosla, 1993 and references cited in these reviews).

There was a massive increase in the extent of plantations (mostly comprising alien species) between 1980 and 1990 (Table 16.1). In Tropical Asia and the Pacific Islands the extent of plantations increased by 190% during this period. In tropical parts of Africa and the Americas, plantation area increased by 74 and 76% respectively over this period.

Rationale behind introductions

The main reasons for using alien trees instead of native trees in afforestation programmes were summarized by Zobel *et al.* (1987) and Mather (1993); these and other perspectives are reviewed briefly here.

1. In many parts of the world, where none now exist or where the indigenous conifers grow poorly or do not respond well to intensive forestry management, there is a need for coniferous trees to produce fibres and solid wood products. The wood of *Pinus* spp. is especially adaptable for a wide range of products.
2. Alien trees are often preferred because they grow much faster than native species.
3. Many of the tree species favoured for commercial cultivation display extreme ecological plasticity, both in terms of the climatic and edaphic conditions under which they flourish, and in that productivity can be greatly increased through basic silviculture. Highly productive plantations of commercial pine species such as *Pinus radiata* owe much

of their productivity to leaf area indices two or three times those found in natural forests (Richardson and Rundel, 1998).

4. Indigenous species are generally more difficult to manage silviculturally than the aliens.
5. Especially in the tropics, the biology of indigenous species is often poorly known (e.g. how to collect, store and germinate seed, how to produce seedlings in a nursery and how to manage them in a forest). Consequently, foresters prefer to work with well-studied alien species.

Table 16.1 The extent of plantations in the tropics, showing the change in area between 1980 and 1990, and the major tree taxa. Data from Pandey (1995). *: Note that several tree species are known in the vernacular as "mahogany", including *Swietenia* spp. (especially *S. mahagoni*), *Khaya*, *Shorea polysperma*, *Calophyllum inophyllum*, *Pentace burmanica*, *Melia azedrach*, *Pterocarpus dalbergiodes*. Regional statistics often do not specify which species are planted.

Region	Area of plantations (1000 ha) 1980	1990	% increase in area	Important tree/shrub taxa
Tropical Africa				
West Sahel	36.6	232.3	535	*Acacia* spp. (incl. *A. albida*), *Anacardium*, *Azadirachta indica*, *Eucalyptus*, *Gmelina*, *Prosopis*, *Tectona grandis*
East Sahel	447.2	762.0	70	*Acacia* spp. (incl. *A. nilotica*, *A. senegal*), *Azadirachta indica*, *Casuarina*, *Eucalyptus*, *Cupressus*, *Pinus*, *Tectona grandis*
Central Africa	64.8	175.0	170	*Acacia* spp. (incl. *A. klaineana*, *A. senegal*), *Azadirachta indica*, *Cassia siamea*, *Eucalyptus saligna*, *Gmelina*, *Khaya ivorensis*, *Pinus*, *Terminalia superba*
West Africa	299.3	440.5	47	*Anacardium*, *Azadirachta indica*, *Cassia siamea*, *Cedrela*, *Eucalyptus*, *Gmelina*, *Tectona grandis*, *Terminalia* spp. (incl. *T. ivorensis*)
Tropical Southern Africa	582.9	1054.5	81	*Acacia* spp. (incl. *A. mearnsii*), *Cupressus*, *Eucalyptus*, *Gmelina*, *Grevillea*, *Pinus*, *Tectona grandis*
Insular Africa	285.0	322.9	13	*Anacardium*, *Casuarina*, *Cryptomeria japonica*, *Eucalyptus*, *Pinus*
Total (Africa)	**1715.5**	**2987.2**	**74**	

Table 16.1 continued

Region	Area of plantations (1000 ha) 1980	1990	% increase in area	Important tree/shrub taxa
Tropical America				
Central America & Mexico	104.6	273.1	161	*Eucalyptus, Cedrela, Cupressus, Gmelina, Pinus* spp. (incl. *P. caribaea*), *Prosopis, Tectona grandis* and "mahogany*"
Caribbean subregion	207.7	441.6	113	*Acacia auriculiformis, Azadirachta indica, Cassia siamea, Casuarina, Catalpa longissima, Cedrela odorata, Eucalyptus, Gmelina, Hibiscus elatus, Leucaena, Pinus* spp. (incl. *P. caribaea, P. occidentalis*), *Tectona grandis* and "mahogany*"
Tropical South America	4594.7	7921.5	72	*Araucaria, Cupressus, Eucalyptus* spp. (incl. *E. globulus*), *Pinus* (incl. *P. caribaea*), *Tectona grandis*
Total (Amer.)	**4907.0**	**8636.2**	**76**	
Tropical Asia & Pacific				
South Asia	4957.0	19758.0	299	*Acacia nilotica, Casuarina, Dalbergia sissoo, Eucalyptus, Gmelina, Leucaena leucocephala; Melia, Morus alba, Pinus* spp. (incl. *P. roxburghii*), *Paraserianthes falcataria, Populus, Shorea, Tectona grandis, Terminalia arjuna* and mangrove species
Continental SE Asia	1795.1	3196.3	78	*Casuarina, Eucalyptus, Mangletia glauca, Melaleuca, Pinus* spp. (incl. *P. kesiya, P. massoniana, P. merkusii*), *Pterocarpus macrocarpus, Styrax tonkinensis, Tectona grandis, Xylia*
Insular SE Asia	4336.0	9155.5	111	*Acacia mangium, Eucalyptus, Gmelina, Leucaena leucocephala, Pinus, Paraserianthes falcataria, Pterocarpus indicus, Swietenia macrophylla, Tectona grandis*
Pacific Islands	88.0	189.2	115	*Agathis, Campnosperma brevipetiolata, Eucalyptus, Gmelina, Pinus* spp. (incl. *P. caribaea, P. elliottii*), *Tectona grandis, Terminalia* and "mahogany*"
Total (Asia/ Pacific)	**11088.1**	**32109.8**	**190**	

6. The availability of seed is a key to success in plantation forestry. Seeds of native species are often difficult to obtain whereas seeds from genetically-improved aliens are readily available.
7. Alien trees are frequently better suited to planting in grasslands or scrublands (marginal forest lands) where most afforestation is required. Also, aliens are frequently particularly successful in degraded forest lands.
8. It is often necessary to develop local forest industries to improve the balance of trade by reducing the need for imported wood products. Knowledge of markets and manufacturing technology currently favours the use of the wood of aliens such as pines or eucalypts.
9. Aliens are sometimes used to replace native species that are very susceptible to diseases or insects and cannot be grown profitably.

Species selected for planting must: grow well (also in plantations); grow on bare ground without shelter; produce desired products; and be physiologically well adapted to take advantage of favourable growing periods in new environments. In recent decades some attention has also been given to selecting species that pose less risk to the environment, but the aforementioned factors almost always override such considerations.

Many alien trees and shrubs targeted for planting in various parts of the world have been hailed as "wonder plants". A few prominent examples are *Azadirachta indica* (neem - "the wonder tree of India"); *Leucaena leucocephala* ("a wonder tree ... in Indonesia"); *Pinus radiata* (one of its common names is "remarkable pine"); and *Prosopis cineraria* ("wonder tree of the Thar desert"). Among the reasons for the admiration expressed for these plants are: their exceptional adaptability to a wide range of sites; their rapid growth; and the multiplicity of uses for their various products (e.g. for *Leucaena leucocephala*: fodder, firewood, building timber, erosion control, "green manures", providing shade and shelter for other crops). If one is to believe all the properties and benefits attributed to neem (e.g. Ruskin, 1993), then one cannot but concur that this is the "tree for solving global problems".

The global scale of introductions

Statistics on the global extent of afforestation are difficult to assimilate, but Gauthier (1991) suggests that the total area under "industrial" plantations amounted to almost 100 million ha in 1987, with 84% of this area under conifers. The regional distribution of these plantations was as follows: Asia (40%), Europe (19%), the former USSR (17%), North America (13%), Latin America (6%), Africa (3%) and Oceania (2%). Only about a quarter of the total area under industrial plantations comprises fast-growing (>12 $m^3 ha^{-1} yr^{-1}$)

plantations, most of which were planted after World War II (Bazett, 1992). More recent assessments of plantation areas (e.g. Pandey, 1995) deal with selected parts of the world (e.g. the tropics). Le Maitre (1998) has documented the exponential growth of plantation areas in southern hemisphere countries, mainly after 1945.

Dominant species being introduced

Pinus and *Eucalyptus* are by far the most important genera used in alien commercial forestry in the tropics and subtropics (*sensu* Zobel *et al.*, 1987). The main pine species are *P. caribaea*, *P. elliottii*, *P. kesiya*, *P. oocarpa*, *P. patula*, *P. pinaster*, *P. radiata* and *P. taeda* (although several other Central American/Mexican species are increasing in importance). Among the eucalypts, the most important species are *E. globulus*, *E. grandis*, *E. camaldulensis*, *E. tereticornis*, *E. urophylla* and *E. deglupta*. Other important taxa listed by Zobel *et al.* (1987) are *Acacia* spp., *Gmelina arborea*, *Tectona grandis* (teak) and *Swietenia* (mahogany) (see also Table 16.1). According to Pandey (1995), the total areas under plantations of the dominant tree taxa in the tropics in 1990 were (ha x 1000): *Eucalyptus* (10 064), *Pinus* (4 636), *Acacia* (3 400) and *Tectona grandis* (2 190). Large-scale plantings in temperate zones include species from the following genera: *Abies*, *Fagus*, *Eucalyptus*, *Larix*, *Quercus*, *Picea*, *Populus*, *Pinus* and *Pseudotsuga*. Species of *Eucalyptus* and *Pinus* are probably the most important taxa widely planted outside their natural ranges in temperate zones (but global statistics are not available).

Hundreds of tree species are fairly widely planted in nonconventional forestry. For example, in June 1996 the MultiPurpose Tree and Shrub (MPTS) database administered by the International Centre for Research in Agroforestry (ICRAF), Nairobi, Kenya (Carlowitz *et al.*, 1991) had records for more than 2 000 species. Legumes are especially favoured; important taxa are *Acacia* (many species), *Albizia* spp. (including *Paraserianthes*), *Leucaena leucocephala*, *Parkinsonia aculeata*, *Prosopis* (many species), *Robinia pseudoacacia* and *Sesbania* spp. (Hughes and Styles, 1989). Table 16.1 also lists some important taxa planted in 13 tropical regions.

Alien forest trees as invaders

All trees that are widely planted in alien environments will become naturalized and spread under certain conditions. Richardson *et al.* (1994) distinguished between naturalized species (those that often regenerate freely, but mainly under their own canopies), and those that frequently recruit seedlings, often in very large numbers, in natural or semi-natural vegetation at distances of more than 100 m from parent plants (often much further). It is the latter category ("invaders") that are considered here. The extent of such invasions and the impacts that they produce depend on many factors. The invasive spread of alien trees from commercial forestry plantations is currently a bigger problem than invasions from agroforestry plantings. The main reasons for this are: 1) the former are generally planted in much greater numbers over larger areas; 2) afforested areas often adjoin natural vegetation (especially in the southern hemisphere); and 3) spreading agroforestry trees are more likely to be kept in check by agricultural practices and utilization by humans, at least at a local scale (although exploitation of invading *Prosopis juliflora* in eastern and central Sudan has not prevented it from becoming a major problem).

Of the **dominant** tree genera used for commercial forestry (see above), all except *Abies*, *Fagus* and *Gmelina* are noted as alien invaders in recent synthesis volumes (although *A. nordmanniana* is mildly invasive around forestry plantings in Canterbury, New Zealand; Webb *et al.*, 1988). Among the most widely planted trees, *Acacia* and *Pinus* species feature most prominently on weed lists and in reviews of trees as invaders. This is hardly surprising, as many taxa in both genera are superb colonizers, with many adaptations that equip them to become invaders in a wide range of habitats (O'Dowd and Gill, 1986; Hughes and Styles, 1989; Fagg and Stewart, 1994; Richardson *et al.*, 1994). For these genera, the importance of species as invaders is usually correlated with the extent of planting (e.g. Richardson *et al.*, 1990 for *Pinus*). The Australian Tree Seed Centre (1993) lists 10 *Eucalyptus* species that are known to pose problems as invaders in South Africa. Eucalypts are also represented on many national or regional weed lists. Despite this, they have not been nearly as successful in invading alien environments as other widely planted trees such as pines and legumes. Many eucalypts produce very large quantities of seeds, so their "lack of real punch" as invaders is rather puzzling.

A search (in June 1996) of the MPTS data base (see above) which contains information on more than 2 000 species used in agroforestry showed that 135 species (c. 7%) were weedy under certain conditions, but that some 25 species (c. 1%) were considered weedy in more than half the records: *Acacia* (8 spp.), *Ailanthus altissima*, *Calotropis procera*, *Casuarina glauca*, *C. littoralis*, *Dichrostachys cinerea*, *Gleditsia triacanthos*, *Guazuma ulmifolia*, *Justicia adhatoda*, *Leucaena leucocephala*, *Melaleuca quinquenervia*, *Parkinsonia aculeata*, *Prosopis* (3 spp.), *Psidium guajava*, *Sesbania bispinosa* and *Ziziphus nummularia*. Hughes and Styles (1989) list 144 species of woody legumes, many of which are used in agroforestry, as being weeds; 29 species (20%) are major pests.

184 species (out of 653) in the Invasive Woody Plant Database (Binggeli, 1996) are considered "highly invasive". Of these, 34 species (18%) were introduced for forestry and another 49 species (27%) for amenity purposes. Forestry and amenity plantings are, thus, very important sources of highly invasive woody plants. Trees (> 15 m) made up 25% of the highly invasive species in the database. These data clearly show that trees and shrubs purposefully introduced to new areas: a) serve a wide range of uses; and b) often become invasive.

Case study: pines as alien invaders in the southern Hemisphere

Although alien pines have spread from planting sites in parts of the northern hemisphere (e.g. Brown and Neustein, 1972 for Britain) and have probably had significant impacts in places, these invasions have, surprisingly, been poorly studied. The spread of pines from planting sites in the southern hemisphere has been much more thoroughly studied, probably because the invading pines are unquestionably aliens, are very conspicuous, and have clear impacts. Probably the earliest record of pines invading natural vegetation in the southern hemisphere is from 1855, when *P. halepensis* was noted to be spreading into South African fynbos (Richardson and Higgins, 1998). At least 19 *Pinus* species are now well established as invaders of natural ecosystems in the southern hemisphere, and eight species are weeds of major importance. Species with self-perpetuating invasive populations over the largest areas are *P. contorta* (mainly in New Zealand), *P. halepensis* (South Africa), *P. nigra* (New Zealand), *P. patula* (Madagascar, Malawi and South Africa), *P. pinaster* (South Africa, Australia, New Zealand, Uruguay), *P. ponderosa* (New Zealand), *P. radiata* (Australia, New Zealand, South Africa) and *P. sylvestris* (New Zealand). *Pinus contorta*, *P. nigra*, *P. ponderosa* and *P. sylvestris* are not important species for commercial forestry, but have been widely planted over large areas in New Zealand for erosion control and other purposes (see Richardson *et al.*, 1994). Richardson *et al.* (1994) and Richardson and Higgins (1998) give a detailed assessment of the factors that control these invasions. They suggest that the extent of invasions can be explained by a model incorporating information on: species attributes, residence time, extent of planting, ground-cover characteristics, locality (latitude), disturbance regime, and the resident biota in the receiving environment. Besides helping to explain the dynamics of pine invasions, the model also illustrates factors that can be managed to reduce the extent of invasions (see Richardson and Higgins,

1998). Richardson *et al.* (1990), Rejmanek and Richardson (1996) and Rejmanek (1998) discuss the suite of life-history attributes that distinguish successful pine invaders from less successful congeners. Based on this information, Richardson and Higgins (1998) suggest that the extent of the problem is set to increase dramatically over the next few decades if left unchecked, both in areas where invasions are already evident, but also in other areas with recent plantings over large areas and/or where changes to landuse practices are altering invasibility.

Ecological and economic effects of using alien species

Afforestation with alien trees was initially driven by the belief that such plantings were beneficial to the environment, and trees were often planted to repair damaged ecosystems. Even until several decades ago, plantations of alien trees were considered to have little impact on the environment. In recent decades, however, there has been growing realization of the significant impacts of this form of land transformation on various ecosystem properties. Most afforestation in the southern hemisphere has transformed grasslands and scrub-brushland habitats. This is because: site preparation in such sites is relatively easy (compared to forests, where tree removal is expensive); much tropical grassland is unsuitable for agriculture or grazing (although such sites are very often very rich in native biodiversity; Huston (1993)); and ecological/political reasons dictate the desirability of planting grasslands and scrub-brushlands rather than tropical forests (Zobel *et al.*, 1987). Clearly, establishment of plantations in such vegetation totally alters vegetation structure, and consequently also biomass distribution, plant density and vegetation height, leaf area index, litterfall and decomposition rates, fire behaviour, nutrient cycling and energy balance (Versfeld and van Wilgen, 1986). Most pine invasions also affect these vegetation types; self-sown stands are similar in most respects to man-made forests and have essentially the same impacts. The invasive spread of alien trees from plantations into adjoining areas of natural vegetation has meant that the damaging effects of forestry with alien species is extending into watersheds and/or areas set aside for conservation. Impacts following the invasion of natural ecosystems in the southern hemisphere by pines were reviewed by Richardson *et al.* (1994); they consider the principal effects to be caused by the shifts in life-form dominance, reduced structural diversity, increase in biomass, disruption of prevailing vegetation dynamics, and changed nutrient cycling that occur when grasslands and scrublands are invaded. Most resultant impacts are detrimental to the invaded systems and threaten sustained productivity and functioning.

Detailed cost-benefit analyses for invasive alien trees that are also important crops are scarce. In probably the most detailed study to date, Higgins *et al.* (1997) produced a dynamic ecological-economic model to aid in conflict resolution when alien plants with value in parts of the landscape in South African fynbos impact on ecosystem services in other parts. Geldenhuys (1986) considered the value of *Acacia melanoxylon*, a valuable timber species in the Western Cape of South Africa which is also an important invader of indigenous forest and riparian zones. Invasive pines (mainly *P. pinaster* and *P. radiata*) and other alien trees have a major impact on watershed hydrology in the fynbos biome of South Africa. Runoff from watersheds with dense stands of alien trees is between 30% and 70% lower than for uninvaded fynbos, depending on the annual rainfall and the age and density of the alien stand. During the dry summer months, when water needs are greatest, runoff in invaded watersheds may be reduced to zero, converting perennial streams to seasonal ones. Besides the obviously detrimental impacts to aquatic biota, the reduced streamflow has serious implications for water use in regions where shortages of water are already limiting development. Large parts of the watersheds that yield about two-thirds of the region's water requirements and more than 90% of the water supply of Greater Cape Town are covered by dense alien stands. A recent modelling study suggests that further invasion could result in an average decrease in water production from fynbos watersheds of 347 cm^3 water ha^{-1} yr^{-1} over 100 years, resulting in an average loss of more than 30% of the water supply to Cape Town (Le Maitre *et al.*, 1996).

Problems caused by the invasive spread of agroforestry species are generally less easy to quantify. An exception is *Prosopis* in arid parts of South Africa, where the dense thickets that covered some 180 000 ha in 1989 render invaded land useless for normal farming practices (Harding and Bate, 1991). For South African fynbos, Higgins *et al.* (1997) describe a model for assessing costs and benefits of *Acacia cyclops* which is an important source of fuelwood, but is damaging to natural ecosystems.

What can be done?

It is necessary to consider actions to deal with the current problems caused by alien trees and shrubs that have spread from planting sites, but also to establish protocols to reduce the magnitude of the problem in the future. In many cases options exist for managing invaded areas by manipulating disturbance regimes (e.g. fire cycles, grazing levels) to impede invasion (e.g. Richardson and Higgins, 1998 for pines). The solution to problems caused by invasive alien trees lies in integrating various control methods.

The situation is generally simpler in agroforestry systems, where decisions need to be made on whether benefits derived from the invasive spread of an alien tree species outweigh the reduced value of ecosystem services, e.g. the loss of grazing land in areas invaded with *Prosopis* (Figure 16.1).

In commercial forestry, the situation is usually more complex, both because multi-national forestry companies frequently wield considerable economic and political power, especially in developing countries, and because the methodology for costing the impacts of invasions (usually in conservation areas) is poorly developed. In a major development in South Africa, a detailed cost-benefit analysis showed that the major costs of clearing invasive alien trees are warranted, since the financial benefits from improved water yield from watersheds outweigh the substantial costs of control. This is because the increased streamflow reduces the need for other actions directed at increasing water supply, e.g. dam building (Van Wilgen *et al.*, 1996, 1997). Results of this analysis, enhanced by the political will to provide jobs for the large corps of unemployed and unskilled people, led, in 1995, to the financing from state coffers of what is probably the largest and most expensive programme of alien plant control ever undertaken (Figure 16.2).

Options for the replacement of invasive with less invasive species seem limited in commercial forestry. For example, with pines, all of the most productive species are invasive. The gradual replacement of known invasive species with more innocuous species has more scope in agroforestry systems. Such undertaking are in progress in several regions. For example, in the Galapagos archipelago, four alien trees are highly invasive in natural systems. Two of these (*Cedrela odorata* and *Cordia alliodora*) were introduced for timber, one for its fruits (*Psidium guajava*) and one for quinine (*Cinchona succirubra*). Besides attempts at controlling the densities of these species, conservation authorities are encouraging people to use other species such as *Centrolobium paraense*, *Juglans neotropica*, *Swietenia macrophylla* and *Tectona grandis*. These species have been on the islands for at least 30 years, grow well and show no signs of invasive spread (A. Mauchamp, pers. comm.). Although the replacement of invasive with noninvasive species is a commendable aim, is it possible to proclaim a species "safe" for planting? For example, at least one of the species recommended for planting in the Galapagos is invasive elsewhere (*S. macrophylla* in Sri Lanka; Cronk and Fuller, 1995) and could well become invasive in the Galapagos in the future.

What can be done to reduce the magnitude of the problem in the future? Firstly, there is an urgent need to predict the nature and magnitude of the problem in the future. Invasions are already a problem, especially in regions with the longest history of afforestation with alien species. Given the massive increase in the extent of plantings over the last few decades (e.g. see Table 16.1 for the tropics), I have no doubt that the scale of the problem will increase rapidly. This is because of the increased size of the propagule pool of known invasive species, the increased contact area between plantations and potentially invadable habitats, but also because so many new species are being cultivated. Experience has shown that, on average, about 10% of introduced species become naturalized, and 1% become "pests" (Williamson, 1996, p. 33). However, these averages are likely to change in the future; sophisticated techniques for species-site matching (e.g. Booth and Pryor, 1991) are likely to increase the proportion of introductions that become naturalized and invasive.

Early-warning systems are needed to identify species with invasive tendencies. The arboreta that have been established in many parts of the world to assess the suitability of many tree species for cultivation in particular environments are important "natural experiments". Such plantings provide a useful source of information regarding the invasive tendencies of different trees (e.g. Richardson and Higgins, 1998, p. 456) (see also Figure 16.3).

An international protocol for species translocations is urgently needed (see Hughes, 1994, 1995 for a thorough discussion). With the emergence of computerized databases of invasive species (Frost *et al.*, 1995; Binggeli, 1996) and global syntheses of plant invasions, ignorance is no longer an excuse for disseminating invasive trees. Afforestation policies need to include clearly stated objectives to reduce impacts outside areas set aside for forestry. But who plays policeman? Organizations such as ICRAF have an important role to play in the case of agroforestry. More information should be disseminated when species are recommended, and restrictions should be placed on the translocation of some species. ICRAF routinely issues a warning when supplying seeds of known weedy species such as *Acacia karroo* and *Mimosa pigra*, but has no means for ensuring that any procedures will be implemented to prevent spread from planting sites. Commercial forestry companies have an important responsibility in this regard, and many organizations are currently implementing the ISO 14001 Environmental Management System which involves regular environmental reviews of their silvicultural practices.

Tree species can be screened for invasive potential in pre-planting tests. *Pinus greggii*, for example, was found to have high seed efficiency in natural stands compared to other pines (c. 63%; Lopez-Upton and Donahue, 1995). This species, which has recently been planted in South Africa, shares many attributes with the closely related, and already invasive, *P. oocarpa* and *P. patula*; it must therefore be flagged as a high-risk introduction. In a test of forty legume species, Aronson *et al.* (1992) were able to exclude weedy candidate species from field trials in favour of similarly performing candidates

Figure 16.1 *Prosopis* in South Africa.
a) Early plantings of *Prosopis glandulosa* var. *torreyana* and *P. velutina* around a farm house near Kenhardt in the Northern Cape Province, South Africa. The two taxa hybridized soon after the initial planting in the early 1900s, and hybrids spread rapidly to cover very large areas of rangeland.
b) A dense stand of *Prosopis*, showing the results of ineffective (mechanical) control in the past. Coppicing has lead to an impenetrable stand of multi-stemmed shrubs. Such stands are totally useless: hardly any pods are produced, wood is too fine for firewood or charcoal, and the dense thickets are impenetrable for livestock and humans. (Photos: D.M. Richardson).

Figure 16.1 *Prosopis* in South Africa.
c) Biological control, using seed-attacking insects can be successful against invasive *Prosopis* in South Africa, if fallen pods are protected from consumption by livestock until the insects have consumed the seeds. Where this occurs, fewer than 1% of seeds remain intact in one-year-old pods on the ground.
d) An innovative approach to managing *Prosopis* involves harvesting fallen pods for providing supplementary forage for sheep. The storage of the pods in the bins shown here protects the pods from livestock until the introduced insects have destroyed almost all the seeds. (Photos: D.M. Richardson).

(e.g. *Prosopis alba*) with less potential of becoming invasive weeds. The power of such screening exercises remains to be tested. Views on the relative invasiveness of different taxa are gleaned from experience in different regions, and the careful (but risky) extrapolation of results to other areas (Tucker and Richardson, 1995).

Biological control offers considerable potential for reducing conflicts of interest where indispensible crops cause problems as invaders. Considerable progress has been made in this regard in South Africa. The histories of control efforts on *Acacia mearnsii*, *A. cyclops* and *Prosopis* spp. should prove particularly instructive in the search for a general model for resolving conflicts. *Pinus* seems an obvious target for biocontrol action. Despite the view that prospects for effective biocontrol of pines are limited, since potentially suitable seed-attacking insects are also vectors of diseases (de Groot and Turgeon, 1998), a pilot study to explore options for using seed-attacking insects on invasive pines in South Africa was launched in 1997 (J.H. Hoffmann and V.C. Moran, University of Cape Town, pers. comm.). Work is also underway in South Africa to produce seedless clones of *P. elliottii*, *P. patula* and *P. radiata* by irradiating seeds. Pilot trials have shown the range of doses that should still allow germination but which will hopefully affect seed production (H. de Lange, National Botanical Institute, Cape Town, pers. comm.). Until major advances are demonstrated using such techniques, it will be necessary to introduce a "polluter pays" policy, possibly through the imposition of a levy on timber products. In cases, such as South African fynbos, where the magnitude of invasions has already reached chronic proportions, levies on the supply of ecosystem services such as water will probably be necessary to fund at least initial clearing programmes.

The growing database of invasion case studies facilitates the development of screening models for given systems (e.g. Tucker and Richardson, 1995, for fynbos). Such tools provide objective means for rationalizing introductions and enabling managers to test the effects of various habitat manipulation options. Recent developments in spatial modelling also allow managers to explore the implications of interactions between invasive organisms, inherent features of the invaded environment and disturbance (e.g. Higgins and Richardson, 1998).

Figure 16.2 The "Working for Water" programme in South Africa encompasses 140 projects (at the end of 1997), employs over 8 300 people, and has an annual budget of about US$ 53 million. The picture shows workers clearing a dense stand of invasive *Pinus pinaster* in the Western Cape province (Photo: D.M. Richardson).

Conclusions

Everyone I approached for comment had a different view of the issue of the invasive spread of commercially important trees. Clearly, every case needs to be evaluated in terms of risks, costs and benefits. The dissemination of information is crucial for ensuring that local policies can be developed and implemented.

The fact remains that the traits that make some tree species highly suitable for forestry (e.g. rapid growth rates) are an integral part of the suite of characteristics of early-successional or "weedy" trees. Important forestry species, therefore, carry with them components of fitness (life-history traits) that make them early-seral plants, and potential invaders. Many species have already realized this potential, and many more will do so over the next few decades.

Figure 16.3. Of the 86 alien tree species that were planted at Hosmer's Grove in Haleakala National Park, Maui, Hawaii, in the early 1900s, only about 20 species survived. Among these are nine *Pinus* species, four of which (*P. jeffreyi*, *P. patula*, *P. pinaster* and *P. radiata*) produce seedlings; three of these (all the aforementioned except *P. jeffreyi*) readily establish seedlings at considerable distances from parent trees. The photograph, taken in 1997, shows a *P. patula* seedling established in subalpine heath scrub > 75m from the nearest planted tree. (Photo: D.M. Richardson).

Acknowledgements

I thank James Ball (FAO, Rome), Bob Boardman (Primary Industries, South Australia), Doug Boland, Trevor Booth (both CSIRO, Australia), Ian Dawson (ICRAF, Kenya), Bill Dvorak (North Carolina State University, USA), Hannah Jaenicke (ICRAF, Kenya), Steve Higgins (IPC, South Africa), John Hoffmann (University of Cape Town), Colin Hughes (Oxford University, UK), Fred Kruger (CSIR, South Africa), André Mauchamp (Charles Darwin Research Station, Galapagos, Ecuador), Alan Pottinger (Oxford University, UK), John Scotcher (SAPPI Forests, South Africa), Tony Simons (ICRAF, Kenya), Kobus Theron and Gerrit van Wyk (both University of Stellenbosch, South Africa), Tim Vercoe (CSIRO, Australia) and Helmuth Zimmermann (Plant Protection Research Institute, South Africa) for sharing useful ideas, commenting on parts of the manuscript, and helping in other ways. None of them should be held responsible for the conclusions I have drawn.

References

Aronson, J., Ovalle, C., Avendano, J., Avendano, R.J. and Ovalle, M.C. (1992) Early growth rate and nitrogen fixation potential in forty-four legume species grown in an acid and a neutral soil from central Chile. *Forest Ecology and Management,* **47**, 225–244.

Australian Tree Seed Centre (1993) *Seeds of Australian Trees Project.* Project design document for Australian Centre for International Agricultural Research. Canberra, CSIRO Division of Forestry.

Bazett, M.D. (1992) *Shell/WWF Tree Plantation Review. Study No. 3. Industrial Wood.* Shell International/WWF, London.

Binggeli, P. (1996) A taxonomic, biogeographical and ecological overview of invasive woody plants. *Journal of Vegetation Science,* **7**, 121–124.

Booth, T.H. and Pryor, L.D. (1991) Climatic requirements of some commercially important eucalypt species. *Forest Ecology and Management,* **43**, 47–60.

Brown, J.M.B. and Neustein, S.A. (1972) Natural regeneration of conifers in the British Isles, in *Proceedings of the Third Conifer Conference*, Royal Horticultural Society, London, pp. 93–116.

Carlowitz, P.G., Wolf, G.V. and Kemperman, R.E.M. (1991) *Multipurpose Tree and Shrub Database – An Information and Decision Support System.* Version 1.0. ICRAF, Nairobi, Kenya.

Cronk, Q.C.B. and Fuller, J.L. (1995) *Plant Invaders. The Threat to Natural Systems.* Chapman & Hall, London.

DeGroot, P. and Turgeon, J. (1998) Insect-pine interactions, in *Ecology and Biogeography of* Pinus (ed. D.M. Richardson), Cambridge University Press, Cambridge, pp. 354–380.

Fagg, C.W. and Stewart, J.L. (1994) The value of *Acacia* and *Prosopis* in arid and semi-arid environments. *Journal of Arid Environments,* **27**, 3–25.

Frost, H.M., Terry, P.J. and Bacon, P. (1996) A feasibility study into the creation of a database on weeds and invasive plant species, in *Weeds in a Changing World*, (ed. C.H. Stirton), BCPC Symposium proceedings No. 64. BCPC, London, pp. 35–69.

Gauthier, J.J. (1991) *Plantation wood in international trade.* Paper presented at Centre for Applied Studies in International Negotiations (CASIN) Conference, Geneva.

Geldenhuys, C.J. (1986) Costs and benefits of the Australian blackwood *Acacia melanoxylon* in South African forestry, in *The Ecology and Management of Biological Invasions in Southern Africa,* (eds I.A.W. Macdonald, F.J. Kruger and A.A. Ferrar), Oxford University Press, Cape Town, pp. 275–284.

Harding, G.B. and Bate, G.C. (1991) The occurrence of invasive *Prosopis* species in the north-western Cape, South Africa. *South African Journal of Science,* **87**, 188–192.

Higgins, S.I., Azorin, E.J., Cowling, R.M. and Morris, M.J. (1997) A dynamic ecological–economic model as a tool for conflict resolution in an invasive alien plant, biological control and native plant scenario. *Ecological Economics,* **22**, 141–154.

Higgins, S.I. and Richardson, D.M. (1998) Pine invasions in the southern hemisphere: modelling interactions between organism, environment and disturbance. *Plant Ecology,* **135,** 79-93.

Hughes, C.E. (1994) Risks of species introductions in tropical forestry. *Commonwealth Forestry Review,* **73,** 243–252.

Hughes, C.E. (1995) Protocols for plant introductions with particular reference to forestry: Changing perspectives on risks to biodiversity and economic development, in *Weeds in a Changing World,* (ed. C.H. Stirton), BCPC Symposium proceedings No. 64. BCPC, London, pp. 15–32.

Hughes, C.E. and Styles, B.T. (1989) The benefits and risks of woody legume introductions. *Monographs in Systematic Botany of the Missouri Botanical Garden,* **29**, 505–531.

Huston, M. (1993) Biological diversity, soils, and economics. *Science,* **262**, 1676–1680.

Khurana, D.K. and Khosla, P.K. (1993) *Agroforestry for Rural Needs.* IUFRO Workshop Proceedings. Indian Society of Tree Scientists, Solan, India.

Lavery, P.B. and Mead, D.J. (1998) *Pinus radiata*: A narrow endemic of North America takes on the world, in *Ecology and Biogeography of* Pinus, (ed. D.M. Richardson), Cambridge University Press, Cambridge, pp. 432–449.

Le Maitre, D.C. (1998) Pines in cultivation: A global view, in *Ecology and Biogeography of* Pinus, (ed. D.M. Richardson), Cambridge University Press, Cambridge, pp. 407–431.

Le Maitre, D.C., van Wilgen, B.W., Chapman, R.A. and McKelly, D.H. (1996) Invasive plants and water resources in the Western Cape Province, South Africa: modelling the consequences of a lack of management. *Journal of Applied Ecology,* **33**, 161–172.

Lopez-Upton, J. and Donahue, J.K. (1995) Seed production of *Pinus greggii* Engelm. in natural stands in Mexico. *Tree Planter's Notes,* **46**.

Mather, A., ed. (1993) *Afforestation. Policies, Planning and Progress.* Belhaven Press, London.

O'Dowd, D.J. and Gill, A.M. (1986) Seed dispersal syndromes in Australian Acacia, in *Seed Dispersal,* (ed. D.R. Murray), Academic Press Australia, Sydney, pp. 87–121.

Pandey, D. (1995) Forest resources assessment 1990. Tropical forest plantation resources. *FAO Forestry Paper,* **128**, FAO, Rome.

Rejmanek, M. (1998) Invasive plant species and invasible ecosystems, in *Invasive Species and Biodiversity Management,* (eds O.T. Sandlund, P.J. Schei and Å. Viken), Kluwer Academic Publishers, Dordrecht, The Netherlands.

Rejmanek, M. and Richardson, D.M. (1996) What attributes make some plant species more invasive? *Ecology,* **77**, 1655–1661.

Richardson, D.M., Cowling, R.M. and Le Maitre, D.C. (1990) Assessing the risk of invasive success in *Pinus* and *Banksia* in South African mountain fynbos. *Journal of Vegetation Science,* **1**, 629–642.

Richardson, D.M. and Higgins, S.I. (1998) Pines as invaders in the southern hemisphere, in *Ecology and biogeography of* Pinus, (ed. D.M. Richardson), Cambridge University Press, Cambridge, pp. 450–473.

Richardson, D.M. and Rundel, P.W. (1998) Ecology and biogeography of *Pinus*: An introduction, in *Ecology and Biogeography of* Pinus, (ed. D.M. Richardson). Cambridge University Press, Cambridge, pp. 3–43.

Richardson, D.M., Williams, P.A. and Hobbs, R.J. (1994) Pine invasions in the Southern Hemisphere: determinants of spread and invadability. *Journal of Biogeography,* **21**, 511–527.

Ruskin, F.R., ed. (1993) *Neem. A Tree for Solving Global Problems.* National Academy Press, Washington.

Tucker, K. and Richardson, D.M. (1995) An Expert System for Screening Potentially Invasive Alien Plants in Fynbos. *Journal of Environmental Management,* **44**.

Van Wilgen, B.W., Cowling, R.M. and Burgers, C.J. (1996) Valuation of ecosystem services. *BioScience,* **46,** 184–189.

Van Wilgen, B.W., Little, P.R., Chapman, R.A., Gorgens, A.H.M., Willems, T. and Marais, C. (1997) The sustainable development of water resources: History, financial costs, and benefits of alien plant control programmes. *South African Journal of Science,* **93**, 404–411.

Versfeld, D.B. and van Wilgen, B.W (1986) Impact of woody aliens on ecosystem properties, in *The Ecology and Management of Biological Invasions in Southern Africa,* (eds I.A.W. Macdonald, F.J. Kruger and A.A. Ferrar), Oxford University Press, Cape Town, pp. 239–246.

Webb, C.J., Sykes, W.R. and Garnock-Jones, P.J. (1988) *Flora of New Zealand. vol. 4. Naturalized Pteridophytes, Gymnosperms, Dicotyledons.* Botany Division, DSIR, Christchurch.

Williamson, M. (1996) *Biological Invasions.* Population and Community Biology Series 15. Chapman & Hall, London.

Zobel, B.J., van Wyk, G. and Stahl, P. (1987) *Growing Exotic Forests.* John Wiley and Sons, New York.

17 Consequences of spreading of pathogens and genes through an increasing trade in foods

EYSTEIN SKJERVE and YNGVILD WASTESON
Norwegian College of Veterinary Medicine, Oslo, Norway

Abstract

An increased volume of international food trade after the establishment of WTO will also speed up the transfer of pathogenic agents and microbial genes around the world.

Food-borne infections are today representing an increasing health problem in many areas. Examples include pandemic spread of *Salmonella* varieties and a continuous dispersion of antibiotic resistant bacteria in fresh foods. The problem is, at least partially, related to trade of biological products and the industrialization of agriculture production.

The molecular mechanisms for establishment and spread of microbial genes are known and described, and the microbes' ability for exchanging gene elements is clear. The complexity of microbial establishment, spread and persistence in natural ecosystems is, however, less understood.

Mathematical risk assessment is today presented as the methods to limit the health and ecological consequences of an increased food trade, but will most likely not accomplish this. Dynamic modeling is an alternative approach to understand the long-term behavior of the complex microbial ecosystems linked to the food production chains.

Introduced microbes or genes may disrupt hitherto stable microbial communities. The introduced varieties may be permanently established, or only slowly disappear from the ecological systems they are introduced into.

Introduction

International trade in foods has throughout history been a part of man's basic activities, and controlling this trade has been a basis for political power and empire building. The present liberalization of international rules for trade of agricultural products is no exception to previous examples from recent and remote history. However, the volume of this trade is so large that the biological and economical consequences are larger than previously seen. The power of empires is now replaced by strong blocks of nations determining the rules for such trade – of course mainly for their own benefit. The underlying assumption behind the political trend is that free trade is *per se* a positive phenomenon.

O. T. Sandlund et al. (eds.), Invasive Species and Biodiversity Management, 259–267.

While many areas of production and trade are typically objects of pure socioeconomic studies, the trade in agricultural products should be viewed as a biological process, modified by socioeconomic factors, and not vice versa. The possible consequences of allowing economists to take the lead on agricultural production and trade might be a major disruption of food production systems in many areas of the world, possibly with irreversible or slowly reversible consequences. Understanding the complex and dynamic structure of regional and local food production systems is a prerequisite for creating agricultural systems able to support an increasing world population, at the same time avoiding major damages to valuable ecological biosystems.

This paper will concentrate on the direct health consequences of an increased food trade, and the possible environmental consequences closely linked to these health issues. The use of risk analysis in order to reduce these consequences is discussed.

World Trade Organization and the new international order

The establishment of the World Trade Organization (WTO) was a new step in the internationalisation of trade and commerce. With its focus on agricultural products, the Uruguay negotiations of General Agreement on Tariffs and Trade (GATT) paved the way for an integration of agricultural production into a common world market. The WTO documents also recognizes the problem of spreading pests and invasives through this trade. More specifically, the so called Sanitary and Phytosanitary Protocol (SPS) of the WTO agreement (Anonymous, 1993) prescribes the use of risk assessments to limit the possible negative effects on human, plant and animal health. Furthermore, the same protocol expresses certain rights for an individual country to protect itself against importation of plant, animal or human pathogens or ecologically harmful organisms. There is, however, a dramatic change from the time before the WTO. Previously each country had its own regulations for the control of imported invasives, based on their history, geographic location and health status of plants, animals and man. Under WTO, the only way to claim protection will be through scientifically based risk assessments, where decisions based upon these analyses can be overruled by WTO organs. So far, only minor effects of the WTO agreement have appeared, due to the high toll barriers replacing the previous legal barriers. Within a few years the full effect of WTO will emerge, due to a reduction in toll barriers, and a major increase in the trade of agricultural products is expected.

Food-borne infections – an increasing problem

Infectious diseases constitute a major health problem today, as in all previous history of man. Of the approximately 17 million persons dying directly from infections every year, respiratory infections claim 5 million lives and enteric infections about 3 million. Most of the latter group consists of childhood infections in poor countries caused by contaminated drinking water or foods. Most of these deaths could easily be prevented at a relatively low cost. The increasing importance of infections is described in detail by several groups of researchers, using the term *emerging infections* to describe the new (?) diseases as AIDS and Ebola and *re-emerging infections* to describe e.g. the return of TB in the western countries (Lederberg *et al.*, 1992; Lederberg, 1993; Bryan, 1998).

The most rapidly increasing problem in the rich countries of the west is the increase of food-borne infections caused by organisms linked to contamination of water or foods by typical zoonotic agents as *Salmonella* spp., *Campylobacter* spp. and verotoxinogenic *E. coli* (VTEC). While most problems linked to contaminated foods and water used to be associated with poor countries, there has been a rapid increase in the incidence of these diseases also in the industrialized countries. Most of the recent increase is linked to the spread of infections through fresh animal products as poultry meat and eggs, pork and to a lesser extent beef (Anonymous, 1994a, b, 1995). These zoonotic agents serve as a good illustration of the effects of changing the ecological basis of systems for bioproduction, combined with a rapid spread of pathogens through international trade. While the problem has been most typically expressed in the intensive animal production in the western countries, also other areas of the world has been hit by spread of pathogens through trade in live animals and fresh foods. The global spread of *Salmonella* Enteritidis PT4 through trade in chicken and eggs is a good illustration on this (Altekruse, 1997).

The health consequences of this increase is not fully acknowledged. In the Netherlands, recent research indicates that up to 60% of the population experience food-borne infections every year (Hoogenbom-Verdegaal *et al.*, 1994). Results from other countries show similar figures. The economic consequences for each country is difficult to assess, but in the US, it is estimated that food-borne infections may cost up to US$ 23 billion every year. The health effects are not only linked to the enteric stage, but also to possibly severe sequelae from intestines, joints, CNS or skin. Up to 5% of the patients might experience sequelae lasting for months or years after an enteric infection (Baird-Parker, 1994; Kvien *et al.*, 1994).

Food chains as ecological niches for pathogenic organisms and microbial genes

There are numerous explanatory factors behind the observed increase of food-borne infections. Accumulation of pathogens in the domestic animal food chains is to a large extent linked to the industrialization of animal production, with large units of animals on small areas creating ideal conditions for pathogens to spread, establish and persist in the production chain. Of special interest is the spread of pathogens through the trade in animal feeds, which is possibly the most important initial factor during the post World War 2 years (Maurice, 1994).

The extensive use of antibiotics as feed additives for domestic animals has contributed to the increasing problem of drug resistant bacteria. The use of avoparcin in European poultry and pig production has lead to an increase in the number of enterococci resistant not only to avoparcin, but also to vancomycin, the drug of choice for treating human infections with multiresistant staphylococci. The possible establishment of the vancomycin resistance in nosocomial staphylococci is a feared outcome of this situation (Barrow, 1989; Klare *et al.*, 1995). Establishment of resistance genes in potentially pathogenic bacteria (bacteria that exist as part of the normal intestinal flora but on some occasions may produce disease) might represent a serious future health problem. During an outbreak of shigellosis caused by Spanish iceberg lettuce, resistance genes not previously observed in the Nordic countries were detected in *E. coli* on the lettuce (Høiby *et al.*, 1995). Some scientists also claim that the intensive use of antibiotics in animal feed creates a selection pressure on the intestinal flora where most pathogens are included, and where antimicrobial resistance is most easily established (Threlfall, 1992).

The consumer trends of fresh food preference has also changed the market. The "fresh" foods with long shelf-life because of technological use of vacuum or modified atmosphere, clearly produces new ecological niches ideal for supporting growth of psychrotrophic pathogens as *Listeria monocytogenes* and *Yersinia enterocolitica*. At the same time, the number of persons with defect or suppressed immunosystem is on the rise. In the US, approximately 20% of the population is immunocompromised, due to diseases as AIDS and cancer, transplantation or therapy for other diseases, high age, pregnancy or infancy (Anonymous, 1995).

Mechanisms for spreading, establishment and persistence of microbial genes in different ecosystems

The molecular mechanisms involved in microbial exchange of essential genes are to a certain extent described. Microbes' ability to interact with their surroundings seems almost unlimited. Bacterial populations should be regarded as parts of one microbial world where genes are expelled, exchanged and included due to basic selection mechanisms and microbial competition. These interactions enable bacteria to adapt to changing environments, and genes harboured by harmless microbes might be transferred to pathogenic or environmentally widespread organisms.

Bacteria communicate within natural ecosystems and can be transported over long distances. The large epidemic of cholera in Latin-America in 1991 was probably caused by *Vibrio cholerae* transported via ocean currents across the Pacific Ocean. It has been shown that resistance plasmids from *V. cholerae* can be easily transferred to aquatic microbes such as *Aeromonas salmonicida* (Kruse *et al.*, 1995). Aquatic environments can probably act as a reservoir of resistance genes, where human pathogenic bacteria and aquatic organisms can exchange resistance genes. The ocean can be regarded as one ecosystem where the use of antibiotics in one location can have an impact on the occurrence of antibiotic resistant bacteria in distant areas.

Other studies of resistance genes also demonstrates communication between bacteria originating from different ecosystems. Similar resistance genes exist in different bacteria from different ecological niches, indicating that different bacteria can utilize genes from a common gene reservoir. Transfer experiments of multidrug resistance plasmids between bacteria of diverse origins performed in natural microenvironments show that transfer mechanisms are environmental phenomena occurring between bacterial strains that are evolutionary or ecologically unrelated (Roberts *et al.*, 1990).

The mechanisms behind the long term behavior of genes in populations and the environment are less understood. Due to the complexity of ecological systems, it has been difficult to describe and model the spread, establishment and persistence of microbial genes. Modeling a single mechanism of exchange between two distinct bacteria is definitely less complex than modeling the fate of a gene being introduced in the intestinal environment of animals or man or in a food chain consisting of animals, man and the environment, all closely interrelated. The mechanistic molecular models used for modeling microbial mechanisms can not cope with this task. Stochastic based models as used in risk assessment can describe isolated phenomena (Vose, 1996), but are of limited value when studying complex microbial interactions. To grasp the full complexity of these systems, dynamic simulation models offer a possible approach (Hannon and Ruth, 1997). To support the establishment of dynamic models, an extensive set of data from population based epidemiological studies are needed.

The spread of genes through trade in foods might be an important factor in the building up the global reservoir of pathogens and genes coding for pathogenicity and drug resistance. So far the primary focus has been on travelers harboring special microbes. However, the massive spread of genes through an increasing food trade might cause larger consequences in the long term. Any human action creating favorable conditions for pathogenic microbes to grow and spread, or unwanted genes to spread and persist, will contribute to a situation where we can lose control over our food production systems. The overall effect of this process is unclear. A possible reduction of microbial diversity through importing new varieties of microbes and microbial genes may not only cause problems for plant, animal and human health, but also induce problems for the agricultural production itself.

Another aspect of the intensification of animal husbandry is the environmental problems linked to excessive amounts of manure. For every Dutch citizen, approximately 6 metric tons of animal manure is produced every year. Most of this manure is produced by animals living on imported animal feed, mainly from Asia, Africa and Latin America. The country is not large enough to properly utilize such a large quantity of manure, and a massive contamination of ground water with nitrogen has been demonstrated. A large number of different pathogens have been (often unnoticed) passengers on this trade (Skjerve, 1994). The Dutch example is not unique, and the same can be found in many regions in Northern Europe and North America.

Risk analysis as a tool do reduce damages to ecosystems caused by an increasing food trade

Within the WTO system, risk analysis is supposed to be **the** way trading on agricultural products will be regulated. An interesting difference can be noted between controlling the spread of infectious animal diseases and the spread of zoonotic agents or agents of environmental importance. Animal diseases have for a long time been controlled by conservative means as testing of individual animals, quarantine and strong legal actions to protect animals. Foods can, however, be traded as long as they are accepted in the country of origin. Thus, it can be claimed that animals have better protection than human beings and the environment.

Methods for risk analysis are well developed for predicting the risk of spreading animal diseases (Ahl *et al.*, 1993; MacDiarmid, 1993). WTO has transferred the responsibility for risk analysis method in this area to Office International d'Epizooties (OIE) in Paris. Under the new rules, OIE will be responsible for establishing permanent guidelines for methods for risk analysis, as well as having the power to accept or reject specific analyses from different countries.

In the area of food, the Codex Alimentarius Commission (CA) has the same responsibility. While risk analysis methods have been established inside the area of chemical food contaminants, no method has been accepted for microbial hazards linked to foods. The process has now started within the CA, and within a few years standards for risk analysis will be presented (Anonymous, 1997a, b). Of interest is to note the discrepancies between the approach used in these papers, and another paper presented at the same meeting, where risk evaluation is used as a term to allow information about the epidemiological status of different geographical regions into the whole process (Anonymous, 1997c).

A major weakness in the main approach used by WTO/CA, is the assumption that microbial risks can be translated into probabilities. This is necessary to establish a stochastic model typically used in risk assessments. As previously stated, dynamic modeling offers a means to study the long term effect of introduction of new microbes or genes. It might be possible to predict the effect of importing a certain number of tons of meats to a country from another country – if sufficient epidemiological data are available. However, the main question will be the long-term effect of an overall increase in the volume of trade. A dynamic model may give us some basic ideas about possible consequences on a global scale, also on a longer term.

A safer approach for the consumer and the environment would be to include a stage with critical evaluation of the overall consequences of an increased food trade before the methods of classic risk analysis are applied. The CA paper on risk evaluation would allow such a process (Anonymous, 1997c). It is already well documented that trade on animal feed, live animals, fresh animal products, fresh seafood and fresh vegetables constitutes a health and environmental problem.

Without describing the long term consequences on health and environment, risk analysis as a tool can not contribute to securing the stability of ecosystems linked to the food production. Aggressive exporters may be able to use risk assessments against weak countries without the means to present analyses of a quality that will be accepted by the ruling bodies inside CA and WTO.

Conclusions

The expected increase in the volume of trade in foods might cause severe disruptions of food chains in many areas in the world. The systematic mixing of ecosystems represented by an increased trade of fresh foods might appear as one of the factors threatening human health and the stability of ecosystems. Once microbial genes are established, they may exist for generations in a close interrelation with humans, animals and environmental organisms. The WTO approach to reducing the damages through risk analysis based trade systems can not eliminate the damages caused by an increased trade, only reduce them. On the contrary, risk analysis may be used as a means to increase trade. There is an urgent need for establishing dynamic risk models for studying the ecological consequences of an increased trade on biological systems.

References

Ahl, A.S., Acree, J.A., and Gipson, P.S. *et al.* (1993) Standardization of nomenclature for animal health risk analysis. *Reviews in Science and Technology*, **12**, 1045-53.

Altekruse, S.F., Cohen, M.L. *et al.* (1997) Emerging food-borne diseases. *Emerging Infectious Diseases*, **3**, 285-93.

Anonymous (1993) *Agreement on the Application of Sanitary and Phytosanitary Measures. Final Act Embodying the Result of the Uruguay Round of Multilateral Trade Negotiations of 15 December, 1993.* GATT Secretariat, Geneva.

Anonymous (1994a) *Report of WHO Working Group on Shiga-Like Toxin Producing* E. coli *(SLTEC), with Emphasis on Zoonotic Aspects.* WHO, Geneva.

Anonymous (1994b) *WHO Surveillance Programme for Control of Food-borne Infections and Intoxications in Europe. Sixth Report 1990-1992.* Robert von Ostertag Institute, FAO/WHO Collaborating Institute.

Anonymous (1995) *Report of the WHO Consultation on Emerging Food-borne Diseases, Berlin, 20-24 March 1995.* WHO, Geneva.

Anonymous (1997a) *Recommendations for the Management of Microbiological Hazards for Foods in International Trade (ICMSF and the United States).* Codex Alimentarius, Washington DC.

Anonymous (1997b) *Proposed Draft Principles and Guidelines for the Conduct of Microbiological Risk Assessment.* Codex Alimentarius, Washington DC.

Anonymous (1997c) *Microbiological Risk Evaluation in Relation to International Trade in Foods and Animal Feed.* Codex Alimentarius, Washington DC.

Baird-Parker A.C. (1994) Foods and microbiological risks. *Microbiology,* **140,** 687-95.

Barrow P.A. (1989) Further observations on the effect of feeding diets containing avoparcin on the excretion of salmonellas by experimentally infected chickens. *Epidemiology and Infections*, **102,** 239-52.

Bryan, R. (1998) Alien species and emerging infectious diseases: past lessons and future implications, in *Invasive Species and Biodiversity Management*, (eds O.T. Sandlund, P.J. Schei and Å. Viken), Kluwer Academic Publishers, Dordrecht, The Netherlands.

Hannon, B. and Ruth, M. (1997) *Modeling Dynamic Biological Systems*. Springer-Verlag, New York.

Hoogenbom-Verdegaal A.M.M., de Jong, J.C., During, M. *et al.* (1994) Community-based study of the incidence of gastrointestinal disease in the Netherlands. *Epidemiology and Infection*, **112**, 481-87.

Høiby A., Kapperud, G. and Rørvik, L.M. (1995) The tip of the iceberg. *ASM News*, **61**, 615.

Klare I., Heier, H., Claus, H. *et al.*(1995) vanA-mediated high-level glycopeptide resistance in *Enterococcus faecium* from animal husbandry. *FEMS Microbiological Letters*, **125**, 165-71.

Kruse H., Sørum, H., Tenover, F.C. and Olsvik, Ø. (1995) A transferable multiple drug resistance plasmid from *Vibrio cholerae* O1. *Microbial Drug Research in Epidemiological Diseases*, **1**, 203-10.

Kvien T.K., Glennas, A., Melby, K. *et al.* (1994) Reactive arthritis – incidence, triggering agents and clinical presentation. *Journal of Rheumatology*, **21**, 115-22.

Lederberg J., Shope, R.J. and Oaks, S.J., eds. (1992) *Emerging Infections*. National Academy Press, Washington DC.

Lederberg J. (1993) Emerging infections: Microbial threats to health. *Trends in Microbiology*, **1**, 43-44.

MacDiarmid S.C. (1993) Risk analysis and the importation of animals and animal products. *Reviews in Science and Technology*, **12**, 1093-108.

Maurice J. (1994) The rise and rise of food poisoning. *New Scientist*, 17 December 1994, 28-33.

Roberts M.C. and Hillier, S.L. (1990) Genetic basis of tetracycline resistance in urogenital bacteria. *Antimicrobial Agents and Chemotherapy*, **34**, 261-64.

Skjerve E. (1994) Meat production and the environment. *Outlook on Agriculture*, **23**, 115-22.

Threlfall E.J. (1992) Antibiotics and the selection of food-borne pathogens. *Journal of Applied Bacteriology*, **73**, 96S-102S.

Vose, D. (1996) *Quantitative Risk Analysis. A Guide to Monte Carlo Simulation Modelling*. John Wiley & Sons, Chichester, UK.

Part 4

Management tools

18 Legal authorities for controlling alien species: A survey of tools and their effectiveness

MICHAEL J. BEAN
Environmental Defense Fund, Washington DC, USA

Abstract

A wide variety of legal authorities at the national, subnational, and international levels exists for preventing or controlling the harmful consequences of the introduction of alien species. Trade controls aimed at preventing the intentional importation of potentially harmful alien species are among the earliest measures tried. A key consideration in the effectiveness of such measures is whether they assume all alien species to be harmful until proven otherwise, or whether they require an affirmative showing of harmfulness before trade may be restricted. Trade restrictions aimed at ensuring that imported products do not carry pests, pathogens, or other species harmful to agriculture or human health are also common and have had significant success. Less obvious pathways for introduction, such as ship ballast water, have been the focus of more recent legal attention. The success of these efforts cannot yet be determined.

Once introduced, alien species have been the subject of a variety of legal measures aimed at controlling their spread or achieving their eradication. Such measures include internal quarantines, requirements that aquaculture facilities be self-contained, and other strategies. The success of such measures has often been less than dramatic. Imposition of legal liability upon parties responsible for introduction of harmful species is a potentially useful though largely untried legal strategy. A tax on classes of activities known to be significant sources of introduced alien species could generate the revenues needed to finance education, inspection, control, or eradication programs.

The international law mandate

Article 8(h) of the Convention on Biological Diversity directs that each contracting party shall "as far as possible and as appropriate ... prevent the introduction of, control or eradicate those alien species which threaten ecosystems, habitats or species." The Convention thus recognizes alien species as a potentially significant threat to the conservation of biological diversity. It imposes upon contracting parties an obligation to do something about that threat. The precise nature of that obligation is not explained, nor is the obligation unqualified: It is to be undertaken only "as far as possible and as appropriate."

O. T. Sandlund et al. (eds.), Invasive Species and Biodiversity Management, 271–282.

International recognition of the conservation threat posed by alien species is reflected in a number of global conservation treaties that preceded the Convention on Biological Diversity. The Convention on the Law of the Sea, for example, provides in Article 196.1 that states "shall take all measures necessary to prevent, reduce and control ... the intentional or accidental introduction of species, alien or new, to a particular part of the marine environment, which may cause significant and harmful changes thereto." The Convention on the Conservation of Migratory Species of Wild Animals (also called the "Bonn Convention") directs the parties to it to "endeavor ... to the extent feasible and appropriate, to prevent, reduce or control factors that are endangering or are likely to further endanger the species, including strictly controlling the introduction of, or controlling or eliminating already introduced, exotic species."

Similar provisions are found in a number of regional conservation treaties. The 1979 Convention on the Conservation of European Wildlife and Their Natural Habitats requires its member states "to strictly control the introduction of non-native species." In Asia, the 1985 ASEAN Agreement on the Conservation of Nature and Natural Resources, which is not yet in force, directs its party states to endeavor to "regulate and, where necessary, prohibit the introduction of exotic species." In the Caribbean, the 1990 Protocol on Specially Protected Areas and Wildlife of the Convention for the Protection and Development of Marine Resources of the Wider Caribbean Region (which is also not yet in force) directs contracting parties to "regulate or prohibit intentional or accidental introduction of non-indigenous or genetically altered species to the wild that may cause harmful impacts to the natural flora, fauna or other features of the Wider Caribbean Region."

From this short summary, it is clear that there is widespread recognition as a matter of international law that one of the obligations of all nations seeking to advance the cause of conserving biological diversity is to address the threat posed by introduced, alien species. The important question that each of these treaties leaves open is precisely how this obligation is to be met. More specifically, what legal measures can or should nations take to carry out this obligation effectively? To date, a wide variety of legal authorities at the national, subnational, and international levels have been put in place for preventing or controlling the harmful consequences of the introduction of alien species. This paper examines the various types of tools that have been tried and attempts to assess their effectiveness.

Import restrictions

The simplest, most direct, and surely one of the earliest approaches to this problem was to prohibit or restrict the importation of harmful, alien species. In the United States, for example, the very first wildlife conservation law enacted at the national level (the Lacey Act of 1900) prohibited the importation of such birds or other animals as the Secretary of Agriculture determined were injurious to the interests of agriculture or horticulture. From its initial, narrow focus on animals injurious to agriculture or horticulture, the law has subsequently been amended to restrict the importation of animals declared injurious to any of several interests, including forestry, wildlife, and human beings (OTA, 1993). Paralleling this broadening of the interests to be protected, the authority to declare animal species injurious was shifted from the Secretary of Agriculture to the Secretary of Interior, who is responsible for wildlife conservation programs at the national level in the United States. Penalties for violations of the Lacey Act are significant. They include the possibility of imprisonment for up to six months as well as monetary fines.

Though it is now nearly a century old, the Lacey Act has generally been judged to have had very little beneficial impact on the problem of introduced alien species. Three shortcomings in particular have hobbled its effectiveness. First, because it is directed only at the intentional importation of animals, it does nothing to address the much larger problem of unintentional introductions, such as when the offending organism enters as a "hitchhiker" or "stowaway" in ballast water, a shipment of lumber, or a truckload of fruit (Bean, 1991). Second, because the Lacey Act primarily restricts the "importation" of injurious species, it implicitly treats the problem as one of "foreign species" versus "native species." It overlooks altogether the fact that in large nations with diverse ecosystems, internal commerce that moves native species from one area where they occur naturally to another where they do not can cause as much ecological disruption as international commerce. Yet, except with respect to shipments between the island state of Hawaii and the U.S. mainland, or between the island possessions of the U.S. and the mainland, this internal commerce lies wholly outside the Lacey Act's purview. Finally, the Lacey Act requires a prior determination that a given foreign species is injurious before its importation can be restricted. The burden is effectively placed on the government to prove that the species to be imported is injurious, rather than on the importer to prove that the species to be imported is safe. As a result, the list of species whose importation is actually restricted under the Lacey Act consists of fewer than two dozen species or genera, most of which have already been established in the United States and for which the feared injury has already been done. Meanwhile, many non-native species, including many fish imported for the pet trade, have become established in the wild in the United States notwithstanding the Lacey Act (OTA, 1993).

Reversing the burden of proof, effectively presuming every foreign species to be injurious until demonstrated otherwise, could substantially improve the effectiveness of the Lacey Act and similar laws that seek to protect against the threat of imported species. In the United States, an effort was made to reverse the burden of proof under the Lacey Act two decades ago. The idea was to create a "clean list" of species that could be imported, a "dirty list" of those that could not, and a "gray list" of everything else. Until a species on the gray list was shown to be safe, no importation of it would be allowed. The effort was blocked politically by exotic pet importers and allied interests who feared a significant impediment to their business. The alternative to reversing the burden of proof is to mount an aggressive effort to expand the list of injurious wildlife to include not only those that have already been imported and have caused problems, but others that have the potential to do so.

Unlike the United States, New Zealand has established a "clean list" approach (Anonymous, 1996). The authority for restricting importation of potentially injurious alien species is contained in 1993 legislation known as the Biosecurity Act. Incoming visitors, merchandise, etc. are subject to rigorous inspection. Plants or animals permitted entry are given a "biosecurity clearance." Such clearances may not be given to any plant or animal not already established in New Zealand unless it is on an approved list (i.e., a "clean" list). A separate "dirty" list identifies organisms that may not be cleared for entry, regardless of whether the species may already be present in the country. The costs of scientific analysis, inspection, enforcement, and related activities are intended to be covered through a "user pays" approach. Like New Zealand, the state of Hawaii also restricts the entry of all major animal groups to species that occur on a clean list.

A further shortcoming of the Lacey Act in the United States is that, while it empowers the Secretary of Interior to prohibit altogether the importation of injurious wildlife, it confers no authority to control those species once they have gotten past customs inspectors and are within the country. By contrast, the Federal Noxious Weed Act is a U.S. law that gives the Secretary of Agriculture extensive authority over "noxious weeds," a term that is defined as a plant of foreign origin that is new to, or not widely prevalent in the United States and that may be injurious to agriculture, wildlife, or other interests. The Secretary of Agriculture can restrict not only the entry into the country of such species but also its movement within the country. His powers include the authority to impose area quarantines and to require the inspection of other articles and products as a means of preventing the dissemination of noxious weeds.

While the Federal Noxious Weed Act confers much broader powers than the Lacey Act, the use of those powers has been criticized as too timid. Fewer than 100 plant species are designated as noxious weeds and therefore prohibited from entry into the United States. About half of these are thought to occur already in the U.S.; small control programs are directed at only eight of these and at least nine are actually sold in interstate commerce notwithstanding their designation as noxious weeds. Only one species is subject to a quarantine order.

Risk-benefit assessments

Placing a species on a clean or dirty list is not simply a matter of determining whether it is inherently dangerous. The same species, when introduced in one country, could pose a great danger to its biological diversity, but could pose no danger at all to another country's biological diversity. Each country will therefore need its own scientifically qualified body to make these sorts of determinations, based on the likelihood that the species in question could become established and do harm. Not all countries will necessarily wish to bar every species that presents ecological risks. Instead, they are likely to want some mechanism for balancing those risks against the economic or other benefits of introducing the species. In New Zealand, risk assessments of this sort are overseen by a central Hazard Control Commission. The Commission's duties are wide-ranging but largely advisory. Among other things, it must "balance the benefits which may be obtained from hazardous substances and new organisms against the risks and damage to the environment and to the health, safety and economic, social and cultural wellbeing of people and communities" (Anonymous, 1996)

Another means of securing at least a measure of risk and benefit assessment, as well perhaps of public disclosure, is through environmental impact assessments. In the United States, the National Environmental Policy Act requires a detailed, written evaluation of the expected environmental consequences of most major actions undertaken by the national government. Most of the 50 states of the United States have similar state-level laws. The requirement to prepare an environmental impact statement in connection with a recent proposal to introduce anadromous fish native to the Pacific coast into streams on the Atlantic coast was instrumental in bringing to light the ecological risks associated with that proposal and ultimately in preventing it from going forward. This example illustrates the power of public disclosure of potential ecological risks as a means of achieving practical influence over intentional species introductions.

The weighing of risks and benefits does not always consist of balancing ecological risks against economic benefits. Sometimes, there may be both ecological risks and ecological benefits associated with alien species introductions. For example, the highly endangered Guam rail no longer survives on its native island of Guam and there is no near-term prospect for successfully reintroducing it there because of the presence on Guam of an introduced non-native snake that decimated the rail and other Guam birds. A few captive rails have been introduced on the nearby island of Rota, where the species is not known to have occurred naturally. This introduction, technically an intentional introduction of an alien species, was seen as a last resort effort to keep the species alive in the wild. Similarly, after a subspecies of peregrine falcon that formerly occurred in the eastern United States had become extinct, a related subspecies was introduced there to restore the species to a significant part of its former range. This example illustrates not only the possibility that alien species introductions can sometimes offer ecological benefits, but also the conceptual complexity that sometimes accompanies the question whether a particular introduction should properly be characterized as one involving an "alien" species. At the subspecies level, the introduced falcons were not native; at the species level, however, they were clearly native. Because the ecological role that the introduced subspecies was expected to play was presumed to be identical to that of the extirpated subspecies, the introduction was approved and has been viewed as highly successful.

Controlling unintentional introductions

Though injurious alien species have often been imported for the horticultural and pet trades, a far more serious problem is that of unintentional introductions of such species. The means by which such unintentional introductions can occur are nearly limitless. The legal measures that have been taken to address this problem typically focus on the categories of activities from which unintentional introductions occur.

A recent example is the emerging set of international and national regulations governing the discharge of ballast water in shipping. The 1992 United Nations Conference on the Environment and Development adopted *Agenda 21*, one provision of which calls upon nations, through the International Maritime Organization (IMO) and other organizations, to consider the "adoption of appropriate rules on ballast water discharge to prevent the spread of non-indigenous organisms." Largely in response to this call, the IMO the following year adopted detailed, voluntary ballast water guidelines and urged governments to adopt them. Australia and New Zealand adopted comprehensive, mandatory ballast water discharge programs, but few other nations have yet done so. The United States, two years prior to the UN conference, adopted the Nonindigenous Aquatic Nuisance Prevention and Control Act of 1990, largely in response to the establishment of zebra mussels in the Great Lakes region as a result of ballast water discharges (ANSTF, 1994). The 1990 Act directed the U.S. Coast Guard (the agency responsible for enforcing maritime laws in the United States) to develop initial voluntary guidelines to reduce the risk of unintentional species introductions into the Great Lakes through ballast water discharges. After two years, these voluntary guidelines were to be replaced by binding regulations. The ballast water guidelines and regulations were to be limited to the Great Lakes because the zebra mussel problem was recognized to be a severe one there, while much less was known about the severity of similar problems elsewhere. New legislation has recently been proposed that would expand the voluntary ballast water management program to all U.S. waters. Rather than making a planned transition from voluntary guidelines to mandatory regulations, the new legislation would authorize regulations wherever compliance with voluntary guidelines is inadequate.

While the threat of unintentional introductions through ballast water discharge is a quite recently recognized problem, a much more longstanding concern has been with the risk of harmful introductions accompanying shipments of agricultural goods. Indeed, nearly a half century ago, the International Plant Protection Convention (IPPC) established a cooperative framework for the regulation of agricultural pests. The IPPC requires signatories to set up central agencies for the certification, inspection and control of the importation of agricultural goods, though it stops short of mandating uniform inspection criteria. The focus of the IPPC is limited to the control of pests potentially injurious to agriculture and horticulture. Prompted by the IPPC, many nations have enacted domestic legislation providing for inspection of agricultural goods, quarantining where appropriate, and issuance of phytosanitary certificates signifying that goods have been inspected and found to be pest or disease free.

In the United States, the purposes of the IPPC are carried out primarily through the Federal Plant Pest Act and the Nursery Stock Quarantine Act. The former confers broad powers on the Secretary of Agriculture to "detect, eradicate, suppress, control, or to prevent or retard the spread of plant pests." The Nursery Stock Quarantine Act requires that all imported nursery stock (including all field-grown florists' stock, trees, shrubs, vines, cuttings, and certain other materials) be accompanied by a certificate of inspection declaring it to be free of injurious plant diseases and insect pests. This law also authorizes broad quarantine powers, both with respect to imported nursery stock and with respect to internal commerce between the various states when necessary to prevent the spread of a dangerous plant disease or insect infestation.

Still further authority useful in reducing the threat of alien species introductions is contained in the Federal Seed Act. This law's primary purpose was to protect the purchasers of seeds from inadequate labeling, false advertising, and other deceptive or fraudulent practices. This purpose was to be achieved by setting standards for seed purity and requiring that the contents of seed packages be accurately identified. The seed purity standards have also tended to reduce the spread of undesirable weeds by requiring that the percentage of weed seeds in a shipment not exceed specified tolerances.

These various authorities over the importation and interstate shipment of agricultural products are far-reaching. The resources to implement these authorities are also considerable. Annual expenditures for agricultural inspection and quarantine have been in excess of US$ 100 million recently (by comparison, the Fish and Wildlife Service carries out port inspections of imported fish and wildlife with resources of about US$ 3 million annually). Nearly 2 000 employees of the Agriculture Department carry out these tasks. In addition, about 500 state seed inspectors carry out the Federal Seed Act. Despite the far-reaching authority and the considerable resources available, the number of introduced agricultural pests continues to grow.

Because many of the pests and diseases that threaten agriculture or horticulture also potentially threaten biological diversity, the considerable array of legal measures prompted by the IPPC undoubtedly aids in reducing the threat to biological diversity stemming from trade in agricultural commodities. However, the narrow focus on risks to agriculture or horticulture limits the potential effectiveness of these measures in reducing ecological threats more generally.

The government as land owner and manager

In many countries, the government itself is a significant owner and manager of land. At a minimum, most countries maintain national parks, wildlife refuges, and other protected areas. Others manage national forests, public rangelands, and a wide variety of other areas with widely varying designations and purposes. The policies that govern the management of these publicly owned lands can either exacerbate or relieve the problem of alien species (Gillis, 1991). This potential can be illustrated with a few examples drawn from the United States, where land owned by the national government comprises nearly a third of the entire land base.

The National Park System in the United States is operated pursuant to a set of formal management policies developed by the National Park Service (NPS, 1985). With respect to the management of biological resources, the general objective of these policies is to perpetuate the native plant and animal life as part of the natural ecosystems of the parks. The Service policies distinguish between "native" and "non-native" species. Native species are those that "as a result of natural processes occur on or occurred on lands now designated as a park. Any species that have moved onto park lands directly or indirectly as the result of human activities are not considered native." In general, the Park Service adopts a "hands-off," "let nature take its course" approach to park management. However, it is Park Service policy that non-native species "will not be allowed to displace native species if this displacement can be prevented by management." Neither will non-native species be introduced into natural areas within parks "except in rare cases where they are the nearest living relatives of extirpated native species [recall the peregrine falcon example earlier], where they are improved varieties of native species that cannot survive current environmental conditions, [or] where they may be used to control established exotic species."

One of the very popular activities in U.S. national parks is recreational fishing. For many years, Park Service policy allowed and encouraged the stocking of park waters with game fish, including native and non-native species. Current policy, however, is much more cautious. Areas that are naturally barren of fish are not to be stocked. In so-called "natural zones," stocking may only be permitted for native species, and then only to reestablish the species within its historic range. Certain park areas commemorate historical or cultural phenomena rather than natural history phenomena. In these areas, stocking of native or non-native fish may be permitted "only where there is a special need associated with the historic events or individual commemorated." However, stocking of non-native fish is allowed in these areas only if the stocked fish "could not spread to natural zones or waters outside the park." Within some national parks, there are highly altered environments such as reservoirs. In such altered environments, fish stocking is allowed, but preference is given to native species and non-natives may be stocked only if they are already present and only where "scientific data indicate that introducing additional exotics would not diminish native plant and animal populations and the exotics could not spread to natural zones or waters outside the park."

Even with respect to the introduction of native species, Park Service policy tries to ensure minimal ecological disruption. Service policy requires that such introductions are to be “accomplished using organisms taken from populations as closely related genetically and ecologically as possible to the park populations, preferably from similar habitats in adjacent or local areas, except where the management goal is to increase the variability of the park gene pool to mitigate past, human-induced losses of genetic variability.”

Separate from the National Park System is an extensive National Wildlife Refuge System administered by the U.S. Fish and Wildlife Service. Whereas Park Service philosophy embraces minimal management to achieve natural diversity, refuge system management has often sought to favor a particular type of wildlife resource. Hence, the policies relating to refuge management are more ambivalent with respect to alien species (Bean and Rowland, 1997). On the one hand, one of the basic goals of the refuge system is to preserve "the natural diversity and abundance of fauna and flora on refuge lands." On the other hand, refuge policies permit specific wildlife management objectives for individual refuges to override the objective of preserving natural diversity. Non-native fish may be introduced wherever native fish cannot be used to support the fish and wildlife objectives of a particular refuge. Non-native grasses can be planted to boost waterfowl populations. No effort need be made to eliminate exotic species that have become established on a refuge for at least 25 years if their presence does not conflict with the specific objectives of that refuge.

Yet a third major public land system in the United States consists of the vast semi-arid rangelands in the western United States that are administered by the Bureau of Land Management. Historically, grazing practices on these lands are believed to have been largely responsible for the transformation of native prairie ecosystem dominated by perennial grasses to a highly altered ecosystem heavily comprised of introduced annual plants. This transformation occurred before the Bureau of Land Management came into existence. That fact presents the Bureau with an interesting challenge in trying to "improve" range conditions. Under the standards the Bureau has traditionally used, range quality is a reflection of the percentage of vegetation in an area that is similar to what is believed to be the climax plant community for that area. However, the Bureau considers the "climax" community to be what would develop there now, without further human interference, rather than what was there prior to European settlement. Thus, under these standards, an area comprised largely of alien species could be considered to be in excellent condition.

The Bureau, in cooperation with the relevant state wildlife agency, may authorize the release of non-native wildlife on Bureau lands. Its policies restrict such releases in wilderness areas, or in areas where they could conflict with endangered species conservation efforts. Bureau policy states that those responsible for unauthorized releases will be held liable for damages and costs of control. Apparently, however, there has yet to be an instance in which the Bureau sought to recover damages or expenses for any unauthorized release.

Each of these federal land managing agencies had no explicit statutory authority or directive to control non-native species on its own lands, until 1990, when the Federal Noxious Weed Act was amended to confer that authority. The amendments require each federal land managing agency to create a program to manage "undesirable plants" on its lands. These include not only harmful alien species, but potentially a much broader array of plants as well. The required management programs are to seek to control or contain targeted species in cooperation with state agencies where the lands occur.

Legal strategies for the future

Two sobering conclusions from the past several decades of effort to address the problems of harmful alien species are that nothing works perfectly and that many things don't seem to work very well at all. Still, it is possible to identify some legal strategies that offer promise for an improved effort.

First, narrowly focused regulatory programs that address a particular industry or set of practices associated with unwanted species introductions (for example, aquaculture, the seed trade, ballast water discharge, etc.) are probably more likely to be successful than are broader, all-encompassing programs that try to address the problem through sweeping regulatory requirements.

Second, by focusing on particular industries or practices, it is feasible to make the costs of inspection and control largely or entirely self-financing. Imposing the cost of inspection upon the party shipping an agricultural commodity is not uncommon. A surcharge on such persons could generate what in effect would be an "insurance fund" to finance control or eradication measures necessitated by the fact that some unintended introductions will occur despite the best inspection efforts.

Liability for damages (to crops, wildlife resources, ecosystems, or other things of value) should be imposed wherever it is possible to find persons responsible for an unauthorized release, an unintended release, or a release that has unanticipated negative consequences. Requiring persons engaged in activities where these risks are significant to post a bond or maintain insurance would make sure that liability was a meaningful deterrent to careless practices.

Shifting the burden of proof in decisions about intentional introductions will ensure that ignorance about the potential effects of a proposed introduction will not be rewarded.

Planning for future contingencies should also be built into regulatory requirements. Exotic game breeders, aquaculturists, and others whose activities present a risk of unintended escape could be required to develop a plan for recapture or control as part of any application for approval of their activities. Such a requirement would ensure that attention to what needs to be done in the event of such contingencies is given before they arise, not afterwards.

Amnesty as a way of encouraging past violators of restrictions on importing harmful organisms to give up those organisms without fear of penalty may be worth pursuing. The state of Hawaii offers this alternative to those who have plants or animals that were illegally brought into the state in the past. The potentially vast quantity of organisms that were illegally acquired but are still within the control of individuals represent a ticking time bomb of potential future introductions. Programs to encourage the voluntary surrender of such organisms are more likely to succeed than trying to track down those who have already eluded the law once.

Government must lead by example. Whether as a landowner, an importer, or in any other role, the efforts of government to influence the behavior of private citizens and businesses will succeed best if the government itself is subject to rules no less strict than those that apply to others and is scrupulous in adhering to them.

References

Anonymous (1996) *Hazardous Substances and New Organisms Bill Information Kit.* Government of New Zealand (available through http://www.mfe.govt.nz/hsno.html).

ANSPT (1994) *Report to Congress: Findings, Conclusions, and Recommendations of the Intentional Introductions Policy Review.* Aquatic Nuisance Species Task Force, Department of Interior, Washington, DC.

Bean, M.J. and Rowland, M.J. (1997) *The Evolution of National Wildlife Law* (3rd edition), Praeger Publishers, New York.

Bean, M.J. (1991) *The Role of the United States Department of Interior in NIS Issues.* Contractor report prepared for the Office of Technology Assessment, Washington, DC.

Gillis, A.M. (1991) Should cows chew cheatgrass on common lands? *Bioscience,* **41**, 668-675.

NPS (1985) *Management Policies.* National Park Service Department of Interior, Washington, DC.

OTA (1993) *Harmful Non-Indigenous Species in the United States.* Office of Technology Assessment, U.S. Government Printing Office, Washington, DC.

19 Strategies for preventing the world movement of invasive plants: a United States perspective

RANDY G. WESTBROOKS and ROBERT E. EPLEE
US Department of Agriculture, Whiteville, North Carolina, USA

Abstract

Over the past several thousand years, Man has moved many plant species far beyond their historical native range. Many introduced plants that have become established outside of cultivation are benign (so far). However, some introduced species, with free living populations, pose a threat to the biodiversity of natural areas and/or diminish the production capacity of managed or agricultural ecosystems. In the United States, 17 federal agencies have formed the Federal Interagency Committee for the Management of Noxious and Exotic Weeds (FICMNEW). FICMNEW has developed a National Strategy for Invasive Plant Management. The goals of the national strategy are weed prevention, weed control, and restoration of degraded lands. Research, education and partnerships are critical to the success of the strategy. Regulatory strategies to protect the United States and other countries from invasive plants include production of weed free commodities in exporting countries; preclearance of risk commodities at foreign ports of export; port of entry inspections; and finally early detection, containment, and eradication of incipient infestations before they spread.

Introduction

Since the dawn of recorded history, Man has intentionally and unintentionally transported thousands of different plants and animals far beyond their natural ecological range to other parts of the world. Most of these species are beneficial to human society or show no early signs of invasiveness. However, hundreds of relocated species now cause serious problems in agricultural and/or natural ecosystems within the United States. In the absence of co-evolved predators and parasites that usually keep them in check in their natural ranges, introduced species that find suitable habitats may thrive and outcompete and/or displace native species. Over the past several decades, serious problems caused by introduced plants and animals have raised concerns over the movement of species around the world (Elton, 1958; Mooney and Drake, 1986; Zamora *et al.*, 1989; Westbrooks 1991, 1993; Schmitz, 1994; Vitousek *et al.*, 1996). While change and disruption in ecosystems have occurred throughout history, the biological invasions that

O. T. Sandlund et al. (eds.), Invasive Species and Biodiversity Management, 283–293.

are now resulting from human commerce are truly different with regard to origins, rate of introduction, types of organisms, abruptness and magnitude of change (Wagner, 1993).

Recognized invasive species that pose a threat to agricultural and managed ecosystems, or threaten the biodiversity of natural ecosystems have been termed **biological pollutants** (McKnight, 1993; Westbrooks, 1993). Unlike chemical pollutants that typically degrade in the environment, biological pollutants have the ability to grow, multiply, adapt and spread, and cause greater problems over time.

Some examples of introduced plants that have become biological pollutants in the United States include witchweed (*Striga asiatica)* in the Carolinas, kudzu (*Pueraria lobata*) throughout the Southeast, mile-a-minute vine (*Polygonum perfoliatum*) in the Northeast, leafy spurge (*Euphorbia esula*) in the Midwest and West, and miconia (*Miconia calvescens*) in Hawaii. Usually, introduced plants that become invasive typically receive little attention until they become major problems (Eplee and Westbrooks, 1990). By the time a problem is recognized, environmental documentation is prepared, and funding is obtained for control, eradication is often impractical. At this point, an invasive plant becomes a permanent, expanding and detrimental component of the invaded area.

Economic impact of introduced weeds in the United States

Weeds cause billions of dollars of losses annually in the United States by competition with crops and by reducing the quality of food, feed, and fiber. During the 1950s, annual losses due to reduced crop yield and quality and costs of weed control were estimated to be about US$ 5.1 billion per pear (USDA, 1965). In 1962, US$ 200 000 000 was spent in the United States on herbicides alone for weed control (Montgomery, 1964). In 1979, it was estimated that 10-15% of the total market value of farm and forest products in the United States was being lost to weeds, a loss of about US$ 10 billion per year (Shaw, 1979). During the 1980s, farmers spent over US$ 3 billion annually for chemical weed control and about US$ 2.6 billion for cultural, ecological, and biological methods of control (Ross and Lembi, 1985). At that time, about 17% of crop value was being lost due to weed interference and money spent controlling them (Chandler, 1985).

In 1994, it was estimated that the economic impact of weeds on the U.S. economy equals or exceeds US$ 20 billion annually. In the agricultural sector, losses and control costs associated with weeds in 46 major crops, pasture, hay and range, and animal health were estimated to be more than US$ 15 billion per year. In non-crop sectors including golf courses, turf and ornamentals, highways rights of ways, industrial sites, aquatic sites, forestry, and other sites, losses and control costs totaled about US$ 5 billion per year. Value of losses was not available for most non-crop sites, but estimates of control costs were determined (Bridges, 1992, 1994). Since introduced weeds account for about 65% of the total weed flora in the United States, their total economic impact on the U.S. economy equals or exceeds US$ 13 billion per year.

Losses due to individual weeds such as leafy spurge can be quite large. In 1991, researchers reported that annual losses due to reduced carrying capacity from leafy spurge were US$ 2.2 million in Montana, US$ 8.7 million in North Dakota, US$ 1.4 million in South Dakota, and US$ 200 000 in Wyoming. Based on studies of direct and secondary impacts on grazing land, wildlife, and the state's economy, North Dakota was estimated to lose in excess of US$ 87 000 000 annually because of leafy spurge infestations (Goold, 1994). Nearly 6% of the untilled land in North Dakota is infested with leafy spurge (Leistritz *et al.*, 1995).

National strategies for invasive plant management.

Invasive plants grow, adapt, reproduce and spread without respect for agency jurisdictions or property boundaries. Therefore, an effective management strategy to thwart alien species necessarily includes a number of participants and activities. Since the biology of a pest is not negotiable, the strategies of action must consider the total biology of the species as well as political and economic issues. There must be a recognition of need to eliminate the alien species, a commitment of will and resources to the national effort, and good, practical science for developing control methodologies.

To be fully successful, any effort that is made to respond to this serious global problem must bring together a complex set of interests that includes private landowners, industry, and government agencies at all levels. One of the first challenges is to raise public awareness about the seriousness of this issue. A further challenge is to focus public and private resources in a partnership approach to deal with specific weed problems while prevention and control remain economically feasible.

In 1997, the Federal Interagency Committee for the Management of Noxious and Exotic Weeds (FICMNEW) developed a "National Strategy for Invasive Plant Management in the United States". Primary goals of the national strategy are to 1) minimize further introductions of foreign invasive plants in the United States; 2) detect, report and assess incipient infestations; 3) prevent the movement of invasive plants from infested to noninfested areas within the United States; 4) eradicate or control weeds that have already become established; and 5) restore degraded agricultural lands, rangelands and other ecosystems to a healthy and productive state. The strategy will serve as a road map to guide the nation in addressing this growing problem.

In 1998, FICMNEW will publish a national weed fact book entitled "Invasive Plants: Changing the Landscape of America". This publication will serve as a resource guide for legislators, policymakers, agency administrators, and others who are responsible for addressing invasive plants in the United States. FICMNEW is also working with a film producer to develop a video documentary on invasive plants to help increase public awareness and understanding of the invasive plant issue.

Role of federal agencies in weed management

A number of federal agencies have a variety of responsibilities for dealing with weeds in the United States. Major areas of responsibility include weed regulation, research, and management. Within the U.S. Department of Agriculture (USDA), foreign weed exclusion is the responsibility of the Animal and Plant Health Inspection Service (APHIS). APHIS also cooperates with state and local agencies, as well as private land owners/managers in eradicating newly introduced weeds on private lands. Natural enemies of introduced weeds are imported under quarantine to control large infestations (biocontrol). Basic research on agricultural weeds is conducted by the Agricultural Research Service (ARS). Weed management and research on federal lands is conducted by the USDA Forest Service, and a number of agencies in the U.S. Department of the Interior including the U.S. Fish & Wildlife Service (FWS), National Park Service (NPS), Bureau of Land Management (BLM); Bureau of Reclamation (BOR); U.S. Geological Survey (USGS); Bureau of Indian Affairs (BIA). The Department of Defense and the Department of Energy also conduct weed management on their installations. Estimated annual expenditures for weed research and control by some federal agencies in fiscal year 1997 are shown in Figure 19.1.

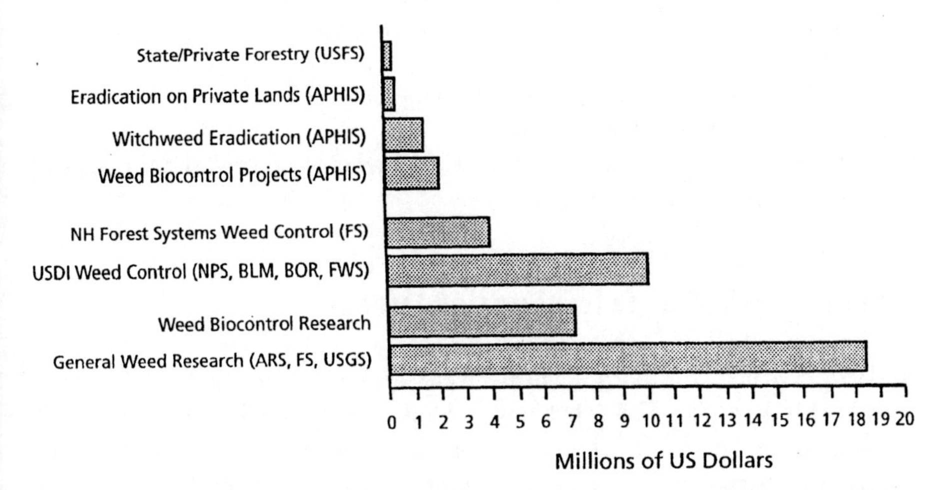

Figure 19.1 Estimated U.S. federal expenditures on invasive plants for selected agencies in the fiscal year 1997. Note that foreign weed exclusion is a part of the APHIS (Agricultural Quarantine Inspection Program), which has a total annual budget of about US$ 150 million. For explanation of acronyms, see text.

Regulatory strategies for exclusion of foreign weeds

One of the missions of APHIS is to prevent the entry of certain foreign pests into the United States. Foreign Pests regulated by APHIS include but are not limited to invasive plants, insects, plant diseases, animal diseases, and molluscs that are of foreign origin. Plant Protection and Quarantine (PPQ) is the operational unit of APHIS that has the responsibility for excluding such foreign pests from the United States. Regulatory strategies for protecting the United States by preventing the entry of foreign pests into the United States include:

- **prevention** (requiring or encouraging the production of pest free commodities in foreign countries to minimize the world movement of recognized pests);
- **preclearance** (inspection/certification of certain commodities at the port of export, prior to being shipped to the United States);
- **exclusion** (port of entry inspections and treatments, designed to detect prohibited pests in imported commodities, and to mitigate pest risk of contaminated shipments);
- **detection** (conducting surveys and communicating with scientists and state agencies for early detection of incipient infestations of prohibited foreign species);

- **containment** (establishment of regulatory rules and programs to prevent the spread of prohibited species from infested areas to non-infested areas);
- **eradication** (total elimination of incipient infestations of prohibited species by appropriate means); and,
- **biological control** (utilizing biological agents to control certain pests if they cannot be eradicated).

Plant taxa listed as federal noxious weeds

In 1976, 26 taxa of foreign weeds were designated as Federal Noxious Weeds (FNWs). The FNW list now includes 94 taxa with 89 species, all species of the parasitic genera *Aeginetia*, *Alectra*, and *Striga*; plus all species of *Cuscuta* and *Orobanche* that are not native to the United States. Melaleuca (*Melaleuca quinquenervia*), a tree in the myrtle family from Australia that is causing major problems in the Florida Everglades, was added to the FNW list in 1992. Tropical soda apple (*Solanum viarum*), a serious new weed of pastures in Florida, was added to the list in 1995. Wetland nightshade (*Solanum tampicense*), a thorny perennial plant that has invaded 500 or more acres of wetlands at five different sites in southwest Florida, is being added to the FNW list at this time.

Detection of noxious weeds at ports of entry

Between 1976 and 1988, resources materials available to APHIS personnel in enforcing the Federal Noxious Weed Act included a list of target species, a short list of high risk commodities, and sampling procedures for inspecting commodities for noxious weeds. At that time, greasy (raw) wool, soil contaminated equipment, aquatic plant shipments, and seed shipments had been recognized as high risk vectors for introducing foreign weeds (Westbrooks, 1989; Westbrooks and Eplee, 1991).

In the late 1980s, a Noxious Weed Inspection System (NWIS) was developed to enhance the ability of PPQ Officers to detect weed contaminants in high risk commodities at ports of entry. The purpose of this system is to provide officers with information on potential associations of target weeds and commodities that originate in habitats where such weeds could be expected to grow. NWIS is based on the principle that certain weeds are likely to be associated with certain commodities from certain countries. NWIS is comprised of a "Federal Noxious Weed Inspection Guide", a "Federal Noxious Weed Identification Guide" with monographs, line drawings, and range maps on all listed species, and a "Noxious Weed Seed Collection". Each PPQ work station at U.S. ports of entry has one set of NWIS materials (Westbrooks, 1989, 1993; Eplee and Westbrooks, 1991; Westbrooks and Eplee, 1991).

Strategies for early detection, reporting, and rapid response

If noxious weeds do enter the United States, in spite of regulatory efforts to exclude them, the next goal is to detect, contain and eradicate incipient infestations before they become established and start to spread. A critical element in this process is early detection. At present, new plant species that are collected in the United States are typically stored at one of the 600+ public or private herbaria that exist around the country. In the past, weed scientists and other plant specialists generally learned about new state and national records through word of mouth or through notes published in botanical journals. Experience has shown that if an infestation is detected early, it can be generally contained and eradicated at a relatively low cost compared to what it will cost for control once it becomes established.

In May 1994, a county extension agent for Mitchell County, North Carolina, submitted an unfamiliar plant to Dr. James Hardin, the Curator of the herbarium at North Carolina State University in Raleigh, for identification. Once Dr. Hardin identified the plant as small broomrape (*Orobanche minor*), he called the authors about the collection. Subsequently, the North Carolina Department of Agriculture and USDA APHIS initiated an eradication project which eliminated the infestation of the new weed (a parasitic plant that is a serious threat to crops such as tobacco, carrots and tomato) before it could become well established and spread throughout western North Carolina.

One way to enhance early detection and reporting of new infestations of new weeds such as small broomrape would be to create a **Plant Detection Network** in each state. Such a network could be established by creating communication links between plant collectors, herbarium curators and appropriate state and federal agencies. Botanists, farmers, county agents, and land managers are just some of the people who need to be encouraged to report new plants that they observe.

To facilitate action on such reports, a **State Interagency Weed Team** could be established in each state. Such a team would be comprised of state and federal officials from agencies and institutions that are involved with weed management and research in a particular state. The goal of a state weed team would be to develop coordinated action plans and to leverage available resources and expertise for dealing with new weeds of common concern. Having one interagency spray crew to cover multiple jurisdictions would be far more efficient and cost effective than having separate county, state, and federal crews in a particular area.

Once a state weed team is informed about a new infestation, it will need input from technical specialists on how to proceed. On a national level this could be accomplished by creating a **Federal Interagency Rapid Response Weed Team**. The purpose of such a team would be to provide technical support to federal, state, and local agencies, in evaluating new infestations of introduced weeds. The national team, which would consist of recognized weed regulatory and control specialists from participating federal agencies, would cooperate with weed scientists, botanists, and state plant regulatory officials in affected states. Such an interagency team would provide a shared pool of expertise that is not normally available to individual agencies. When this or a similar system is adopted nationwide, the United States will be in a much better position to detect new weeds and to respond to them appropriately. Early detection, reporting, and rapid response to new invasive plants are three major goals of numerous federal agency and interagency strategic plans including the APHIS Noxious Weed Policy Implementation Plan, the U.S. Department of Agriculture Strategic Plan for Weeds, the National Strategy for Invasive Plant Management, and the U.S. National Campaign Against Invasive Species.

Federal/state noxious weed eradication projects

Currently, about 45 species of the 94 taxa that are listed as FNWs are known or reported to occur in the United States to a limited degree. Over the past 40 years, APHIS and the agencies that preceded it have been involved in cooperative federal/state efforts to eradicate a number of these species from the United States. Some of these include:

- **Witchweed** (*Striga asiatica*). 177 000 ha infested in NC and SC; now reduced to about 6 000 ha in North Carolina and South Carolina.
- **Branched broomrape** (*Orobanche ramosa*). 283 ha infested in Karnes County, Texas; nearing completion.
- **Goatsrue** (*Galega officinalis*). 16 000 ha infested in Cache County, Utah; nearing completion.
- **Mediterranean saltwort** (*Salsola vermiculata*). 550 ha infested in San Luis Obispo County, California; ongoing effort.
- **Hydrilla** (*Hydrilla verticillata*). 310 km of canals infested in the Imperial Irrigation District, Imperial Valley, California; now 99% eradicated.

- **Japanese dodder** (*Cuscuta japonica*). 1 ha infested in the SC Botanical Garden, Clemson, South Caroina; nearing completion.
- **Small broomrape** (*Orobanche minor*). Spot infestations eradicated in Washington County, Virginia; Mitchell County, North Carolina; Pickens, Abbeville, and Aiken Counties, South Carolina; and in Baker County, Georgia.
- **Catclaw mimosa** (*Mimosa pigra* var. *pigra*). 405 ha infested in Martin and Palm Beach Counties, Florida; ongoing effort.
- **Asian common wild rice** (*Oryza rufipogon*). A rhizomatous red rice; 0.5 ha infested in the Everglades National Park, FL; nearing completion.
- **Wild sugarcane** (*Saccharum spontaneum*). A rhizomatous wild sugarcane; 13 spot infestations along the southeastern shore of Lake Okeechobee in Martin County, FL, totaling less than 1 ha; ongoing effort.
- **Tropical soda apple** (*Solanum viarum*). A perennial herb with thornlike prickles that has invaded about 250 000 ha in Florida. All documented infestations in other southeastern states have been linked to interstate movement of contaminated cattle (that eat the fruit), composted manure, and bahiagrass seeds from Florida.

Most of the early weed eradication projects (e.g., witchweed, goatsrue, and hydrilla) involved large acreages. However, in recent years, there has been a new emphasis on early detection. As a result, most new projects are smaller in scope and duration (1-2 ha; 3-5 yr). This measure of success is mostly due to increased networking between weed scientists and botanists in recent years. Weeds detected early can be eliminated for less money in less time, with less damage to natural and managed ecosystems.

Strategies to prevent the spread of established invasive plants.

The first line of defense against introduced invasive plants is early detection of new infestations. As already noted, the work of amateur and professional field botanists is critical in early detection and reporting of new plant species as they are observed.

The second line of defense against invasive plants is to contain and eradicate incipient infestations as soon as they are detected.

The third line of defense against invasive plants is to prevent movement into noninfested areas.

The fourth line of defense against invasive plants is to develop effective and environmentally sound methods and procedures for control of large infestations.

Summary

Harmful non-indigenous plants are biological pollutants that threaten agricultural production and the biodiversity of natural ecosystems in the United States. Federal agencies in the United States, through FICMNEW, are developing a coordinated national strategy for dealing with invasive plants. One role of USDA APHIS in biological protection of ecosystems is to prevent the introduction of foreign invasive plants into the United States. APHIS also cooperates with affected states to combat incipient infestations of Federal Noxious Weeds before they become widespread. The most effective way to deal with invasive plants is to prevent their introduction from other countries, to detect incipient infestations at an early stage, and to implement an effective eradication program before they begin to spread to other farms, counties, and states. Money spent on weed prevention is a wise investment that will help to minimize future losses and control costs that are typically associated with widespread weeds.

References

Bridges, D., ed. (1992) *Crop Losses due to Weeds in the United States.* Weed Scence Society of America, Champaign, Ill.

Bridges, D. (1994) Impact of weeds on human endeavors. *Weed Technology,* **8**, 392–395.

Chandler, M. (1985) Economics of weed control in crops, in *The Chemistry of Allelopathy Biochemical Interactions Among Plants*, (ed. A. Thompson), ACS Symposium Series, No. 268, American Chemical Society, pp. 9–20.

Elton, C. (1958) *The Ecology of Invasions by Plants and Animals.* Methuen and Co., Ltd. London, UK.

Eplee, R. and Westbrooks, R. (1990) Federal Noxious Weed initiatives for the future. *Proceedings of the Weed Science Society, NC*, pp. 76–78.

Eplee, R. and Westbrooks, R. (1991) Recent advances in exclusion and eradication of Federal Noxious Weeds. *WSSA Abstracts* 31, 31 pp.

Goold, C. (1994) The high cost of weeds, in *Noxious Weeds: Changing the Face of Southwestern Colorado,* San Juan National Forest Association, Durango, Colorado, pp. 5–6 .

Leistritz, F., Bangsund, D. and Leitch, J. (1995) Economic impact of leafy spurge on grazing and wildland in the northern Great Plains, in *Alien Plant Invasions: Increasing Deterioration of Rangeland Ecosystem Health.* BLM/OR/WA/PT-95/048+1792. Proceedings of a symposium by the Range Management Society, Phoenix, Arizona, pp. 15–21.

McKnight, W., ed. (1993) *Biological Pollution: The Control and Impact of Invasive Exotic Species. Proceedings, Symposium on Biological Pollution.* Indiana Academy of Sciences. Oct. 25–26, 1991.

Montgomery, F. (1964) *Weeds of Canada and the Northern United States.* Ryerson Press, Toronto, Canada.

Mooney, H. and Drake, J., eds (1986) *Ecology of Biological Invasions of North America and Hawaii.* Springer-Verlag, New York.

Ross, M. and Lembi, C. (1983) *Applied Weed Science.* Burgess Publishing Company, Minneapolis, MN.

Schmitz, D. (1994) The ecological impacts of non-indigenous plants in Florida, in *An Assessment of Invasive Non-Indigenous Species in Florida's Public Lands,* (eds D. Schmitz and T. Brown), Florida Department of Environmental Protection, Technical Report TSS–94–100, 10–28.

Shaw, W. (1979) National Research Program 20280, weed control technology for protecting crops, grazing lands, aquatic sites, and noncropland. *Weeds Today,* **10**(**4**), 4.

USDA (1965) *A Survey of Extent and Cost of Weed Control and Specific Weed Problems.* Agriculture Research Service, Report 23–1, U.S. Department of Agriculture, Washington, DC.

Vitousek, P., D'Antonio, C., Loope, L. and Westbrooks, R. (1996) Biological invasions as global environmental change. *American Scientist,* **84**, 468–478.

Wagner, W.H., Jr. (1993) Problems with biotic invasives: A biologists viewpoint, in *Biological Pollution: The Control and Impact of Invasive Exotic Species*, (ed. B.N. McKnight), Proceedings, Symposium on Biological Pollution, Indiana Academy of Sciences, Oct. 25–26, 1991, pp. 225–241.

Westbrooks, R. (1989) *Regulatory exclusion of Federal Noxious Weeds from the United States.* PhD dissertation, Department of Botany, North Carolina State University, Raleigh, NC.

Westbrooks, R. (1991) Plant Protection Issues. I. A commentary on new weeds in the United States. *Weed Technology,* **5**, 232–237.

Westbrooks, R., and Eplee, R. (1991) USDA APHIS Noxious Weed Inspection System. 1991 update. *WSSA Abstracts,* **31**, 29.

Westbrooks, R. (1993) Exclusion and eradication of foreign weeds from the United States by USDA APHIS, in *Biological Pollution: The Control and Impact of Invasive Exotic Species.* (ed B.N. McKnight), Proceedings, Symposium on Biological Pollution, Indiana Academy of Sciences, Oct. 25–26, 1991, pp. 242–252.

Zamora, D., Thill, D. and Eplee, R. (1989) An eradication plan for plant invasions. *Weed Technology,* **3**, 2–12.

20 Managing insect invasions by watching other countries

MICHAEL J. SAMWAYS
University of Natal, Scottsville, South Africa

Abstract

Many insects are highly mobile, regularly reaching remote areas. This movement is further promoted by the huge amount of human traffic. Colonization does not necessarily mean invasion. Invasion is the establishment, population spread and increase, leading to a keystone role in a new community. Some invasions benefit agroforestry, but most are detrimental, especially to natural ecosystems. Local invasiveness gives a glimpse of potential intercontinental invasiveness. Prediction based on taxonomy, behaviour, ecology, etc. is unreliable. The best approach is to be aware of which species have been local, regional or intercontinental invasives elsewhere, and then be highly vigilant for their first appearance in a new area. On first discovery, a new invasive should be eradicated. If this opportunity is missed, an invasive can only be suppressed at best. As the cost of damage by invasives and their control is so high, strict quarantine implemented by taxonomically-trained and highly perceptive officers is a very sound investment.

Natural insect invasiveness

Insects are master travellers, colonists and survivors. The fossil record suggests that they have responded positively to changing climatic conditions by shifting their ranges (Coope, 1995) and have shown very low rates of extinction throughout their history (Labandeira and Sepkoski, 1993).

Insect variety virtually ensures that some, somewhere, will be on the move. Besides their great number of species (Stork, 1993), they can be extremely abundant. Together with other arthropods, temperate lands may support 1 000 kg fresh weight ha^{-1} (Pimentel *et al.*, 1992), while the world biomass of ants (about 9 000 species) is probably four times that of all land vertebrates (Wilson, 1991). This large number of individuals means wide genetic variation (Brakefield, 1991) including polymorphisms of selective advantage in colonizing new habitats (Samways, 1993).

Many insects are migratory, and some of these are major pests (Drake and Gatehouse, 1995; Pedgeley, 1982). Significantly, many of these migrants are the most difficult pests to mitigate. Migrations are directional movements, honed by natural selection (e.g. Monarch butterfly movements;

O. T. Sandlund et al. (eds.), Invasive Species and Biodiversity Management, 295–304.

Brower, 1977). But many insect movements are undirected dispersal. Insects on wind currents regularly land on snowfields (Ashmole *et al.*, 1983), colonize recently-cooled lava flows (New and Thornton, 1988) and travel on flotsam arriving serendipitously on islands (Peck, 1994a). Some small insects regularly travel high up on wind currents (Berry and Taylor, 1968) and colonize islands (Peck, 1994b). Even first-instar moth larvae can be lifted 800m by atmospheric turbulence and travel 19 km at a time on wind currents (Taylor and Reling, 1986). Thornton *et al.* (1988) recorded densities of around 20 arthropod individuals m^{-2} day^{-1} landing on Anak Krakatau. This is equivalent to about 50 million individuals arriving each day on this tiny island. Even large insects can traverse the ocean, with the sweepstakes effect resulting in the most distant locales being colonized by at least some species (Coope, 1986). Movements however, do not always equal invasiveness. There have been more insect invasions from Europe elsewhere than into Europe.

Human spread of insects

In addition to the natural showers of insects across the world, there is now major dispersal by human agency. This is partly because insects are excellent stowaways, particuiarly Diptera, but also Hemiptera, Hymenoptera, Lepidoptera and Coleoptera on aircraft (Russell *et al.*, 1984), and Lepidoptera and Odonata on ships (Holdgate, 1986). These are free insects, while others are the bane of quarantine officers by travelling on plants.

The number of new colonists can be high. Hawaii accumulates 20–30 new insect species per year (Beardsley, 1991) and Guam 12–15 new species, facilitated by the huge amount of human traffic (Schreiner and Nafus, 1986). Although these figures are impressive, they are probably only a small percentage of insect species raining on foreign lands (Williamson, 1996). Some arrivals however, establish and remain local and relatively rare, such as the moth *Phyllonorycter messaniella* around cities in southeast Australia (New, 1994).

It is difficult to predict which insect taxa and behavioural types will colonize (Lawton and Brown, 1986). Also, those that travel as free agents or as stowaways are not necessarily those that make up the colonizing fauna. Exotic Thysanoptera, for example, are over-represented in Hawaii (Simberloff, 1986). There has also been debate on which ecosystems are susceptible to invasion (Simberloff, 1986). But generalizations about invasives must avoid the circular argument that success succeeds (Holdgate, 1986; Williamson and Brown, 1986).

What constitutes an insect invasion?

It may take only a few vagrants or stowaways to eventually build up to a keystone invasion. The ant *Solenopsis invicta* problem in the USA developed from a few individuals in ship ballast dirt which arrived in Mobile, Alabama in the 1930s (Schmidt, 1995).

The important point is not the colonization but the huge impact. An invasion is therefore best described as the establishment, population spread and increase, leading to a keystone role in a new community. This new community may be only metres away, but generally it is at a distance of hundreds of kilometres. An invasion may be to a natural ecosystem, agroforestry, urban environment or be of medical or veterinary significance.

Not all insect invasives to agroecosystems are economically detrimental. The arrival of the ladybird *Chilocorus nigritus* on citrus in South Africa has had a major positive economic impact, but possibly an adverse competitive effect on the native *C. wahlbergi* (Figure 20.1). The ecological point however, is that *C. nigritus* is a keystone species in the citrus agroecosystem, especially when its host population levels are high (Samways, 1988). Similarly, the exotic Argentine ant, *Linepithema humile,* has had a major keystone adverse impact on the fynbos ecosystem (Bond and Slingsby, 1984).

Local versus continental effects: glimpses of prediction

Invasions can be highly localized, as well as happening on a global scale. This is more than curiosity, as it gives us a glimpse of the types of organisms and the circumstances that are invasives. The ant *Pheidole megacephala,* besides being a highly noxious invader of Pacific islands, has also invaded Australia (New, 1995). This insect is native to Africa where its biology is well documented (Broekhuysen, 1948). Its population increases rapidly in response to physical disturbances such as the construction of a new road and subsequent road kills for food (Samways *et al.*, 1997) and to mutualistic honeydew-producing homopteran populations (Samways, 1981). As disturbance and food availability increase, this species then dominates and excludes all others. But when the habitat becomes unfavourable for it, or there is balanced biological control of its host scales and mealybugs, it becomes less dominant and even non-dominant (Samways, 1983a), with distinct amensalism in favour of other ant species (Samways, 1983b).

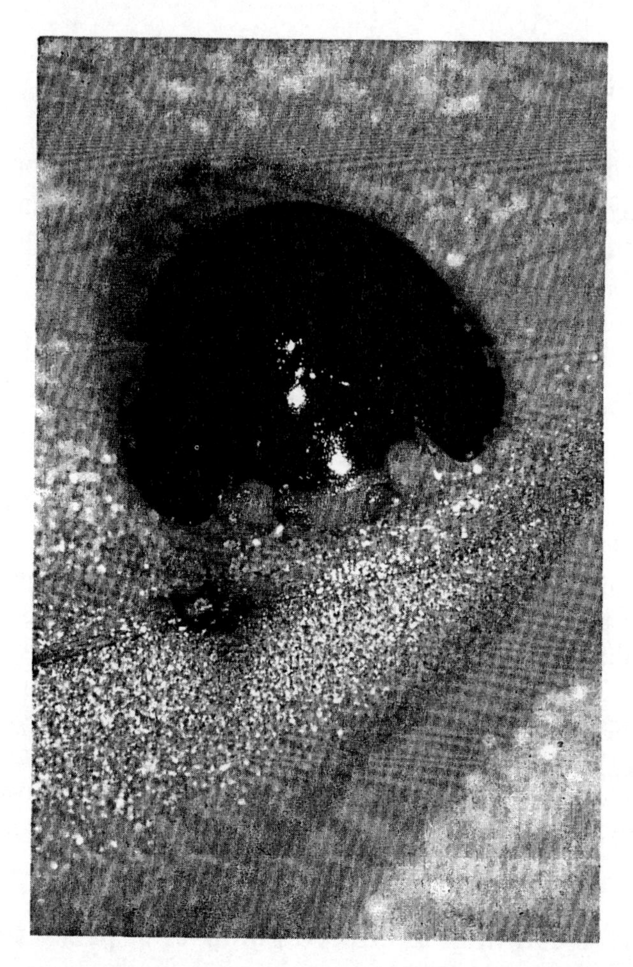

Figure 20.1 One invasive alien species, the ladybird *Chilocorus nigritus*, that has benefitted South African agriculture but may have reduced the population level of the native *C. wahlbergi.* (Photo: M.J. Samways).

The important point is that we **know** that *P. megacephala* is a highly important inter-continental invasive. We also know it is a highly significant local invasive. This gives us a clue as to potential risks of other species To name one, the ant *Anoplolepis custodiens* behaves and responds in a similar way to *P. megacephala* (Samways, 1986c) and can reach enormous and totally dominating numbers in southern Africa (Samways, *et al.*, 1981; Figure 20.2). As yet, this species has not reached other areas. But the chances are that this could be a devastating invasive of new regions with a similar climate. Of course, we cannot risk testing this hypothesis.

In the medical, veterinary and urban contexts, an invasive is a pest when it poses a health or hygiene risk, or is a considerable nuisance. Pharaoh's ant, *Monomorium pharaonis*, and recently the ghost ant, *Tapinoma melanocephalum* in buildings such as hospitals are examples (Coghlan, 1996). In natural ecosystems, an invasive is a pest when native populations are threatened, especially if the end result could be species extirpation or extinction.

There is an important corollary. An invasive colonises and then spreads. In other words, the **potential** impact must also be considered. This might also be associated with a change in its behaviour and even its genotype (Schmidt, 1995).

Prediction of insect pest invaders

Even though the South African ant *Anoplolepis custodiens* could be a potential invader to other countries (Figure 20.2), in the Cape fynbos it can be outcompeted by the exotic *Linepithema humile*. What does this tell us? One of the clearest patterns to emerge on the types of insect invaders that become pests, is that colonization and high impact at one location, can mean similar adverse effects elsewhere, even if indigenous dominants (and potential invasives) are present. Although there are contentions that island ecosystems are inherently more susceptible in this regard (Carlquist, 1965), there is little firm evidence that this is actually so (D'Antonio and Dudley, 1995).

There is a wide range of taxa that are noxious invasives. Some of the serious invaders of Australia, for example, (New, 1995), are also invaders of South Africa, e.g. the wasp *Vespula germanica*, the wood wasp *Sirex noctilio*, the codling moth *Cydia pomonella*, the peach–potato aphid *Myzus persicae*, the green vegetable bug *Nezara viridula* etc. A natural extension of this line of thinking is that crops in similar climates in different parts of the world eventually accumulate the same pest complex. Indeed, *M. persicae* is a world pest. Citrus orchards in Australia and South Africa have very similar pest complexes with some notable exceptions. The native South African psyllid *Trioza erytreae* and *Scirtothrips aurantii* have not yet reached Australia. In turn, South Africa, in the absence of sufficient quarantine, is highly likely to receive the wasp *Vespula vulgaris*, now a well-established pest in New Zealand (Barlow *et al.*, 1996).

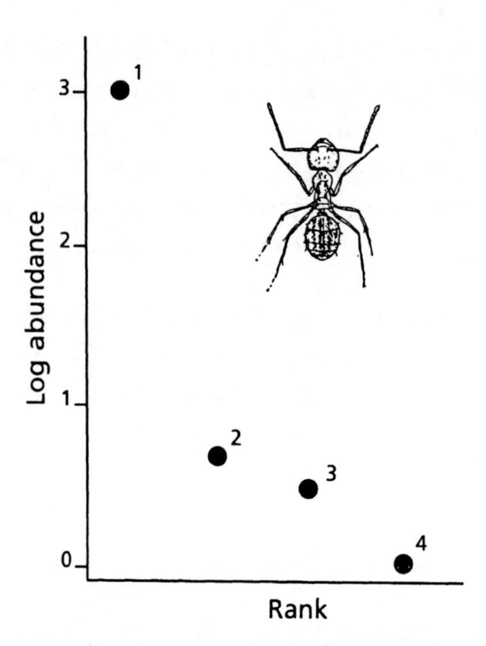

Figure 20.2 The extreme local dominance of the pugnacious ant *Anoplolepis custodiens* has caused impoverishment in abundance and a reduction in species richness (to only four species, 1—4) of the local ant fauna in South Africa. This ant can be an "alien invasive" at the local geographical scale. Once it gains a foothold it can change local community structure. Such an insect should be known to quarantine officers worldwide, as it has the potential to be a significant alien invasive. 1 = *A. custodiens*; 2 = *Pheidole sculpturata*; 3 = *Tetramorium mossamedense*; 4 = *Solenopsis punctaticeps*. (Redrawn from Samways *et al.*, 1981).

Development of a protocol

Although our knowledge of exactly which organisms are potential invasives and which ecosystems are highly invasable is still scant, we can nevertheless develop a protocol for containing this global homogenization of invasive pests. Accurate prediction looks to be far off and possibly lies more in genetic nuances of individual species than in generalization about origin, size, life-history trait etc. Nevertheless, we can develop a protocol:

1) Although totally undervalued and unacknowledged, it is essential that a country has experts on exotic invasive insects. Recognition is the first step to intelligent control of a pest (Coppel and Mertins, 1977). Early

recognition, followed by expert knowledge of what to do, led to instant and highly successful control of the citrus blackfly *Aleurocanthus woglumi* in South Africa (Bedford and Thomas, 1965). Such experts would also be knowledgeable on potential serious invaders. In other words, the experts would be aware of invasive insects elsewhere in the world. This has recently been exemplified by the response to the Eucalyptus snout beetle *Gonipterus scutellatus* which has established in California (Cowles and Downer, 1995). Such experts would also be familiar with trade routes into their particular area or country so as to be primed on which species may arrive.

2) There should always be extremely strict quarantine. The potential impact and cost of an invading insect can be enormous. The fire ant, *Solenopsis invicta*, moving into a new area can reduce the native ant species by 70% and arthropods in general by 30% (Schmidt, 1995) causing millions of dollars of damage despite millions of dollars being spent on control (Thompson, 1990). Prevention is indeed far cheaper than cure. Quarantine can be highly effective, with the example set by Australia (New, 1995) a valuable model for other countries. Yet still, this is not sufficient, with the recent invasion of the poinsettia whitefly (*Bemisia argentifolii*) from California and set to cost Australia perhaps A$ 300million yr^{-1} (Anderson, 1996).
3) There should be early detection and eradication (or suppression where eradication is not possible) when a very new invasive appears. There should be an immediate and all-out response, as is being attempted in California against the newly-arrived exotic termite *Coptotermes formosanus* (Haagsma *et al.*, 1995). In other words, besides having expertise on the taxonomy, behaviour, monitoring and control methods of potential or actual invasive insects, it is also essential to have the logistic response capability.
4) Utilization of eco-climatic modelling (Sutherst and Maywald, 1985) can be very effective in determining to where a new invasive might spread. This has been done very effectively for the wasp *Vespula germanica* in South Africa (Tribe and Richardson, 1994). There appears to be even more suitable localities for this wasp beyond its initial establishment site. Every effort should be made to contain the wasp. An important corollary too, is that pest scenarios will change with global climate change (Sutherst, 1995).

In summary, the ideal approach to invasive insect management is that set up by the Center for Exotic Pest Research (CEPR) at the University of California, Riverside (Metcalf, 1995). The long-term goal of the CEPR is to develop a systematic methodology for coping with exotic pests by:

- Conducting risk assessment and analysis of invasion biology.

- Improving techniques for early detection of pest invasions through lures, traps and monitoring.
- Developing control / eradication measures employing biological, cultural, genetic and chemical techniques.
- Exploring potential transgenic manipulations to control exotic pests.
- Evaluating the economic and sociological implications of exotic pest introductions.

Acknowledgements

I thank Ms Åslaug Viken and The Trondheim Conference on Biodiversity for supporting this presentation. Ms Pamela Sweet kindly processed the manuscript.

References

Anderson, I. (1996) Aliens dodge Australia's pest police. *New Scientist,* **150,** (no. 2027), 12.

Ashmole, N.P., Nelson, J.M., Shaw M.R. and Garside, A. (1983) Insects and spiders on snowfields in the Cairngorms, Scotland. *Journal of Natural History,* **17**, 599–613.

Barlow, N.D., Moller, H. and Beggs, J.R. (1996) A model for the effect of *Sphecophaga vesparum vesparum* as a biological control agent of the common wasp in New Zealand. *Journal of Applied Ecology,* **33**, 31–44.

Beardsley, J.W. Jr. (1991) Introduction of arthropod pests into the Hawaiian islands. *Micronesia,* **(Suppl.) 3**, 1–4.

Bedford, E.C.G. and Thomas, E.D. (1965) Biological control of the citrus blackfly, *Aleurocanthus woglumi* (Ashby) (Homoptera: Aleyrodidae) in South Africa. *Journal of the Entomological Society of Southern Africa,* **28,** 117–132.

Berry, R.E. and Taylor, L.R. (1968) High altitude migration of aphids in maritime and continental climates. *Journal of Animal Ecology,* **37,** 713–722.

Bond, W.J. and Slingsby, P. (1984) Collapse of an ant–plant mutualism: the Argentine ant (*Iridomyrmex humilis*) and myrmecochorous Proteaceae. *Ecology,* **65**, 1031–1037.

Brakefield, P.M. (1991) Genetics and the conservation of invertebrates, in *The Scientific Management of Temperate Communities for Conservation,* (eds I.F. Spellerberg, G.B. Goldsmith and M.G. Morris), Blackwell Scientific Publications, Oxford, UK, pp. 45–79.

Broekhuysen, G.J. (1948) The brown house ant (*Pheidole megacephala* F.). *Bulletin, Department of Agriculture, Union of South Africa,* No. **266**.

Brower, L.P. (1977) Monarch migration. *Natural History,* **86**, 40–53.

Carlquist, S. (1965) *Island Biology: A Natural History of the Islands of the World.* Natural History Press, Garden City, New Jersey, USA.

Coghlan, A. (1996) Call in the ghost-ant busters. *New Scientist,* **150**, (no. 2027), p.12.

Coope, G.R. (1986) The invasion and colonization of the North Atlantic islands: a palaeoecological solution to a biogeographic problem. *Philosophical Transactions of the Royal Society of London, Series B.*, **314,** 619–635.

Coope, G.R. (1995) The effects of Quaternary climatic changes in insect populations: lessons from the past, in *Insects in a Changing Environment,* (eds R. Harrington and N.E. Stork), Academic Press, London, pp. 30–48.

Coppel, H.C. and Mertins, J.W. (1977) *Biological Insect Pest Suppression.* Springer-Verlag, Berlin.

Cowles, R.S. and Downer, J.A. (1995). Eucalyptus snout beetle detected in California. *California Agriculture,* **49**, 38–40.

D'Antonio, C.M. and Dudley, T.L. (1995) Biological invasions as agents of change on islands versus mainlands, in *Islands: Biological Diversity and Ecosystem Function,* (eds P.M. Vitousek, L.L. Loope and H. Andersen), Springer-Verlag, New York, pp. 103–121.

Drake, V.A. and Gatehouse, A.G. (1995) *Insect Migration: Tracking Resources through Space and Time.* Cambridge University Press, Cambridge.

Haagsma, K., Rust, M.K., Reierson, D.A., Atkinson, T.H. and Kellum, D. (1995) Formosan subterranean termite established in California. *California Agriculture,* **49**, 30–33.

Holdgate, M.W. (1986) Summary and conclusions: characteristics and consequences of biological invasions. *Philosophical Transactions of the Royal Society of London, Series B,* **314,** 733–742.

Labandeira, C.C. and Sepkoski, J.J., Jr. (1993) Insect diversity in the fossil record. *Science,* **261**, 310–315.

Lawton, J.H. and Brown, K.C. (1986) The population and community ecology of invading insects. *Philosophical Transactions of the Royal Society of London, Series B,* **314**, 607–617.

Metcalf, R.L. (1995) Invasion of California by exotic pests. *California Agriculture,* **49**, 2.

New, T.R. and Thornton, I.W.B. (1988) A pre-vegetation population of crickets subsisting on allochthonous aeolian debris on Anak Krakatau. *Philosophical Transactions of the Royal Society of London, Series B.,* **322**, 481–485.

New, T.R. (1994) *Exotic Insects in Australia.* Gleneagles, Adelaide.

Peck, S.B. (1994a) Sea-surface (pleuston) transport of insects between islands in the Galápagos Archipelago, Ecuador. *Annals of the Entomological Society of America,* **87**, 576–582.

Peck, S.B. (1994b) Aerial dispersal of insects between and to islands in the Galápagos Archipelago, Ecuador. *Annals of the Entomological Society of America,* **87,** 218–224.

Pedgeley, D.E. (1982) *Windborne Pests and Diseases: Meteorology of Airborne Organisms.* Horwood, Chichester, UK.

Pimentel, D., Strachow, U., Takacs, D.A., Brubaker, H.W., Dumas, A.R., Meany, J.J., O'Neil, J.A.S., Onsi, D.E. and Corzilius, D.B. (1992) Conserving biological diversity in agricultural/forestry systems. *BioScience,* **42**, 354–362.

Russell, R.C., Rajapaksa, N., Whelan, P.I. and Langsford, W.A. (1984) Mosquito and other insect introductions to Australia aboard international aircraft, and the monitoring of disinfection procedures, in *Commerce and the Spread of Pests and Diseases,* (ed. M. Laird), Praeger, New York, pp. 109–141.

Samways, M.J. (1981) Comparison of ant community structure (Hymenoptera: Formicidae) in citrus orchards under chemical and biological control of red scale, *Aonidiella aurantii* (Maskell) (Hemiptera: Diaspididae). *Bulletin of Entomological Research,* **71**, 663–670.

Samways, M.J. (1988) A pictorial model of the impact of natural enemies on the population growth rate of the scale insect *Aonidiella aurantii. South African Journal of Science,* **84**, 270–272.

Samways, M.J. (1983a) Community structure of ants (Hymenoptera: Formicidae) in a series of habitats associated with citrus. *Journal of Applied Ecology,* **20**, 833–847.

Samways, M.J. (1983b) Asymmetrical competition and amensalism through soil dumping by the ant, *Myrmicaria natalensis. Ecological Entomology,* **8**, 191–194.

21 Biological pest control for alien invasive species

GEORGE I. ODUOR
CAB International, Africa Regional Centre, Nairobi, Kenya

Abstract

The movement of man across the globe has greatly increased the rate at which large numbers of living organisms are moving from one ecosystem to another. In man's endeavour to achieve increased agricultural productivity, aesthetic satisfaction, etc., increased numbers of alien species are invading natural communities and threatening ecosystems, habitats and species. To manage these alien pests, governments and different organisations are increasingly turning to biological control because it is a strategy which has the potential to be self-sustaining and inexpensive, as well as ecologically non-disruptive, thereby protecting biological diversity. Successful and safe biological control requires thorough ecological studies and research to demonstrate that agents to be introduced are effective and do not pose undesirable risks to non-target species. Biological control usually begins with an exploratory programme in the area of origin of the pest, followed by research on the natural enemies that are found, careful safety testing and quarantine, and finally introduction and impact assessment. Management problems that may be encountered when a natural enemy is imported into an ecosystem to suppress the population of an alien (or indigenous) pest organism, and how they can be resolved, are discussed.

Introduction

As a result of increased trade and movement of man between countries, more problems associated with alien pests continue to cause concern. World-wide, arthropod pests, weeds and plant pathogens are acknowledged as major impediments to agricultural production and a system of control that does not require continuous expenditure or inaccessible technology, is ecologically non-disruptive and maintains local biodiversity, is of increasing appeal to developing as well as developed countries. The use of natural enemies to control pest species, i.e. biological control, is such a strategy.

Although biological control principles can be traced back to the domestication of cats and their employment to protect stored food from the ravages of mice and rats, the modern, scientific practice of biological control began in 1888 with the first deliberate transfer of beneficial insects from one geographical area to another to help in the control of an alien insect pest. In this programme, the Australian cottony cushion scale, *Icerya*

O. T. Sandlund et al. (eds.), Invasive Species and Biodiversity Management, 305–321.

pattersoni, a devastating alien pest of citrus in California, was permanently suppressed by introduction of the specific Australian ladybird beetle, *Rodolia cardinalis*. A century later, the pest persists in California maintained at low, non-damaging levels by its natural enemy, and the control of introduced insect pests by the introduction of specific natural enemies such as parasitoids, predators and pathogens from their native homes has become a recognised and widely used control method. A few decades after the control of cotton cushion scale, in 1925, this method was first applied to the biological control of weeds, with the control in Australia of cactus (*Opuntia* spp.) by the South American moth *Cactoblastis cactorum*, achieving permanent suppression of this weed over a vast area (see below).

Biological control aims at attaining a balance between the population of a pest species and that of its natural enemies ensuring that neither becomes too abundant and thus maintaining species diversity. In this paper, biological control agents and strategies are described, with particular emphasis on the "classical" approach applied to alien pest species. Case studies which illustrate this method are presented, and finally risks and conflicts of interest and their resolution are discussed.

Terminologies

The management of alien invasive species over the past century has focused on species affecting agriculture, fisheries, forestry, livestock production and other human activities. Management of alien invasive species which threaten the conservation of indigenous species or habitats is only now becoming popular. This explains why so much alien species management is couched in the language of "pest control". In the broadest sense, a pest can be considered any organism (plant or animal) which causes harm or damage to man, his animals, crops or possessions, or simply annoys him (Hill, 1983). The most important aspect of a pest species is the damage (to species, habitats or ecosystems) caused and the values placed upon these consequences by man. In agricultural terms, when loss in yield or quality of production reaches a certain level, with concomitant reduction in profit to the farmer, a species can be considered an economic pest and control measures may be implemented if their cost is less than the value of losses due to the pest. In an agricultural context, biological control has today become a desirable alternative to chemical pesticides where these carry a high social cost and where biological control can, by virtue of its self-renewing nature, be free to the farmer.

In this paper biological control shall be defined as "the use of living organism as pest control agents". However, biological control is sometimes used to describe as well control methods which rely on the biology of the pest or its host plant and include use of pheromones, genetic manipulation, production of sterile males for release, and the breeding of plants resistant to pests.

Biological control agents

Living organisms which are natural enemies of pest species can be conveniently classified as either parasites, parasitoids, pathogens, predators, insects for weed control, antagonists or competitors. These natural enemies can maintain a population balance with their hosts or prey, suppressing their numbers through parasitism or predation, but becoming rare themselves when pests are rare, thereby allowing pests to persist at low levels. Indeed, this process has been identified as one of the key mechanisms which permits species coexistence in nature, and hence contributes to the maintenance of ecosystems rich in biodiversity. And natural enemies are a substantial component of biodiversity. If, for instance, we include as potential biological control agents of invasive weeds and insects all of the species of insects which feed on plants or, as predators and parasitoids, on plant pests, we will have in our potential biological control agent category over half of the species of living organisms on this planet – adding pathogens of these pests would increase this further (Waage, 1991)

The major natural enemy groups utilised in the management of alien arthropods and weeds are now briefly described.

Parasites: A parasite is an organism living at the expense of another organism and deriving all its nourishment from it. Individual parasites tend to weaken and reduce the reproductive success of their host (the pest) rather than kill it. Parasites usually have a short free living stage solely concerned with dispersal between the hosts. With respect to the management of insect pests, a good example is a parasitic nematode, which find and penetrate into an insect where it reproduces. Some parasitic nematodes of insects kill their hosts through the action of symbiotic bacteria, and hence are used in biological control.

Parasitoids: This group of parasitic insects, which attack other invertebrates, has a distinctive life history which set them apart from other parasites. Parasitoids develop parasitically in one host which is eventually killed i.e. individual parasitoids consume only one prey (or host) during their lifetime, and hosts always die. This unique lifestyle is exhibited by a remarkably diverse group of insects mostly belonging to the orders

Hymenoptera (wasps) and Diptera (flies), comprising about 300 000 species. Although the larval stage is parasitic, the adults are free living and highly mobile so that they are able to search actively for hosts in which to lay eggs. Many are monophagous, that is, specific to only one species of host.

Pathogens of insect pests: Unlike parasitic nematodes and parasitoids, pathogens are parasitic micro-organisms and are usually either bacteria, viruses, fungi or protozoa. Dead hosts liberate millions of individual microbes which are passively transmitted between insect hosts especially by ingestion. Because of their minute size and rapid reproduction in the host, pathogens are easily mass produced and can be released against pests using equipment developed for application of chemical pesticides.

Predators: Important characteristics of predators are that an individual consumes a number of prey during its lifetime and that they are free living active organisms which seek their food. Some predators feed on a wide range of prey species (polyphagy), some on a few species (oligophagy) while others are extreme specialists (monophagy). Of particular interest to biological control are invertebrate predators (e.g. insects) since they are more specialised and more able than vertebrates (e.g. birds and mammals) to reproduce rapidly to keep up with the growth of pest populations. Ladybird beetles, phytoseiid mites, predatory bugs, and larvae of hover flies are common predators of insects in different habitats.

Insects and pathogens for weed control: Biological control of weeds is commonly effected by use of specific plant-feeding insects or fungal pathogens. Here, as in all cases, the host range of the insect or pathogen must be thoroughly studied, to ensure that it does not have the potential to be a pest itself.

Antagonists and competitors: These are organisms which exclude the pest by either preventing the pest from establishing at a particular site of infection (antagonist) or out-competing it (competitor). These natural enemies are particularly important in the biological control of plant pathogens.

Biological control strategies

Different approaches may be used to utilise the organisms described above in the management of pests. The biological properties of these natural enemies determine the appropriate strategy.

Introduction or "classical biological control" (so-called because of its first application in 1888) involves the introduction of an alien natural enemy from the pest's area of origin into a new area where it becomes permanently established. Alien pests often reach sustained outbreak levels in the absence of specific natural enemies, and introduction is intended to address this problem, as an alternative to the need for recurrent and extensive use of pesticides or other controls. When introductions succeed, the pest's generational mortality is increased to the extent that the pest becomes of negligible importance, a biological balance is achieved and control continues indefinitely.

Inoculation. Where an effective indigenous natural enemy is absent from a particular area, or an introduced natural enemy cannot persist from one cropping season to another, repeated releases can be made at the beginning of a season to colonise the area for the duration of the season, and so prevent pest build up.

Augmentation. Where indigenous natural enemies are effective, but their numbers inadequate in the chosen environment, e.g. because their numbers do not build up sufficiently rapidly, or because they are unable to maintain adequate densities under current agricultural practices, their numbers may be usefully increased by release of laboratory reared individuals.

Inundation. This is where large numbers of natural enemies, indigenous or alien, are cultured and applied at critical periods for short-term suppression of pest numbers, in much the same way as chemical pesticides. This is a frequent approach for the use of pathogens, because pathogens frequently have difficulty persisting and spreading in the crop environment, and because they are often inexpensive to produce. Inundation is popular as an alternative to environmentally damaging chemical pesticides.

Conservation. This is an indirect method in that measures are taken to conserve and enhance the impact of natural enemies already present in the crop environment e.g. increasing food or shelter in crops, for instance encouraging flowers around crop margins, or modifying pesticide use to cause less mortality to important natural enemies. Conservation is important to maintaining the impact of introduced natural enemies as well, and forms the basis of sustainable pest management for farmers around the world.

Development and implementation of a classical biological control programme

Alien organisms frequently become pests because they have been freed from natural enemies which keep their populations below injurious levels in their areas of origin. Identifying and introducing such specific co-evolved natural enemies into the new expanded range of the pest can reduce the abundance of the invading alien organism, the area it infests and its competitive abilities. Before undertaking an introduction programme, it should be established that the pest problem justifies a control effort and that introduction is an appropriate approach. An assessment should be made of the extent of the pest problem, the nature and amount of damage, the value of losses being incurred and the cost to the nation as a whole and/or the farmers.

- **Preliminary studies.** It is imperative that the identity of the alien pest is accurately determined. For this purpose, well prepared and labelled specimens should be sent to an appropriate taxonomist or museum for an authoritative identification. Inadequate taxonomy led to delays in the biological control of pests such as the Kenya mealybug *Planococcus kenyae* and the cassava mealybug *Phenacoccus manihoti* (Homoptera: Pseudococcidae) and weeds (e.g. water fern *Salvinia molesta* in Africa). After accurate identification the following information on the pest can be gained from the scientific literature. The biology, natural enemies and probable origin of the pest, experience with the pest elsewhere including control attempts, and related species to the pests and their pest status.
- **Development of a project proposal.** When successful, a classical biological control programme can be highly cost effective, because established control is permanent with no recurrent costs. However, implementation of a programme can have a high initial cost. Therefore, a proposal is usually required to secure funding. It is important to prepare a proposal in phases so that an evaluation can be made at the end of each stage to ensure that funds are committed only when it can be seen that it's worth continuing to the next phase. It is logical to start with surveys and exploration which will show the likelihood of finding promising natural enemies.
- **Surveys and exploration.** Surveys are made to obtain an inventory of natural enemies, pest incidence and seasonality so as to select sites where the life history, incidence and the host range of the natural enemies are studied. Preferred areas of studies should be areas where the pest is scarce and probably under the control of natural enemies, where the greatest species diversity of the pest genus occurs and where the climate and other ecological conditions match the area where control is required.
- **Selection and screening of natural enemies.** Following identification of candidate natural enemies from surveys, detailed biological studies in the field and laboratory should be undertaken to establish the natural enemy's biology, host specificity, searching efficiency, etc. Characteristics of a good natural enemy include high searching capacity and dispersal rates, specificity to the host, ability to survive in

all habitats occupied by the host, short life cycle and high reproductive rate. Studies will also need to consider risks to non-target organisms and environments (see below).

- **Quarantine.** Like all other alien organisms, biological control agents from other countries are usually subjected to quarantine restrictions. Whereas quarantine can be done after the agent is imported into a country, many countries prefer that it be done outside to minimise risk (third country quarantine). There are two basic requirements of quarantine for a biological control agent, i.e. the agent must not pose a risk to agriculture or the environment after it is introduced, and the agent must not bring in with it undesirable organisms e.g. plant pathogens and natural enemies of the agent itself.
- **Release and evaluation of agent.** Releases should be of adequate numbers of well fed mated individuals in calm weather preferably in the evenings so as to minimise the risk of extreme temperatures and of wind spreading the introduced population too thinly to reproduce and establish. Following first release, progress should be monitored for as long as possible to obtain an assessment of the introduction on the pest and the environment. Evaluation may involve studies of pest population before and after releases, with measurements of predation or parasitism after release, life table studies of single pest generations or cohorts at specific times before and after releases and exclusion studies after establishment of the agent using cages or an insecticide check (in the case of insect agents).

The three main factors which influence the success of classical biological control programmes are (i) the ecology of the pest, (ii) the kind of damage it does, and (iii) the cropping system in which it occurs. Introductions against insect pests are most successful against pests which have low to moderate reproductive rates, tend to form local non-migratory populations, are not vectors of diseases and are "indirect" pests i.e. do not damage the part of the plant that is harvested, and inhabit permanent or stable agricultural systems.

Not all biological control programmes have led to the suppression of pest populations. Rates of establishment (i.e. persistence following their release) of biological control agents are 30% for programmes against alien insects (Greathead and Greathead, 1992) and 64% against alien weeds (Julien *et al.*, 1984), estimates of the percentage of establishments leading to substantial levels of control are about half of these figures. The introduction into Togo in 1987 of the parasitoid *Gyranusoidea tebygi* (Hymenoptera: Encyrtidae) from India led to the control of the mango mealybug *Rastrococcus invadens* (Homoptera: Pseudococcidae). Exploratory and research work together with quarantine and provision of natural enemies cost US$ 175 000 and the country now earns US$ 3.9 million per year from the sale of mangoes (Voegele *et al.*, 1991). Putting a monetary value on the biological control of alien pests threatening conservation of species and habitats is more difficult as the benefits may be entirely environmental.

As global trade has increased, so have the chances of alien pests invading natural communities, thereby threatening agriculture and local biological diversity. This has led to an increase in recent years in biological control projects and the exchange of natural enemies between countries. At least 98 countries world-wide have been the source of biological control agents for one or more programmes, and 121 countries have made at least one introduction. International co-operation is essential, as one country's rare indigenous insect or plant may someday be another's invasive pest problem. Access to potential biological control agents is an important justification for conservation of indigenous biodiversity, and for mechanisms for the fair and equitable exchange of genetic resources between countries, as specified in the Convention on Biological Diversity.

Risk management in biological control

Any programme to introduce alien species to control other alien species carries risks that the control agent may itself have negative effects on managed or natural habitats. Indeed, the first, non-scientific attempts at biological control illustrate the potential for control agents to become pests, where generalist vertebrate predators such as mongooses and birds were introduced on islands against insects and vertebrates.

The use of highly specific biological control agents in programmes against insects and weeds has dramatically reduced biological control risks, and increased success, because specific agents are also more effective. Safety testing of biological control agents involves collection of field data on specificity as well as laboratory studies. Biological control of weeds, because of risks to crops, has a particularly extensive testing procedure, which involves testing of the feeding and reproductive stages of the agent against taxonomic groups increasingly distant from the target species (centrifugal screening) to establish host range, accompanied by tests against agricultural plants and against indigenous plants of conservation interest. Screening a single weed insect or pathogen agent in this way may take several years and involve over 100 test plants. The great majority of natural enemies of potential target pests are rejected in this process. For insect pests, until quite recently, decisions about safety were based on the biology of the natural enemy, often with specific laboratory tests against beneficial insects (e.g. bees, other beneficials and butterflies). Today, this process is becoming more restrictive, with growing reluctance to introduce anything but the most specific, effective agent available.

Because of the overriding need for safety, and increasing interest in biological control by countries with little experience in the procedure outlined above, a Code of Conduct on the Import and Release of Biological Control Agents was developed and ratified by the FAO Council in 1995. Guidelines for the Code have been prepared by CABI Bioscience. The Code is voluntary and recommends that importers submit detailed dossiers on intended introductions for scrutiny by the relevant national authority.

The largely agricultural history of biological control has meant that safety testing has been the responsibility of national plant quarantine authorities. For this reason, risks of impact of introduced control agents on indigenous species in natural habitats have often not received the attention that they now deserve in an increasingly environmentally-conscious society. Moreover, if biological control is going to realise its potential as a means of managing invasive species affecting conservation, these non-target effects will be a primary concern.

The outcome of past biological control programmes are too poorly known to assess their impact on non-target species. A number of cases have been cited (Howarth, 1991), some of which are the subject of ongoing scientific debate (e.g. Sands, 1997). It is possible to say that over the approximately 5 000 introductions against insect pests and 1 000 against weeds, only a handful of programmes **record** any unexpected non-target effects, but this does not preclude the existence of impact on non-targets which were not apparent or not investigated.

Biological control in future will require increased attention to non-target effects in natural habitats. It is important to note, however, that a great deal of risk in biological control is deliberate, and the result of decision making and not a lack of good science. For instance, much attention has been drawn recently to the non-target effects of insects introduced to control alien thistles in North American grasslands, which are now infesting some indigenous thistle species (Louda *et al.*, 1997; CABI, 1997). This is not an accident of bad science. The risk to indigenous thistles was stated clearly in the safety studies undertaken in the 1970s for this introduction. At the time, however, decision makers viewed the threat of the weed to pastures as of greater significance than non-target effects on native thistles. Perhaps today that judgement would be different. In a more recent example, Australian authorities, renowned for their concern for conservation of unique flora and fauna, approved introduction of insects and fungi for control of the alien asclepiad, rubber vine or *Cryptostegia grandiflora*, despite the determination in safety testing that these agents pose some, albeit very minor, risk to rare Australian asclepiads. Here, the much greater effect on

Australian biodiversity of extensive rubber vine infestations supported the decision for introduction.

Case studies

The following are three examples of cases in which alien pests threatened local species, habitats or ecosystems, and how these threats were averted using classical biological control.

Control of the water hyacinth, Eichhornia crassipes *(Pontederiaceae) in Sudan*

The Neotropical aquatic weed, water hyacinth, causes more serious and more extensive problems than any other floating aquatic weed (Harley, 1994). Its high rates of growth and reproduction, high competitive abilities relative to other floating aquatic plants, ability to move by wind and water currents and spread by man due to its attractive flowers all combine to make this plant a particularly noxious weed. By accelerating water loss through evapo-transpiration and reducing water flow by siltation, the weed can reduce the depth of water and eventually convert open water into shallow marshes. By shading, competition for essential resources and depletion of oxygen levels in the water, the water hyacinth mat may cause high mortality of phytoplankton and other aquatic flora, and also fish and zooplankton which depend on the phytoplankton. This leads to a reduction in the biodiversity of the aquatic flora and fauna (cf. Ogutu-Ohwayo, 1998). Although chemical control by use of herbicides has been successful in controlling isolated water hyacinth mats, control on a large scale has not been completely successful. Residues of herbicides affect the aquatic environment and may kill fish directly, further reducing the biodiversity. Like chemical control, physical removal of the weed is restricted to small strategic areas and the need for repeated removals and expensive specialized equipment makes this control strategy prohibitive on a larger scale. Biological control offers the only long term environmentally sound solution to this weed problem.

Water hyacinth was first recorded in Africa in 1879 in Egypt. However, it was only in the last decade or so that its population in some water bodies has increased to constitute a pest problem. In Sudan large floating mats of up to 113 km^2 used to accumulate seasonally behind the Jabel Aulia Dam on the River Nile at Khartoum necessitating large scale herbicidal applications (Beshir and Bennett, 1985). Between 1978 and 1981 two weevil species

Neochetina bruchi and *N. eichhorniae* and a moth *Sameodes albiguttalis* all from South America, were released into the Nile and adjacent wetlands around Juba and established. While civil strife has prevented extensive surveys at release sites, vast mats of the weed have not been seen coming down the Nile from Juba since 1982. Chemical control of this weed has ceased. Similar control of water hyacinth has been achieved in other countries including Australia (Wright, 1981), U.S.A. (Center *et al.*, 1989), India and recently Zimbabwe (G. Phiri, pers. comm).

Control of the prickly pear, Opuntia inermis *and* O. stricta *(Cactaceae) in Australia*

Prickly pear is a succulent perennial plant of New World origin which have been introduced into various countries for their edible fruits, drought resistance and emergency forage value for certain spineless forms, garden ornamental and for use as hedges. In several countries, several species have escaped from cultivation and have become serious agricultural pests (Dodd, 1940). *Opuntia* reproduces from seeds, as well as vegetatively by the rooting of pads that become separated from their parent plants. Seeds are disseminated primarily in the faeces of birds and other animals and partly by wind, surface water and animals. Its aggressive ability to spread naturally, up to 0.5 million ha/year (Dodd, 1940), has made it a range weed, crowding out native vegetation and rendering grazing land inaccessible to livestock. Prickly pear is composed of many species in the genus *Opuntia* and much publicised success in control was achieved only in some environments colonised by this weed. One of the most documented successes in the biological control of this weed occurred in Australia, into which it was introduced, probably from Mexico in 1839. By 1925, the weed had invaded an area of 24 million hectares necessitating its biological control by the introduction of specific natural enemies including the moth *Cactoblastis cactorum* (Lepidoptera: Pyralidae) from Argentina. By 1928 most of the large pure stands of prickly pear had been effectively reduced to scattered plants (Dodd, 1940). It is important to note that both the moth and weed have continued to persist in low numbers and there is no evidence to suggest any adverse effects on the native ecosystems (White, 1981). Another natural enemy, *Dactylopius* sp. (Homoptera: Dactylopidae) is also a minor control factor of the weed, killing fruit and young growth, especially in the dry season. Zimmermann and Malan (1980) also reported that *C. cactorum* and *D. opuntiae* played an important role in suppressing the population of *Opuntia* in South Africa.

Control of the cassava mealybug, Phenacoccus manihoti *(Homoptera: Pseudococcidae) in Africa*

Cassava, *Manihot esculenta* (Euphorbiaceae) was imported into Africa from South America in the 16th century and now serves as a major staple food crop to over 200 million people in Africa (Herren and Bennett, 1984). It is also an important source of carbohydrates to many people in South America and Asia. Being exotic in Africa and having high contents of poisonous cyanogenic glucoside, cassava has been relatively free of major pests until the first records of the cassava green mite and cassava mealybug in the early 1970s. The cassava mealybug originates from South America and now occurs in almost the entire cassava belt in Africa causing yield losses of up to 84% (Nwanze, 1982). Farmers in areas recently invaded by this pest frequently abandoned the cultivation of cassava leading to fears of famine in areas which are too dry for any other crop to grow. The mealybug's high reproductive rate through parthenogenesis, efficient dispersal rates by wind and man on infested cuttings and specificity to plants in the genus *Manihot*, ensured that this pest out-competed the native herbaceous fauna on cassava plants thereby reducing biological diversity. Weed and erosion problems after plant growth had been crippled sometimes led to total destruction of the crop. Chemical control of this pest is not practicable since cassava is a low income crop grown mainly by peasant subsistence farmers and the cassava leaves are often used as a vegetable. Foreign explorations in South America and subsequent biological studies by the International Institute of Biological Control (now CABI Bioscience), International Institute of Tropical Agriculture (IITA) and other organisations led to the identification of a number of candidate natural enemies and in particular the specific wasp parasitoid *Epidinocarsis (= Apoanagyrus) lopezi* (Encyrtidae). This was to lead to the largest tropical biological control programme ever undertaken which was co-ordinated by IITA and included over 25 African countries. *E. lopezi* was first released in Nigeria in 1981 (Herren and Lema, 1982) and later to a number of other African countries. In areas where it has established (an area of 1.5 million km^2 by 1988), *E. lopezi* is usually the only natural enemy found associated with the mealybug all the year round, with parasitism rates which alternate with the pest's density, indicating a regulatory effect by this parasitoid (Hammond *et al.*, 1987). Populations of the mealybug have fallen from the pre-release numbers of over 60 per tip to below 10 per tip after release (Hammond and Neuenschwander, 1990). The biological control of the cassava mealybug in Africa is one of the cases where relationships between indigenous insects and the introduced pest and its natural enemies were studied. No primary parasitoid was found on the pest in Africa, but *E. lopezi* was found to be attacked by 4 species of indigenous hyperparasitoids. Local coccinellid predators of the mealybug e.g. *Exochomus* sp., *Hyperaspis pumila, H. delicatula* etc. were not driven into extinction but continued to persist albeit at lower populations than that of *E. lopezi*, thus maintaining species diversity (Hammond and Neuenschwander, 1990). Biological control has thus suppressed the mealybug population and indirectly conserved native arthropod fauna associated with the previously threatened cassava habitat.

Challenges for biological management of invasive pests

Biological control is a method intended to solve alien invasive pest problems with a minimal disturbance of the ecosystems. It can be argued that biological control should be given a chance to naturally suppress an exploding population of an alien pest before opting for alternative, more disruptive control strategies. However, biological control still remains in many alien pest problems the method of last resort, and often follows failures with extensive intervention with pesticides or other measures. Lack of awareness of the nature and value of biological control is one reason for this. The requirement that biological control leaves continuing pest populations is also difficult for many pest managers. The objective of biological pest control is not the eradication of pests but the reduction and long term stabilisation of pest density at a sub-economic level. Further, impatience often works against the decision for biological control. Biological control programmes are long term, and a time lapse of a few or many years should be anticipated before results become apparent.

Biological control programmes are complex actions, and some aspects should be kept in mind during planning and implementation:

Biological control may not be evenly effective over all areas infested by the invasive species. Even after the release and establishment of a natural enemy in a given large area (e.g. the biological control of the cassava mealybug just cited), the pest may still be spreading at the periphery of the release area. There may be particular areas with high infestations (a consequence of bad farming practices e.g. pesticide applications), and people may fail to appreciate the continuing and underlying benefits of established control.

Biological control often raises conflicts of interest, particularly with respect to invasive weeds. Whenever a weed is thought to have one or more redeeming features there is usually hesitation about introducing free ranging weed-feeding organisms to control it. These conflicts of interest may be economic, ecological or aesthetic. An example of economic conflict is the biological control of gorse, *Ulex europaeus*, in New Zealand where beekeepers benefited by harvesting profits from the weed whereas ranchers lost out by expending resources on mechanical and chemical control (Hill, 1987). In Britain, the prolific bracken fern, *Pteridium aequilinum*, is fast displacing a diverse native flora, creating an ecological conflict as to whether it should be controlled or conserved, since it also serves as an important habitat of a rare bird and butterfly species (Pakeman and Marrs, 1991). Aesthetic conflicts of interest arise from attractiveness of a plant which may be an invasive weed. *Rhododendron ponticum* in Britain is an example of an invasive in national parks whose attractiveness and importance in gardens will probably prevent any efforts for classical biological control.

Government mechanisms for implementing classical biological control are often lacking, with responsibilities resting in different ministries. This is particularly true for invasive weeds of conservation, where weed control is the responsibility of the Ministry of Agriculture and conservation that of the Ministry of Environment. The biological control of *Salvinia molesta* and *Eichhornia crassipes* in Kenya involved at least 6 different government ministries, some with different perceptions of biological control. This often leads to organisational problems and concomitant delays in research activities.

Suggested solutions and recommendations

To increase awareness and understanding of biological control, scientists, extension officers and farmers should be educated on the mechanism of pest impact and biological control and what biological control can and can not achieve. Further, farmers and other land managers could be advised to abstain from engaging in practices that jeopardise successful biological control e.g. indiscriminate pesticide application after release of natural enemies.

Conflicts can be resolved by forming fora at which communication between concerned interest groups can be effected and improved. It is important that all such groups have an opportunity to participate in the discussions and contribute to the decision on whether or not to release the biological control agents and that all are informed on the nature, benefits and risks of biological control. Countries like Australia and New Zealand, with their long history of biological control, have well established mechanisms for conflict resolution and decision making on biological control introductions through public consultation and consensus building

(Hill, 1987; Harris, 1988). Other countries which are struggling with this issue, notably the USA, are investigating similar procedures.

To ensure efficient development and implementation of biological control, a national policy should be established and mechanisms for inter-ministerial co-operation. All concerned parties should be involved right from the start of the programme and specific responsibilities assigned to each. A committee should be formed to co-ordinate all research activities and policy makers should be actively involved e.g. in the release of natural enemies, and regularly kept informed about the progress of the programme. For instance, water hyacinth in Lake Victoria is a regional problem since the lake is shared between Kenya, Uganda and Tanzania (Ogutu-Ohwayo, 1998). The Lake Victoria Environment Management Programme was therefore formed to ensure inter-ministerial and international co-ordination, among other responsibilities.

Countries should be made more aware of the unexploited potential for utilising their own biodiversity for biological control, as an alternative to environmentally unfriendly methods of pest management. Exchange of natural enemies should be facilitated according to the Convention of Biological Diversity. Countries should also be encouraged to use integrated pest management (IPM) with more emphasis on the use of local natural enemies as a solution to environmental problems caused by exotic pests.

Quarantine programmes need to be strengthened and applied to slow the rate of pest introductions (e.g through food aid and importations in wooden containers which harbour pests) and reinforce efforts to change the cultural bias that favour the introductions of alien species. Further, quarantine efforts should be directed at all species, not just pests and deliberately introduced natural enemies. If there are fewer accidental introductions, there will be less need for programmes to correct problems caused by alien species.

References

Beshir, M.O. and Bennett, F.D. (1985) Biological control of water hyacinth on the White Nile, Sudan, in *Proceedings of the VI International Symposium on Biological Control of Weeds*, (ed. E.S. Delfosse), 19-25 Aug. 1984, Vancouver, Canada, pp. 491–496.

CABI, (1997) Weevils are a thorny issue. *Biocontrol News & Information,* **18**, 99–100.

Center, T.D., Cofrancesco, A.F. and Balciunas, J.K. (1989) Biological control of aquatic and wetland weeds in Southeastern United States, in *Proceedings of the VII International Symposium on Biological Control of Weeds*, (ed. E.S. Delfosse), 6–11 Mar. 1988, Rome, Italy, pp. 239–262.

Dodd, A.P. (1940) *The Biological Campaign Against Prickly Pear.* Commonwealth Prickly Pear Board, Brisbane, Australia.

Greathead, D.J. and Greathead, A.H. (1992) Biological control of insect pests by insect parasitoids and predators, the BIOCAT database. *Biocontrol News and Information*, **13**, 61–68.

Hammond, W.N.O., Neuenschwander, P. and Herren, H.R. (1987) Impact of the exotic parasitoid *Epidinocarsis lopezi* on cassava mealybug (*Phenacoccus manihoti*) populations. *Insect Science and its Applications*, **8**, 887–891.

Hammond, W.N.O. and Neuenschwander, P. (1990) Sustained biological control of the cassava mealybug *Phenacoccus manihoti* [Homoptera: Pseudococcidae] by *Epidinocarsis lopezi* [Hymenoptera: Encyrtidae] in Nigeria. *Entomophaga*, **35**, 515–526.

Harris, P. (1988) Environmental impact of weed-control insects. *Bioscience*, **38**, 542–548.

Harley, K.L.S. (1994) *Eichhornia crassipes* (Martins) Solms-Lambauch, in *Weed Management for Developing Countries*, (eds R. Labrada, J.C. Casseley and C. Parker), FAO Plant Protection and Production Paper 120, FAO Rome, pp. 123–134.

Herren, H.R and Lema, K.M. (1982) Cassava mealybug, first successful releases. *Biocontrol News and Information*, **3**, 185.

Herren, H.R. and Bennett, F.D. (1984) Cassava pests, their spread and control, in *Advancing Agricultural Production in Africa. Proceedings of CAB's First Scientific Conference*, (ed. D.L. Hawksworth), 12–18 Feb. Arusha, Tanzania, CAB Slough, UK, pp. 110–114.

Hill, D.S. (1983) *Agricultural Insect Pests of the Tropics and their Control*. Cambridge University Press, London/New York.

Hill, R. (1987) *The Biological Control of Gorse* (Ulex europaes L.) *in New Zealand: an Environmental Impact Assessment*. Department of Scientific and Industrial Research, Christchurch, New Zealand.

Howarth, F.G. (1991) Environmental impacts of classical biological control. *Annual Review of Entomology*, **36**, 485–509

Julien, M.H., Kerr, J.D. and Chan, R.R. (1984) Biological control of weeds: an evaluation. *Protection Ecology*, **7**, 3–25.

Louda, S.M., Kendall, D., Connor, J. and Simberloff, D. (1997) Ecological effects of an insect introduced for biological control of weeds. *Science*, **277**, 1088–1090

Nwanze, K.F. (1982) Relationships between cassava root yields and infestations by the mealybug, *Phenacoccus manihoti*. *Tropical Pest Management*, **28**, 27–32.

Ogutu-Ohwayo, R. (1998) Nile perch in Lake Victoria: the balance between benefits and negative impacts of aliens, in *Invasive Species and Biodiversity Management*, (eds O. T. Sandlund, P. J. Schei and Å. Viken), Kluwer Academic Publishers, Dordrecht, The Nederlands.

Pakeman, R.J. and Marrs, R.H. (1991) *An Environmental Balance Sheet for Bracken and its Control*. Institute of Terrestrial Ecology, Huntingdon, UK.

Sands, D.P.A. (1997) The "safety" of biological control agents: assessing their impact on beneficial and other non-target hosts. *Memoirs of the Museum of Victoria*, **56**, 611–615.

Voegele, J.M., Agounke, D. and Moore, D. (1991) Biological control of the fruit tree mealybug *Rastrococcus invadens* Williams in Togo: a preliminary socio-ecological and economic evaluation. *Tropical Pest Management*, **37**, 382–397.

Waage, J.K. (1991) Biodiversity as a resource for biological control, in *The Biodiversity of Macroorganisms and Invertebrates: Its Role in Sustainable Agriculture*, (ed. D.L. Hawksworth), CAB International, Wallingford, UK, pp. 149–163.

White, G.G. (1981) Current status of prickly pear control by *Cactoblastis cactorum* in Queensland, in *Proceedings of the V International Symposium on Biological Control of Weeds*, (ed. E.S. Delfosse), 22–29 Jul 1980, Brisbane, Australia, pp. 609–616.

Wright, A.D. (1981) Biological control of water hyacinth in Australia, in *Proceedings of the V International Symposium on Biological Control of Weeds*, (ed. E.S. Delfosse), 22–29 Jul 1980, Brisbane, Australia, pp. 529–535.

Zimmermann, H.G. and Malan, D.E. (1980) The role of imported natural enemies in suppressing re-growths of prickly pear, *Opuntia ficus-indica*, in South Africa, in *Proceedings of the V International Symposium on Biological Control of Weeds*, (ed. E.S. Delfosse), 22–29 July 1980, Brisbane, Australia, pp. 375–378.

Part 5

Country case studies

22 Invasive species in Mauritius: examining the past and charting the future

WENDY STRAHM
IUCN – The World Conservation Union, Gland, Switzerland

Abstract

Introduced plants and animals have had a tremendous impact on the native vegetation of Mauritius. Approximately 730 species of vascular plants are now naturalised on the island, compared to about 685 native species. Of the introduced species, more than 50 can be considered highly invasive. While there are now more introduced and naturalised plant species growing on Mauritius than native species, in terms of biomass, the amount of introduced species is greater by many orders of magnitude. The situation is similar on the two other Mascarene Islands; Rodrigues and Réunion. Although the islands are related, they have different histories, problems, and floras, and a comparison is made of species which have become invasive on all three islands. Attempts have been made to control selected species, and case studies of specific nature reserves where alien species have been excluded and/or removed are discussed. The Convention on Biological Diversity should play a significant role as it directs signatories to "prevent the introduction of, control or eradicate those alien species which threaten ecosystems, habitats or species". Alien species are the principal cause of biodiversity loss in the Mascarene Islands today.

Introduction

The Mascarene Islands, situated in the Western Indian Ocean some 900 km west of Madagascar, include the islands of Mauritius, Réunion and Rodrigues (Figure 22.1). Mauritius covers 1 865 km² and attains 828 m in altitude. Réunion (a dependency of France) lies 150 km southwest of Mauritius and is slightly larger (2 512 km²) and much higher (3 069 m). Rodrigues, politically part of the Republic of Mauritius, is located 574 km east of Mauritius, and is much smaller and lower, covering 109 km² and reaching 393 m in height. All the islands are of volcanic origin and have never been united or connected to any other land mass. Their climate, influenced annually by cyclones, is tropical to subtropical, with rainfall ranging from 1 000 to 8 000 mm/year.

The Mascarene Islands are heavily populated. Mauritius, with over a million people, is one of the most densely populated places in the world.

O. T. Sandlund et al. (eds.), Invasive Species and Biodiversity Management, 325–347.

The Mascarenes were uninhabited until the late 16th and early 17th centuries. The rapid population growth and subsequent clearing of land for agriculture has been the major cause of forest loss.

Although largely altered by introduced vegetation, good tracts of native forest still exist on all three islands, with the best examples found on Réunion. Forests on Mauritius are more heavily degraded although this island, probably due to its greater age, has a richer flora. Mauritius is also surrounded by several offshore islets which are very important as many of the non-native species which have caused the decline of the mainland flora were not introduced to the islets (e.g., Round Island and Ile aux Aigrettes; see below).

The unique biodiversity of the Mascarene Islands is seriously threatened. The colonisation by people has decimated the native plant populations and caused much extinction, and the future of the remaining native flora is uncertain due to the uncontrollable effects of introduced plants and animals. This paper reviews the principle species which are invading native forests, presents a few case histories on management measures to attempt to control the invasions, and provides information on which species must be prevented from being introduced to the other Mascarene islands if they are not yet present.

Introduced animals

A major cause of alien plant invasions in the Mascarenes can be attributed to introduced animals. In addition to destroying native species directly, they disturb native habitats and disseminate alien seeds. Each island has a different history of animal introductions. It is no coincidence that Réunion has had fewer animals introduced than Mauritius, and a better conserved flora.

On Mauritius a number of animals have been introduced which damage native vegetation through browsing, soil disturbance, and seed predation (Table 22.1). Many of them also serve as ready dispersal agents for exotic plants. Clearly all *in situ* conservation efforts to conserve the flora on Mauritius must exclude as many of these animals as possible, either by eradication or exclosures.

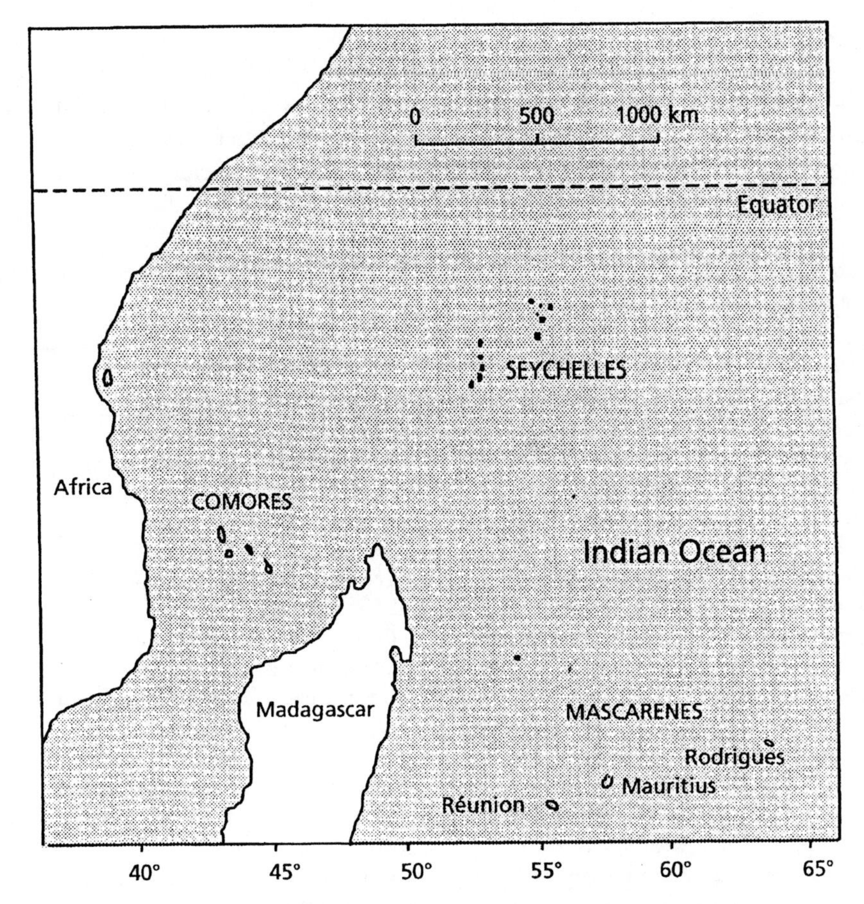

Figure 22.1 The position of the Mascarene Islands in the Western Indian Ocean.

The other islands also suffer greatly from introduced animals. Although deer were eradicated from Réunion by 1793 after the first introduction in 1761 (Cheke, 1987), they have been reintroduced during the past two decades and are damaging native vegetation in the best remaining forests. On Rodrigues deer were introduced in 1862 and had increased to 1 500–2 000 individuals by 1892 (Kennedy, 1893; Cheke, 1987). Subsequently, the population decreased so that there were very few left by 1937 and the last one was shot in 1956 (Cheke, 1987). They probably had a considerable effect on the vegetation, and unfortunately there have been proposals to reintroduce them to this island.

Feral pigs are common in upland wet forests and present although rare in dry areas in Mauritius. They cause extensive damage and are attracted to recently turned soil. Freshly sown seeds often get uprooted by pigs. Pigs also serve as active disseminators of exotic plant species. In a single pig dropping one can find hundreds of germinating guava (*Psidium cattleianum*) seedlings. Feral pigs are not yet present on Réunion or Rodrigues, which is rather extraordinary, and vigilance must be kept to ensure that pigs do not escape into the forests (especially in Réunion). Tenrecs, introduced from Madagascar and present on Mauritius and Réunion, cause similar, although less extensive, damage than pigs. They damage gardens and nurseries by digging up seedlings while searching for invertebrates to eat.

Table 22.1 Introduced animals with documented or assumed detrimental effects on the native flora of the Mascarene Islands

Common name	Latin name	Effect on native vegetation
Deer	*Cervus timorensis*	Browse
Pig	*Sus scrofa*	Soil upheaval, exotic seed dispersal
Monkey	*Macaca fascicularis*	Damage, exotic seed dispersal
Goat	*Capra hircus*	Browse
Black rat	*Rattus rattus*	Seed and seedling predation
Norway rat	*Rattus norvegicus*	Seed and seedling predation
Tenrec	*Tenrec ecaudatus*	Soil upheaval
Mongoose	*Herpestes auropunctatus*	Destroy native avifauna which in turn is needed by native plants
Cat	*Felis catus*	Destroy native avifauna which in turn is needed by native plants
Hare	*Lepus nigricollis*	Browse
Rabbit	*Oryctolagus cuniculus*	Browse (eradicated from Round Island in 1986)
Mouse	*Mus musculus*	Seed predation
Indian house shrew	*Suncus murinus*	Competes with native reptiles which are needed by native plants
Indian mynah	*Acridotheres tristis*	Exotic seed disperser
Red-whiskered bulbul	*Pycnonotus jocosus*	Exotic seed disperser
Agama lizard	*Calotes versicolor*	Effect on native vegetation unrecorded
Wolf snake	*Lycodon aulicum*	Detrimental to native avifauna
Giant African land snails	*Achatina panthera* & *A. fulica*	Browse
Numerous other invertebrates		Seed and plant predation, competition with pollinators and dispersal agents

Monkeys cause extensive damage in both low and upland forests. They selectively choose certain species and pull off branches, flowers and fruits and will sometimes totally defoliate or break off every fruit on a tree. Due to the very slow growing and infrequent flowering strategies of many native tree species, this may negatively influence reproduction. Monkeys have also been observed tearing epiphytic orchids and ferns off trees, and they serve as dispersal agents for exotic species such as guava. Up to now, monkeys have not escaped in Réunion or Rodrigues, but some baby monkeys are each year smuggled in from Mauritius as pets. When these animals become unmanageable, they are either released or given to the zoo, which in 1986 had at least 50 monkeys. It seems to be only a matter of time before these animals become established in the forests unless stricter measures are taken to stop introductions, and to remove monkeys which are already present on Réunion (including those at the zoo).

The most unfortunate recent animal introduction to Réunion has been the red-whiskered bulbul. This species was first reported to have been released on Réunion in 1972 (Cheke, 1987). It is the most common bird on Mauritius, and on Réunion it is now very common around the forest of St. Philippe and is spreading. The species has not yet been introduced to Rodrigues. Bulbuls, besides competing with the native birds, are extremely effective vectors for the dissemination of noxious weeds.

Rat eradication will never be possible on the mainland and many rat-damaged fruits and trees have been observed both on Mauritius (Strahm, 1994) and elsewhere (Campbell, 1978). Hares are common in cultivated areas in Mauritius and are also found in forests, probably having an undesirable effect. However they do not occur on Réunion and Rodrigues. Finally *Achatina* snails are found everywhere despite control efforts by the introduction of predaceous snails *Euglandina rosea* and *Gonaxis quadrilateralis* (Greathead, 1971). The predaceous snails have helped exterminate the native land snails instead. The snails eat young, tender seedlings and shoots as well as cuttings in nurseries. In addition to these exotic animals, the damage caused by introduced insects and various diseases to the native flora must be considerable.

The destruction of the flora of Rodrigues may be largely attributed to the introduction of livestock. Goats, sheep, cattle and pigs were introduced to the island in the 1700's and all, except perhaps sheep, seem to have been feral for some time, although not today. Domestic goats are still destroying the tiny relicts of native vegetation that exist on the island, although recently large areas have been fenced to try to exclude them.

Introduced animals directly and indirectly damage native habitats. Introductions to islands where they have not yet arrived should therefore be avoided. Established populations must be controlled, and eradicated where possible. In addition, programmes to eradicate species only recently introduced need to be undertaken.

Introduced plants

Introduced plants are probably an even greater threat to the native flora than the animals, although both act synergistically to the detriment of the native vegetation. As soon as a gap opens up in the forest, the space is usually filled with fast-growing, aggressive exotic species.

Today there are more introduced than native plant species growing in the Mascarene Islands, although their invasiveness varies greatly. Using the "*Flore des Mascareignes*" (Bosser *et al.*, 1976), and criteria to determine whether a species is native or introduced (Webb, 1985), a provisional list of all native and introduced species has been compiled (Strahm, 1994; see Table 22.2). The introduced species which grow in native habitats and prevent native regeneration are ranked as "very invasive" or "invasive", depending on study plot criteria. The remaining introduced species are designated as "naturalised" if the species is regenerating and growing by itself, "sub-spontaneous" if the species is naturalised but only found just outside of where it has been planted, or "cultivated" if the species do not seem to have escaped. The listed number of cultivated plants is low as only the most commonly cultivated species have been included. Finally there are some species for which data from Réunion was not available. These are listed as "unknown", but were counted as "naturalised" in the summary (i.e. "Total naturalised" in Table 22.2). Although this exercise is still being revised due to recent taxonomic revisions, broad generalisations can be made from the existing data.

Table 22.2 Preliminary table of introduced vascular plants to the Mascarenes in total and to the individual islands.

Locality	Very invasive	Invasive	Naturalised	Sub-spontaneous	Old collections	Cultivated	Unknown	Total listed	Total naturalised
Mauritius	18	29	683	51	38	270	1	1090	731
Réunion	15	17	592	58	24	226	70	1002	694
Rodrigues	7	6	290	19	4	140	2	468	305
Mascarenes	24	45	771	67	40	289	3	1239	843

The naturalised flora on both Mauritius and Réunion is almost the same (731 and 694 species respectively) although 98 species naturalised on Réunion are not naturalised on Mauritius, and 132 species naturalised on Mauritius are not naturalised on Réunion. Despite the islands' close proximity, due to their very different histories (e.g., one island was a French and the other a British colony), different introductions have been made. However, 72% of the total Mascarene naturalised flora is still common to the two islands (coincidentally the same figure as the percentage of the native flora endemic to the Mascarene Islands). On the other hand every introduced species on Rodrigues is also found on Mauritius, probably since the main route to Rodrigues has been via Mauritius.

Since Rodrigues is much smaller and more isolated, fewer species have been introduced and become naturalised. However, the island has a higher percentage of introduced species of its total flora (305 naturalised compared to 134 native species), as well as more introduced species per land area. The introduced species also dominate the flora in terms of number of individuals, and the native flora of Rodrigues has almost entirely disappeared (Strahm, 1989a).

Comparison with the native flora and fauna

Compared with what is believed to be "native", there are now more introduced plant species (731) on Mauritius than native (685). In addition, of the native species, at least 80 are extinct, and a further 478 species are threatened with extinction (Table 22.3). The situation is similar for the other two Mascarene islands. Therefore the proportion of introduced to native species is set to increase unless the extinction processes can be reversed.

Table 22.3 Conservation status of the Mascarene flora, based on the old IUCN conservation categories (IUCN, 1980). Ex = Extinct; E = Endangered; V = Vulnerable; R = Rare, Nt = Not threatened.

Island	Ex	E	V	R	Nt	Total threatened	Total no. of native spp.
Mauritius	80	185	104	189	127	478	685
Réunion	30	78	88	78	272	244	546
Rodrigues	18	36	12	15	53	63	134

The extinction and threat record is as least as bad for the native vertebrates. On Mauritius at one time there were 23 taxa of endemic landbirds, 12 reptiles including giant land tortoises, skinks, geckos and snakes, and two fruit bat species (Cheke, 1987). Today only 9 endemic landbirds (of which 8 species are threatened), four geckos, one skink, and one fruit bat species remain on the mainland. The situation is even worse for Rodrigues, with only 2 endemic landbirds and a fruit bat persisting (all threatened), from a fauna which included 12 endemic landbirds and 4 species of reptiles (Cheke, 1987). The loss of native pollen and seed dispersers must also have had a significant impact on the native flora, and it is likely that the number of plant extinctions is much higher than recorded.

The worst weeds

A list of the worst weeds on the Mascarene Islands (mainly Mauritius and Rodrigues) is given in Table 22.4. Six of the 18 species listed are very invasive on all three islands. Five species are particularly aggressive in the uplands of Mauritius, whereas seven species invade mainly lowland forests. By 1900 the tree species tecoma was being used as a timber tree (Brouard, 1963) and is now forming monotypic, self-seeding plantations in the lowlands.

On Mauritius wet forest is more highly invaded than dry, although both are degrading into exotic secondary thicket. Réunion is much better off than Mauritius, and the principal pest on Réunion is the bramble, or "vigne maronne", which seems to have invaded every type of native forest on Réunion apart from the high altitude ericoid vegetation. Landslips on steep mountain slopes are very common, and when this occurs, bramble forms monotypic stands. Gradually most of the still good forest growing on the steep slopes will probably give way to this species. “Chinese” guava and jamrosa are also invasive in the hot humid forest (Table 22.4), and the longose forms thick undergrowth in the forest which prevents native regeneration. Aloès is also very invasive in dry areas, as well as liane cerf and an introduced woody legume (*L. leucocephala).* Other species also found on Mauritius, such as bois d'oiseau, vieille fille and poivre marron, are equally weedy on Réunion. Macdonald *et al.* (1991) give a more complete list of weedy species on Réunion. Unfortunately privet, not even mentioned in Cadet (1980), is today very common in the forest above Cilaos as well as in that above Hell-Bourg in the Cirque de Salazie, and seems to be spreading rapidly.

The little forest which remains on Rodrigues has been completely invaded by “Chinese” guava, jamrosa, bois d'oiseau, aloès, vieille fille, l'acacia and ravenal, similar to low altitude forests on both Mauritius and Réunion. A few plants of privet and poivre marron have also been found on Rodrigues but it is hoped that these have been eradicated. Tecoma has been introduced and is already invading one of the last relicts of native vegetation found at Mt. Cimitière.

Noteworthy is that several species which are very invasive on Mauritius and Rodrigues seem not to have been introduced to Réunion or Rodrigues (Tables 22.4 and 22.5). All of these have disastrous potential and should not be introduced to the islands where they do not occur. There are also a few cases where certain species have not yet become naturalised or totally invasive (i.e. privet and poivre marron on Rodrigues, *Fuchsia* on Mauritius), and it would be wise to eliminate these species from the island while it is still possible. In addition, although privet is widespread on Réunion, it has not yet reached all native forests there, and control of this species is still possible even if total eradication may no longer be feasible.

Several species not listed as very invasive for Mauritius and Rodrigues turn out to be so on Réunion. This includes both *Hedychium* species, both *Fuchsia* species, and lisandra (Table 22.5; Macdonald *et al.*, 1991). This may be due in part to the absence until recently of the very invasive privet (Table 22.4). However this species is rapidly expanding and in the next 50 years may well be the worst weed on Réunion as it is on Mauritius today.

Table 22.4 The 18 most invasive species on the Mascarene Islands. On Mauritius some species are aggressive invasives mainly in the upland (Uf) or lowland (Lf) forests. * indicates that the species is not yet introduced to Réunion.

Species	Common name	Habitat	Comments
Agavaceae			
Furcraea foetida	aloès	Lf	Very invasive on all three islands in dry areas
Anacardiaceae			
Schinus terebinthifolius	poivre marron	Lf	Very invasive on Mauritius and Réunion, only one record from Rodrigues
Bignoniaceae			
**Tabebuia pallida*	tecoma	Lf	Invasive on Mauritius and Rodrigues, unknown if introduced to Réunion
Euphorbiaceae			
**Homalanthus populifolius*		Uf	Very invasive on Mauritius, should not be introduced to Réunion or Rodrigues
Flacourtiaceae			
Flacourtia indica	prune malgache	Lf	Very invasive on Mauritius, invasive on Réunion and naturalised on Rodrigues
Lauraceae			
Litsea glutinosa	bois d'oiseau		Very invasive on all three islands
Litsea monopetala	yatis	Uf	Very invasive on Mauritius, invasive on Réunion and not on Rodrigues

Table 22.3 continued

Species	Common name	Habitat	Comments
Malphighiaceae			
Hiptage benghalensis	liane cerf	Lf	Very invasive on Mauritius and Réunion, not on Rodrigues
Melastomataceae			
**Ossaea marginata*		Uf	Very invasive on Mauritius, not on Réunion or Rodrigues
Mimosoideae			
Leucaena leucocephala	l'acacie	Lf	Very invasive on all three islands
Musaceae			
Ravenala madagascariensis	ravenal	Uf	Very invasive on Mauritius, invasive on Réunion and Rodrigues
Myrsinaceae			
Ardisia crenata	l'arbre à noël	Uf	Very invasive on Mauritius and Réunion, not on Rodrigues
Myrtaceae			
Psidium cattleianum	"Chinese" guava	Uf	Very invasive on all three islands
Syzygium jambos	jamrosa	Lf	Very invasive on all three islands
Oleaceae			
Ligustrum robustum var. *walkeri*	privet	Uf	Very invasive on Mauritius and Réunion, recently found on Rodrigues
Rosaceae			
Rubus alceifolius	bramble, vigne maronne	Uf	Very invasive on Mauritius and Réunion, not on Rodrigues
Thymelaeaceae			
**Wikstroemia indica*	l'herbe tourterelle	Uf	Very invasive on Mauritius and Rodrigues, not on Réunion
Verbenaceae			
Lantana camara	vieille fille	Lf	Very invasive on all three islands

In addition to the list of 18 very invasive species, at least 30 species may be classified as invasive (Strahm, 1994; Table 22.5). Therefore, even if we were able to remove the 18 most invasive species, there are others which may be able to become very invasive in the absence of this competition. It can also be expected that some of the remaining 680 "naturalised" species could become invasive if more serious competitors were removed.

Speed of colonisation

For planning of future management measures, it is interesting to see how quickly new species have become naturalised, and whether newly introduced species are still able to become naturalised today. Data on when introduced species became naturalised, and then developed into invasives, are sparse, although Baker (1877) and Balfour (1877) did make an attempt to note which species were naturalised and which were merely in cultivation In addition Johnston (1895) and Vaughan (1937) listed species not recorded by Baker, so some idea of the progression of introduced species can be made.

For Mauritius, Baker (1877) or earlier collectors recorded some 410 species out of the 730 species listed as naturalised on the island today. 38 of the 410 species may no longer be on the island. By 1937 Vaughan (1937 and herbarium collections) had recorded at least 170 more naturalised species. This means that some 150 species have been introduced and naturalised on the island in the fifty years since the late thirties. While some may have been overlooked earlier, it is interesting that new species are still being introduced and becoming naturalised, rapidly increasing the number of species found on the island.

On Rodrigues, some 300 species are now naturalised (Strahm, 1989a; 1994). Balfour (1877) listed approximately 135 species as naturalised on Rodrigues in 1874, and although some more were probably present, we may note that twice as many species are now naturalised some hundred years later. This is similar to the development on Mauritius. In addition invasive species which are locally very common such as vieille fille (first recorded in 1938), l'herbe tourtelle, tecoma, *Ageratina riparia* and *Hippobroma longiflora* (see Tables 22.4 and 22.5) were not recorded by Balfour (1877). Moreover, he did not record ubiquitous herbaceous plants such as *Asystasia gangetica*, *Elephantopus mollis*, *Mimosa pudica*, *Oxalis debilis*, *Passiflora suberosa* and *Vernonia cinerea* which are so much a part of the Rodrigues flora today. Other species which are potentially very invasive but which are not that common, or just localised on Rodrigues are prune malgache, *Flacourtia jangomas*, *Gomphocarpus* spp. and *Asclepias curassivica.* Poivre marron and privet may have been eradicated from the island. Very few species recorded as naturalised by Balfour (1877) have disappeared from the island today.

The speed with which weeds move is exemplified by *Tridax procumbens*, said to have been introduced from Mexico in 1840 to Pamplemousses Botanic Gardens where it subsequently escaped. By 1859 it was a great pest in sugar cane plantations (Clark, 1859). Balfour (1877) had already recorded this species on Rodrigues, and it is a common weed in coastal regions today. Clark (1859) noted that it seemed to "impoverish" the soil where it grew, and this species could have had a very detrimental effect on the coastal lowlands of Rodrigues, which are now mostly invaded by vieille fille.

Several species listed above have not been listed in Tables 22.4 and 22.5 as invasive, as they do not grow substantially in native plant associations today. However they may well have been "invasive" in the past.

Table 22.5 Other invasive and naturalised species in the Mascarenes. Mau. = Mauritius; Rod. = Rodrigues; Réu. = Réunion. Preferred habitat is indicated as upland (Uf) and lowland (Lf) forest. From Strahm (1994).

Family/species	Common name	Habitat	Comments
Acanthaceae			
Justicia gendarussa	nitchoulli	Uf	Invasive on Mau. and Rod., naturalised on Réu.
Strobilanthes hamiltonianus	strobilanthes	Uf	Invasive on Mau., naturalised on Réu.
Araceae			
Zantedeschia aethiopica	arum		Invasive on Réu., cultivated on Mau., not on Rod.
Asclepiadaceae			
Cynanchum calliata			Invasive on Mau.
Balsaminaceae			
Impatiens flaccida	balsamine	Uf	Invasive on Mau., naturalised on Réu. and Rod.
Begoniaceae			
Begonia cucullata	bégonia	Uf	Invasive on Réu., and a Begonia sp. is potentially invasive on Mau.
Boraginaceae			
Cordia curassavica		Lf	Invasive but controlled on Mau., not present on Réu. or Rod.
Campanulaceae			
Hippobroma longiflora	lastron blanc	Uf	Invasive on Mau. and Rod., naturalised on Réu.
Caprifoliaceae			
Lonicera confusa		Uf	Invasive on Mau., naturalised on Réu.
Casuarinaceae			
Casuarina equisetifolia	filao		Invasive on Réu., naturalised on Mau. and Rod.
Compositae			
Ageratina riparia	orthochifon		Invasive on Mau., naturalised on Réu. and Rod.
Bidens pilosa	herbe villebague		Naturalised on all three islands and sometimes invasive.
Chromolaena odorata			Invasive on Mau.
Erigeron karvinsianus	marguerite marron		Invasive on Mau. and Réu.
Mikania micrantha	liane margoze		Invasive on Mau., naturalised on Réu.

Table 22.5 continued

Family/species	Common name	Habitat	Comments
Crassulaceae			
Kalanchoe pinnata	soudefafe		Invasive on Mau., naturalised on Réu. and Rod.
Euphorbiaceae			
Phyllanthus (?urinaria L.)			Invasive in the Macabé plot on Mau., naturalised on Réu. and Rod.
Gramineae			
Arthraxon quartinianus		Uf	Invasive on Mau.
Cenchrus echinatus	herbe à cateaux	Lf	Invasive on Mau., naturalised on Réu. and Rod.
Paspalum conjugatum	herbe créole	Uf	Invasive on Mau., naturalised on Réu. and Rod.
Paspalum paniculatum	herbe duvet	Uf	Invasive on Mau., naturalised on Réu. and Rod.
Phalaris arundinacea	herbe mackaye	Uf	Invasive on Mau.
Lauraceae			
Cinnamomum verum	cannelier	Uf	Invasive in parts of Mau., cultivated on Réu.
Melastomataceae			
Clidemia hirta		Uf	Becoming invasive on Réu., naturalised in Mau.
Tibouchina viminea	lisandra	Uf	Invasive on Réu., naturalised in Mau.
Mimosoideae			
Acacia farnesiana	cassie	Lf	Invasive on Mau., naturalised on Réu. and Rod.
Acacia nilotica	cassie à piquants blancs	Lf	Invasive on Rod., naturalised on Mau.
Desmanthus virgatus	petit acacia	Lf	Invasive on Mau., naturalised on Réu. and Rod.
Myrsinaceae			
Ardisia elliptica		Uf	Invasive on Mau., naturalised on Réu. (?)
Onagraceae			
Fuchsia boliviana	fuchsia à grandes fleurs	Uf	Invasive on Réu., not on Mau. or Rod.
Fuchsia magellanica		Uf	Invasive on Réu. not on Mau. or Rod.
Palmae			
Livistona chinensis	latanier de Chine	Uf	Invasive on Mau., naturalised on Réu. and cultivated on Rod.
Papilionoideae			
Ulex europaeus	ajonc épineux	Uf	Invasive on Réu., naturalised on Mau., not on Rod.

Table 22.5 continued

Family/species	Common name	Habitat	Comments
Pontederiaceae			
Eichhornia crassipes	jacinthe d'eau		Invasive on Mau. and Réu., naturalised on Rod.
Rosaceae			
Rubus rosifolius	framboisier	Uf	Invasive on Mau., naturalised on Réu. and Rod.
Rubiaceae			
Paederia foetida	liane lingue		Invasive on Mau., naturalised on Réu. and not on Rod.
Rutaceae			
Murraya paniculata	bois buis	Lf	Invasive on Mau., naturalised on Réu. and cultivated on Rod.
Sapotaceae			
Mimusops coriacea	pomme jacot	Lf	Invasive on Rod., naturalised on Mau. and Réu.
Solanaceae			
Solanum mauritianum			Invasive on Réu., naturalised on Mau.
Urticaceae			
Boehmeria macrophylla	grande ortie; ortie rouge	Uf	Invasive on Réu., naturalised on Mau.
B. penduliflora	bois chapelet	Uf	Invasive on Réu., naturalised on Mau.
Verbenaceae			
Clerodendron serratum			Invasive on Mau., naturalised on Réu.
Zingiberaceae			
Hedychium flavescens	longouze à fleurs jaunes, gingembre jaune	Uf	Invasive on Réu., cultivated on Mau., not on Rod.
Hedychium gardnerianum	longose	Uf	Invasive on Réu., cultivated on Mau., not on Rod.

Conservation measures

Steps have been taken to halt the decline in native vegetation by the establishment of 16 nature reserves on Mauritius covering 2.5% of the island (Vaughan, 1968; Anonymous, 1983), including the Black River National Park established in 1994 on Mauritius, a small (35 ha) nature reserve on Réunion, and two small nature reserves on Rodrigues. However, due to the invasion by alien species, establishing reserves is not sufficient. Smaller, intensively managed areas must be maintained in representative plant communities to conserve both habitat and plant genetic resources. This should be supplemented with artificially propagating species which for various reasons have ceased reproducing in the wild.

Pilot reserves have been set up in various habitats in Mauritius (Figure 22.2). To date, eight small plots on Mauritius ranging from 1.5-15 ha (Macabé, Brise Fer, Perrier, Pétrin, Florin, Bel Ombre, Mondrain and Mt. Cocotte) and two islets (Ile aux Aigrettes and Round Island) are being managed with a view to control alien plant species, and this work is being expanded. A few examples of the success incurred by excluding alien animals and removing alien plants are given below.

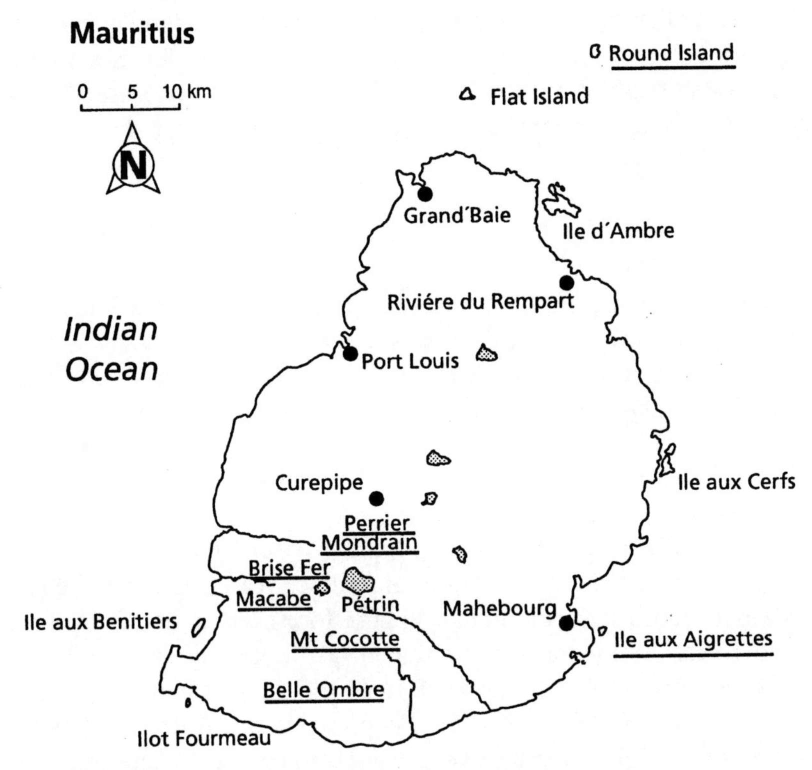

Figure 22.2 Map of Mauritius. Pilot nature reserves cited in the text are underlined.

Perrier

Perrier nature reserve is the model upon which all the other small fenced and weeded reserves on Mauritius are based. Covering only 1.44 ha., Perrier is the classic example of *Sideroxylon* thicket as defined by Vaughan and

Wiehe (1937). Located at 550 m altitude and receiving about 3 000 mm rainfall per year (Padya, 1984), about 140 species of flowering plants and some 30 species of pteridophytes have been recorded from this small reserve (Vaughan, 1980; Strahm, 1994). The vegetation is very dense: Vaughan and Wiehe (1937) counted approximately 34 600 individuals/ha (over 50 cm tall), and 32 550 individuals/ha were estimated to grow in the study plot in 1988 (Strahm, 1994). In this reserve trees tend not to grow over 8 m in height.

Although officially created in 1944, the reserve was not fenced and weeded until 1969 (Vaughan, 1969; Owadally, 1971). Since then two full-time labourers have been employed by the Forestry Service to manually uproot alien species, and this management has permitted the survival and regeneration of a substantial number of species, some unique to this reserve, which would otherwise have disappeared. The reserve is largely surrounded by agricultural or degraded land, and the possibilities for expansion are limited without major restoration work. However, Perrier serves as an example of what can be accomplished by fencing and weeding.

Mondrain

The Mondrain nature reserve, situated on the crest of the Vacoas ridge overlooking the Magenta valley, was established in 1979 by the Royal Society of Arts and Sciences, Mauritius. Covering five hectares of hillside at an altitude of 500-530 m, the area is owned by the Medine Sugar Estate which has leased the reserve to the Royal Society for management and conservation of its unique flora. Some of the species are today almost entirely restricted to this small reserve.

The vegetation is low ridge *Eugenia*/*Sideroxylon* thicket growing under a fairly dry regime of about 1 700 mm rainfall per year. Species in the reserve include both those from the humid upland forest and a few from the drier lowland forest. Many species are stunted at the Mondrain reserve. While some species may elsewhere grow up to 30 m tall, the canopy in the reserve reaches less than 6 m in height. This is probably due to the shallow soil (a thin layer of rich, black humus over laterite), and to wind exposure.

The Mondrain reserve is of great interest because it contains species that have become extremely rare or extinct elsewhere on Mauritius. It also forms one of the best preserved examples of low ridge mid-altitude vegetation. Elsewhere along the Vacoas ridge the native forest has almost entirely disappeared, principally due to degradation by alien species and woodcutting for fuelwood as well as opening areas for hunting grounds.

In only 5 ha inside the Mondrain reserve, 121 species of native vascular plants and over 100 species of introduced plants have been recorded as of 1992 (Strahm, 1994). The reserve has been effectively and consistently weeded since 1985 by uprooting alien species rather than by cutting them down as was done in 1979. This has resulted in a major change in the vegetation, with native species again dominant and regenerating in the reserve. It is apparent that the seed bank of introduced species has been steadily diminishing since weeding, and the number of person-hours required to weed the entire reserve has decreased with each year. Since the reserve is surrounded by mostly alien scrub, there is still constant immigration of alien species into the reserve, and if regular weeding is not continued, the reserve would rapidly revert back to its degraded state. Plant density is also very high in the reserve, with an estimated 28 000 individuals over 50 cm tall found per hectare (Strahm, 1989b; 1994).

Regeneration studies in weeded and non-weeded areas have been undertaken, and a pattern where new seedlings were established during the wet summer months, but then died off during the dry season, became apparent. However, in the weeded plots both the number and survival of native seedlings increased dramatically, and after two years the overall increase in native seedlings was 60% greater in the weeded plots than in the controls (Strahm, 1989b; 1994). The regeneration study plots were set up under varying degrees of shade. Under native canopy, weed colonisation was very slow, although native seedlings also grew more slowly in the shade than in plots that were partially shaded. However, in open, sunny areas colonisation by weeds was great and native regeneration very poor. It appears that most native seedlings need a fair amount of shade in order to survive the dry period. They can colonise small gaps, whereas large gaps favour the alien species.

Round Island

Round Island lies NE of Mauritius separated from the main island by some 22 km of sea less than 80 m deep. The island covers 151 ha in area, attains 281 m in height, and has very steep, eroded slopes, preventing easy access. The last remaining example of "palm savannah" which once covered the northern coastal plain of Mauritius (Vaughan and Wiehe, 1937) is found on the island (apart from two very degraded examples present on Gunner's Quoin and Flat Island).

Round Island has never been inhabited by people and consequently did not suffer the same massive destruction that occurred on the mainland. The major factor which has allowed relict plant and animal populations on the island to survive is probably that rats were never introduced. However, both rabbits and goats were introduced. These species have caused massive erosion as well as plant extinctions.

The goats were shot out from the island between 1976 and 1982 (Bullock *et al.*, 1983), and the rabbits were eradicated in 1986 through an intensive poisoning campaign (Merton, 1987; Merton *et al.*, 1989). Released from grazing and browsing pressure, the vegetation has changed remarkably, although introduced species seem to have profited more than the few native species which still exist on the island. The number of alien species recorded on the island has been increasing rapidly, possibly due to the more frequent management trips made by biologists to the island (despite great precautions taken not to accidentally introduce seeds). In 1994 the number of alien species of vascular plants recorded on the island surpassed the number of native species. However, since all recent introductions have been removed, the development of these introduced species will not be studied.

One potentially invasive weed on the island is *Desmanthus virgatus*, which was first recorded in 1982 as very rare, but which rapidly increased following rabbit eradication. Control began at the same time as rabbit eradication, with annual weeding from 1986 to 1989, followed by more frequent weeding (see Strahm, 1987–1993). Although this species seemed to be under control at the end of 1993, it is not certain that it will remain controlled (with a view to eventual eradication), unless regular weeding is continued.

Round Island is another example showing that even when the factors which have caused the change in the vegetation have been removed, the only way original vegetation can be restored is through active management. This includes controlling alien species as well as planting native species to provide cover, which is an essential element in controlling alien species.

None of the "very invasive" weeds on Mauritius have been introduced to Round Island. Great care must be taken that the most aggressive species (Table 22.4) are never introduced to the island.

Ile aux Aigrettes

Ile aux Aigrettes, located within the lagoon of Mauritius, is a calcarenitic, low islet attaining 12 m altitude, covering 25 ha, and situated 900 m from its nearest point to Mauritius. The islet is of exceptional botanical importance since it harbours a relict of the eastern coastal dry forest which is now virtually extinct on the mainland. Several endemic species are now almost completely confined to this small islet which was declared a nature reserve in 1965.

The islet receives about 1 400 mm rainfall per year (Padya, 1984). The forest is a typical dry forest, not very densely packed, with a surprisingly low number of native species per area. This is probably due to a great number of extinctions on the island. What remains is a very degraded vegetation type with just a few remnants of the original flora. However, species diversity is probably also low due to the calcarenitic substrate, producing very different soils to those on the mainland.

The island has been extensively degraded by deforestation and subsequent erosion. In addition the introduction of goats, black rats, Indian house shrews, cats, giant African land snails and a variety of alien plants have severely damaged the native flora and fauna. However, as no monkeys, deer, mongooses, pigs, and some potentially invasive plants have been introduced to the island, it is still better off than similar sites on the mainland.

Although the island was declared a nature reserve in 1965, once the private lease of the island was terminated in 1975, illegal wood-cutting (for firewood) increased and was quickly decimating the little native vegetation that was left. Not only were woodcutters taking native trees to burn, but they were also trampling any regeneration when removing the wood. However, in November 1985 the Government leased the island to the local conservation NGO, the Mauritius Wildlife Appeal Fund, and two watchmen were placed on the island. Since then illegal woodcutting has stopped and the island quickly became much more densely wooded, although mostly by alien species.

Due to the island's small size, it is however possible to control and eventually eradicate introduced species in order to restore the area to a nearly pristine condition. A programme to remove the most invasive plant species was started in 1986. The ultimate goal of this project is to remove all alien species and encourage native species to regenerate. Most alien species were weeded, although the main target species were prune malgache, bois d'oiseau, l'acacia and aloès (Table 22.4), as well as smaller weeds including l'herbe tourterelle, *Turnera ulmifolia* and *Stachytarpheta indica*. The tecoma tree, which was forming small monotypic stands, was also removed.

After six years of continuous weeding, areas still under native canopy were virtually weed-free. Previously these areas were virtually impenetrable, being densely invaded by a spiny thicket of various invasives (prune malgache, bois d'oiseau, l'acacia). Some suckers from prune malgache and bois d'oiseau were occasionally sprouting, but in general very little weeding was needed, and native species were regenerating well. However, removal of large prune malgache trees from continuous areas created large gaps which take much longer to recover. Very few native plants grow in such areas, and constant weeding is necessary before the occasional native seedling manages to grow. In these cases, supplemental planting of native species can accelerate the restoration process.

The Ile aux Aigrettes project demonstrates that restoration is indeed possible on an island through eradication of introduced species and intensive management of native species. However, the island has been so extensively degraded and early records are so few, that it is impossible to know what we should be restoring the island to. Although in some ways it is tempting to just remove the introduced species and see what happens to the little native vegetation remaining on the island, in practice this will not be feasible. Other species, which have probably been lost from the island, are needed to fill in the gaps and shade out introduced species. Therefore a controlled programme of re-introduction of elements of the Mascarene vegetation which quite possibly once occurred on the island is being undertaken.

Intensive management over the past decade has greatly altered the physiognomy of the island, although until the major invasive species present on the island are eradicated, regular weeding will be necessary. Provided management is consistently undertaken, Ile aux Aigrettes will become a nature reserve which will serve two purposes. First, it will conserve the known lowland coastal vegetation that existed on the island before human colonisation, and second, it will provide a safe haven for numerous other species which may or may not once have occurred on the island.

Summary

Although the long-term solution for the invasive weed problem in the Mascarenes will probably lie in biological control, solutions will take time to develop, and in the meantime those species which are down to critically low levels need to be conserved using other methods. On Mauritius, due to a source of relatively cheap labour and concerns for safety, manual weeding (cutting down and uprooting undesirable species) was found to be most effective, although treating stumps with herbicides was also undertaken. It has been demonstrated that native species will start regenerating again once a native canopy is re-established, and the alien species seed bank diminished.

It is emphasised that Réunion has a significantly better conserved flora due to fewer introductions, both animals and plants, although management is also needed. Introductions which could have been prevented in the last few decades (e.g. deer, bulbuls and privet) have occurred, and there is a risk that pigs, monkeys and several plant species may become feral if stringent measures are not taken.

The Convention on Biological Diversity, of which both Mauritius and France are signatories, could provide the impetus for ensuring:

- that no further harmful introductions are undertaken in the Mascarenes (especially species known to be invasive on other islands);
- that eradication of invasive species from the smaller islets, or where the introduction has been recent and eradication is still possible, is undertaken;
- that continuous and intensive management of invasive species is continued;
- that research on the biological control of the most invasive species is undertaken, to provide new and improved management techniques.

The only hope for the future of the unique flora (as well as fauna) of the Mascarenes will be in continual and intensive management, and prevention of further deleterious introductions. However management costs money, and signatory countries must be committed to providing funds to combat invasive species, which is the only hope for maintaining the level of biological diversity that we enjoy today.

Acknowledgements

I thank the World Wide Fund for Nature (WWF), IUCN – the World Conservation Union, and the Mauritius Wildlife Fund for supporting my work on Mauritius as well as contributing to ongoing alien species eradication efforts. Special thanks also go to the many "weeders" on Mauritius and Rodrigues, both professional and volunteer, and in particular to Mr. Gabriel d'Argent.

References

Anonymous (1983) *The Forest and Reserves Act*. Act No. 41 of 1983. Government of Mauritius.

Baker, J.G. (1877) *Flora of Mauritius and the Seychelles. A Description of the Flowering Plants and Ferns of those Islands*. London, Reeve & Co.

Balfour, I.B. (1877) Aspects of the phanerogamic vegetation of Rodriguez, with description of new plants from the Island. *Journal of the Linnean Society (Botany)*, **16,** 7–24.

Bosser, J., Cadet, Th., Guého, J., Julien, J. and Marais, W., eds (1976–cont.) *Flore des Mascareignes: La Réunion, Maurice, Rodrigues*. Mauritius Sugar Industry Research Inst., l'Office de la Recherche Scientifique et Technique Outre-Mer, Paris and Royal Botanic Gardens, Kew; Port Louis, Mauritius.

Brouard, N.R. (1963) *A History of Woods and Forests in Mauritius*. Government Printer, Port Louis, Mauritius.

Bullock, D.J., Greig, S. and North, S.G. (1983) *Round Island Expedition 1982, Final Report*. Unpub., lodged with the University Libraries, St Andrews and Durham, UK.

Cadet, Th. (1980) *La végétation de l'Ile de la Réunion: Etude Phytoécologique et Phytosociologique*. Imprimerie Cazal, St. Denis, La Réunion.

Campbell, D.J. (1978) The effects of rats on vegetation, in *The Ecology and Control of Rodents in New Zealand Nature Reserves*, (eds P.R. Dingwell, I.A.E. Atkinson and C. Hay), Department of Lands and Survey, Info. ser. 4, Wellington, New Zealand, pp. 99–120.

Cheke, A.S. (1987) A review of the ecological history of the Mascarene Islands, with particular reference to the extinctions and introductions of land vertebrates, in *Studies of the Mascarene Avifauna,* (ed. K.W. Diamond), Cambridge University Press, pp. 5–89.

Clark, G. (1859) A ramble round Mauritius with some excursions in the interior of the sland. By a country Schoolmaster. *Mauritius Almanac, 1859,* 1–43 (reprinted in 1945 in *Revue Agriculture de Ile Maurice,* **24,** 34–51, 96–114).

Greathead, D.J. (1971) *A Review of Biological Control in the Ethiopean Region*. Technical Comm. 5, Commonwealth Inst. of Biological Control. Lamport Gilbert, Reading, UK.

IUCN (1980) *How to use the IUCN Red Data Book Categories*. Threatened Plants Unit, IUCN, Kew, UK.

Johnston, H.H. (1895) Additions to the Flora of Mauritius as recorded in Baker's Flora of Mauritius and the Seychelles. *Transactions of the Botanical Society of Edinburgh,* **20,** 91–407.

Kennedy, W.R. (1893) Notes on a visit to the islands of Rodriguez, Mauritius and Réunion. *Journal of the Bombay Natural History Society,* **7(4),** 440–446.

Macdonald, I.A.W., Thébaud, C., Strahm, W.A. and Strasberg, D. (1991) Effects of alien plant invasions on native vegetation remnants on La Réunion (Mascarene Islands, Indian Ocean). *Environmental Conservation,* **18,** 51–62.

Merton, D.V. (1987) Eradication of rabbits from Round Island, Mauritius: A conservation success story. *Dodo, Journal of the Jersey Wildlife Preservation Trust,* **24,** 19–43.

Merton, D.V., Atkinson, I.A.E., Strahm, W., Jones, C., Empson, R.A., Mungroo, Y., Dulloo, E. and Lewis, R. (1989) *A Management Plan for the Restoration of Round Island Mauritius*. Jersey Wildlife Preservation Trust, UK.

Owadally, A.W. (1971) *Annual Report of the Forestry Service for the Year 1969*. Government Printer, Port Louis, Mauritius.

Padya, B.M. (1984) *Climate of Mauritius*. 2nd ed. Government Printer, Mauritius.

Strahm, W.A. (1987–1993) *Reports on Round Island*. 14 unpublished reports submitted to WWF/IUCN. Available at WWF, Gland, Switzerland and MWAF, Mauritius.

Strahm, W.A. (1989a) *Plant Red Data Book for Rodrigues*. Koeltz, Germany.

Strahm, W.A. (1989b) The Mondrain nature reserve and its conservation management. *Proceedings of the Royal Society of Arts and Sciences, Mauritius,* **5(1),** 139–177.

Strahm, W.A. (1994) *The Conservation and Restoration of the Flora of Mauritius and Rodrigues*. PhD Thesis, Reading University, UK. (2 vols).

Vaughan, R.E. (1937) Contributions to the flora of Mauritius. *Journal of the Linnean Society, Bot.*, **51,** 285–308.

Vaughan, R.E. and Wiehe, P.O. (1937) Studies on the vegetation of Mauritius. I. A preliminary survey of the plant communities. *Journal of Ecology,* **25,** 289–343.

Vaughan, R.E. (1968) Mauritius and Rodrigues, in Conservation of vegetation in Africa South of the Sahara, (eds I. and O. Hedberg), *Acta Phytogeogrica Suecia,* **54,** pp. 265–272.

Vaughan, R.E. (1969) The Mauritius Herbarium. 10th Annual Report 1969, in *Annual Report of the Mauritius Sugar Industry Research Institute. Reduit, Mauritius*. Pp. 157–169.

Vaughan, R.E. (1980) *Notes on some aspects of the plant ecology of Mauritius with a census of the Perrier Reserve*. Unpub. report, 34 pp.
Webb, D.A. (1985) What are the criteria for presuming native status? *Watsonia*, **15**, 231–236.

23 Biodiversity conservation and the management of invasive animals in New Zealand

MICHAEL N. CLOUT
School of Biological Sciences, University of Auckland, Auckland, New Zealand

Abstract

The New Zealand archipelago has been subject to extensive invasion by alien species since human colonisation began c. 1 000 years ago. Since then (and especially in the past 200 years) many species have been introduced and have invaded natural ecosystems, causing massive ecological change and biodiversity loss. Known extinctions have been particularly high amongst endemic vertebrates, including birds, and several of these extinctions are attributable to predation by introduced mammals.

The conservation of native biodiversity in New Zealand is now dominated by the management of invasive species. Introduced mammals are being eradicated from increasingly large islands, and intensive mammal control has yielded some measurable conservation benefits at selected mainland sites. Despite these successes, many endemic species and natural ecosystems remain threatened by invasives. Invaded ecosystems have often been irreversibly altered by a series of biological invasions which interact to produce cascading effects on natural food webs. The long-term consequences are uncertain.

Prevention of further biological invasions remains an important priority in New Zealand. New laws (the 1993 Biosecurity Act and the 1996 Hazardous Substances and New Organisms Act) reflect this, although they are targetted mainly at the prevention and control of economic rather than ecological pests.

Introduction

As in most parts of the world, the natural ecosystems of New Zealand have been greatly diminished and modified by direct human activity. Since the human settlement of this archipelago commenced, with the arrival of Polynesian people c. 1 000 years ago, native forests have been reduced to only 23% of their original area (King, 1990), and much of what remains is in mountainous country. Natural lowland habitats have mostly been cleared or drained for farmland, especially within the past 150 years of European settlement. Although the rate of loss of natural habitats has slowed in recent years, with the passage of environmental legislation, the landscape has been changed irrevocably over much of the country.

O. T. Sandlund et al. (eds.), Invasive Species and Biodiversity Management, 349–361.

Although native biodiversity has been lost through the depletion and fragmentation of habitats, the most severe impacts on New Zealand's native flora and fauna have undoubtedly been caused by introduced species, with the fauna suffering particularly high losses through predation. In their original state, New Zealand ecosystems contained no land mammals apart from some small bats. Biological invasion commenced with the introductions of food plants, dogs (*Canis familiaris*) and kiore (*Rattus exulans)* by Polynesian settlers. The kiore seems to have eliminated several species of small birds, flightless insects, and reptiles (Atkinson and Moller, 1990). Overall, at least 35 bird species became extinct following Polynesian settlement, including several species of large flightless birds (notably the moas; *Dinornithidae*) which were hunted to extinction by people (Anderson, 1989).

Following European settlement, the trickle of alien species became a flood. In the past 200 years Europeans have introduced over 80 species of alien vertebrates, including 34 mammals. Among these were three further species of rodents, three mustelids, six marsupials, and seven deer species. Predatory European mammals (e.g. ship rats, *R. rattus*, stoats, *Mustela erminea*, and cats, *Felis catus*) have caused the extinction of a further nine endemic bird species and threaten several more. Herbivorous mammals (e.g. brushtail possums, *Trichosurus vulpecula*, red deer, *Cervus elaphus*, and goats, *Capra hircus*) have altered the structure and composition of native plant communities through their selective browsing (King, 1990). Known losses of plant species have been less numerous than those of animals, but at least three endemic plants have become extinct since 1840 and a further 45 are highly threatened (Cameron *et al*., 1993).

Mammals have been the most damaging invaders of New Zealand, but they are not the most numerous. Several species of insects, birds and fish, and many species of plants, have also been introduced in the past 200 years (Atkinson and Cameron, 1993). Establishment of new plant species continues at a rate of four per month (E. Cameron pers. comm.). The New Zealand archipelago is now one of the most invaded places on earth, with large sectors of its terrestrial biota dominated by introduced species (Table 23.1).

A recent analysis concludes that there are 403 New Zealand taxa (species, subspecies and forms) which can be classed as threatened (Molloy and Davis, 1994). This total includes 159 plants, 98 invertebrates, and 146 vertebrates. No country has a higher proportion of its surviving avifauna classed as threatened. Of 287 bird species (150 of them endemic), 45 are classed as threatened in the IUCN Red List (IUCN, 1996). Forty-one of these threatened birds are endemic, and many now persist only on mammal-free islands, or in dwindling mainland populations. For most threatened animals and many threatened plants, invasive species now pose the most significant remaining threat to their continued survival. Competition or predation by introduced animals, in particular, threatens many endemic species.

Table 23.1 Numbers of native and alien species in New Zealand. After Atkinson and Cameron (1993).

Group	Native	Alien	%Alien
Plants			
Dicots	1 591	1 199	43
Monocots	621	380	38
Conifers	24	24	50
Ferns and allies	213	20	9
Animals			
Land mammals	2	34	94
Land birds	77	36	31
Seabirds	69	-	0
Reptiles	60	1	2
Amphibians	4	2	33
F/w fish	27	19	43
Insects	c. 18 500	c. 1 500	8

In this paper, I firstly present some examples of the conservation impacts of invasive species in terrestrial ecosystems in New Zealand, focusing on the effects of introduced animals. I then discuss some recent management responses, which have attempted to protect native biodiversity by eradicating or controlling introduced mammals. Finally, I briefly review recent New Zealand legislation which seeks to control introductions and manage existing invasive species.

Throughout this paper the term "introduced species" is taken to mean a species which has been transported beyond its natural range by human agency (either deliberately or accidentally). The terms "exotic", "alien" and "introduced" are essentially interchangeable. An "invasive species" is an introduced species which has invaded natural ecosystems and acts as an agent of change within them.

Conservation impacts of invasive animals: some New Zealand examples

Brushtail possums: the multiple effects of an invasive species

Brushtail possums are herbivorous Australian marsupials which were deliberately introduced to New Zealand in a series of liberations between 1858 and 1900 to establish a fur trade. Their spread was aided by many further releases of New Zealand-bred progeny, leading to colonisation of most parts of the North and South Islands by the 1980s (Cowan, 1990a). Possums were initially valued (and even protected by law), but by the 1940s they were officially recognised as pests, because of the damage caused to crops and native forests. Research over the past 40 years has revealed the scale of their impacts. They are now recognised as a major conservation pest in New Zealand.

In New Zealand, possums have fewer competitors, fewer predators and a smaller range of parasites than in their native Australia (Presidente, 1984). They have also colonised a much wider range of habitats. In New Zealand rainforests they commonly occur at population densities exceeding 10/ha: an order of magnitude greater than in their native Australian dry sclerophyll forests (Cowan, 1990). New Zealand plants have evolved without mammalian herbivores and are generally more nutritious and palatable to possums than Australian *Eucalyptus*, which have toxic secondary compounds and may limit possum populations where alternative food is scarce (Freeland and Winter, 1975).

The primary conservation impact of possums in New Zealand is through the damage which they cause to native forests. Possum browsing changes the composition and structure of forests through the defoliation and progressive elimination of favoured food plants, including *Metrosideros*, *Fuchsia*, *Weinmannia* and many other palatable species (Meads, 1976; Campbell, 1984). Some endemic plants (e.g. *Dactylanthus taylori*, *Peraxilla* spp.) are now threatened with local extinction because of possum damage.

Possums also feed on flowers and fruit, reducing fruit crops of native plants and hence the food supplies of native birds. Some plant species (e.g. *Dysoxylum spectabile, Beilschmiedia tawa, Rhopalostylis sapida*) produce virtually no ripe fruit in the presence of dense possum populations (Cowan, 1990b). The implications of this for forest regeneration and the reproduction and survival of endemic fruit-eating birds which depend on such plants are likely to be severe. Some frugivorous birds (e.g. the New Zealand pigeon, *Hemiphaga novaeseelandiae*) fail to breed in years of poor fruiting (Clout *et al.*, 1995) and possum damage may increase the frequency of such events. Other competitive impacts of possums on endemic birds include competition with hole-nesting species such as kiwi (*Apteryx* spp.) for hollows. Both possums and kiwis are nocturnal and require similar hollows for daytime shelter. Recently, a further serious impact of possums on native wildlife has been revealed. It is now known from video surveillance of bird nests that possums, although primarily herbivorous, also prey on the eggs and chicks of birds, including those of threatened species such as kokako (*Callaeas cinerea*) (Brown *et al.*, 1993). These findings have provided further conservation impetus for control of possums.

Millions of dollars per annum are now spent in New Zealand on possum control to protect native forests and wildlife, and to reduce economic impacts, which include the transmission of bovine tuberculosis by possums to cattle and deer. The possum control methods most commonly used are widespread aerial distribution of toxic baits containing 1080 poison (sodium monofluoroacetate) over large areas of forest, or placement of toxic baits containing anticoagulant poisons (e.g. brodifacoum) in bait dispensers within smaller forest areas. There is concern at the other risks (e.g. secondary or non-target poisoning) associated with widespread toxin use, but no there are currently no other viable options for large-scale possum control. In summary, possums have multiple ecological and economic effects as an invasive species, and it is now clear that their introduction to New Zealand was an unmitigated disaster for the natural ecosystems and native biota of this archipelago.

The combined effects of invasive animals on a forest food web

Most New Zealand ecosystems now contain large numbers of invasive species, which have altered ecosystem processes and in some cases caused extinctions. An example of an ecosystem in which some of these effects have been well-studied is lowland *Nothofagus* forest, which covers large tracts of land in the northern South Island. In these forests native scale insects (*Ultracoelostoma assimile*) naturally infest *Nothofagus* trees and excrete honeydew, which is fed on by a variety of nectarivorous insects and birds, including the threatened kaka (*Nestor meridionalis*), a forest parrot. The honeydew is now also consumed by invasive vespulid wasps.

The German wasp (*Vespula germanica*) colonised New Zealand in the 1940s, followed by the common wasp (*V. vulgaris*) late in the 1970s (Harris *et al.*, 1991). Following the latter invasion, the German wasp was completely replaced in *Nothofagus* forests by the common wasp, which reaches much higher densities (Harris *et al.*, 1991, Thomas *et al.*, 1990). The common wasp now removes over 95% of honeydew during the autumn peak of wasp density, resulting (among other effects) in the abandonment of lowland *Nothofagus* forests by kaka at this time of year. The kaka only return from higher altitude forests in winter when wasp numbers have declined and honeydew is replenished (Beggs and Wilson, 1991). The impacts of the common wasp invasion on long-term survival of kaka and other nectarivorous birds in these forests are still unknown, as are the effects on populations of native invertebrates on which wasps prey. Attempts to control the two vespulid wasps have included local use of toxins (including 1080 poison) in wasp baits and the deliberate introduction of a parasitic wasp (*Sphecophaga vesparum*) for biological control, but there is no evidence as yet of any significant widespread success.

The kaka and other nectarivorous birds in *Nothofagus* forests are also affected by nest predation by introduced mammals, especially stoats (*Mustela erminea*). Stoat populations increase markedly following irruptions of introduced mice after episodic mast seeding of *Nothofagus*, which occurs every few years (King, 1983), greatly raising the predation risks for vulnerable native birds and causing population declines and range contractions for such species. Declines in the populations of pollinators and seed dispersers may have further consequences for reproduction of plants which depend on the services of vulnerable birds. It has been suggested that reduction in nectarivore populations is now limiting the pollination of mistletoes (*Peraxilla* spp.) in *Nothofagus* forests (Ladley and Kelly, 1995). The mistletoes themselves are also directly threatened by browsing possums, which severely defoliate them. Mistletoes are seasonally important nectar and fruit sources for several native birds, including kaka. The invasion of *Nothofagus* forests by introduced mammals and wasps provides a simple but vivid example of how invasive species can produce cascading effects on a natural food web (Figure 23.1), potentially resulting in the extinction of native species..

Figs and wasps: the synergistic effects of introduced species

The potential for sequential invasions to have a synergistic effect is illustrated by the recent naturalisation of banyan figs (*Ficus* spp., Moraceae) and their pollinating wasps (Hymenoptera: Agaonidae) in New Zealand (Gardner and Early, 1996). The Moreton Bay fig (*Ficus macrophylla*) was widely planted last century as an ornamental tree. Its fruits are commonly consumed by a variety of introduced birds, but it did not became invasive. No viable seeds were set in the absence of its specific pollinating wasp (*Pleistodontes froggattii*). Within the past five years this wasp has arrived in New Zealand by unknown means, and the pollination rates of figs have rapidly risen from zero to 100% (Gardner and Early, 1996). With both its specific pollinator and a range of bird dispersers in place, the potential may now exist for this previously benign ornamental fig to invade natural forest communities.

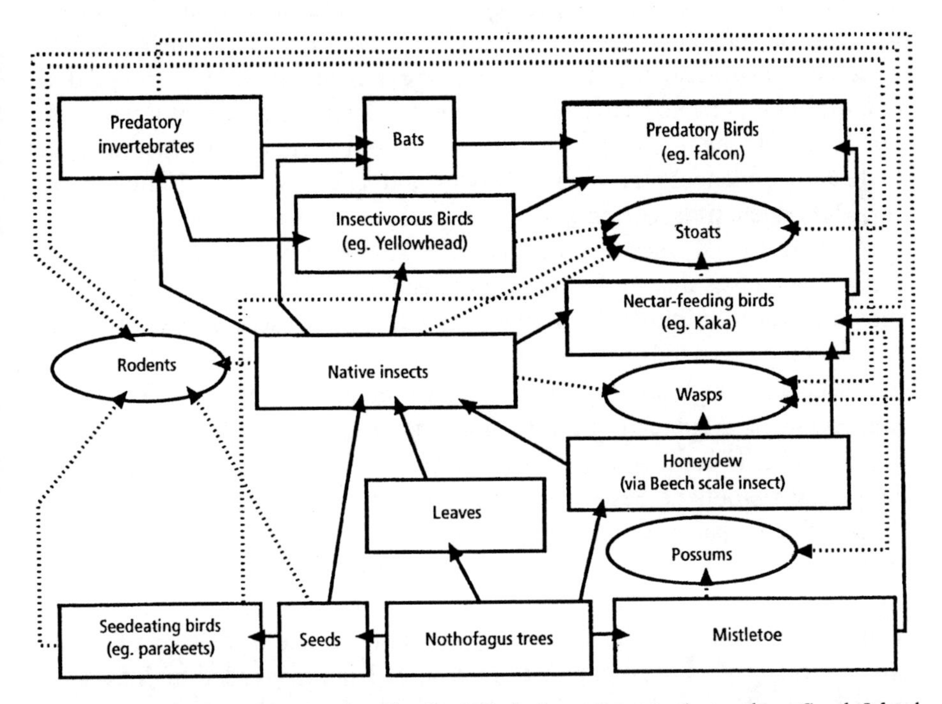

Figure 23.1 Simplified food-web of lowland *Nothofagus* forest in the northern South Island of New Zealand, illustrating the impacts of invasive animals (circled). Direction of energy flow is shown by arrows, with solid lines connecting native elements and dotted lines showing predation on native biota by the invasive animals.

Management responses to invasive animals

The eradication of mammals from islands

The eradication of introduced mammals from islands has been a major advance in New Zealand conservation practice in recent years. Given the potential of mammal-free islands for conservation of threatened species, conservation managers have increasingly attempted to eradicate introduced mammals from islands where there is little prospect of their unaided recolonisation.

In the past decade, in particular, there has been a series of successful eradications of introduced mammals from New Zealand islands (Veitch and Bell, 1990; Veitch, 1994). Successes on large islands include eradication of cattle (*Bos taurus*) and sheep (*Ovis aries*) from Campbell Island (11 216 ha), goats (*Capra hircus*) from Raoul Island (2 938 ha), brushtail possums, Norway rats (*Rattus norvegicus*), and kiore from Kapiti Island (1 970 ha), and rabbits (*Oryctolagus cuniculus*) and mice (*Mus musculus*) from Enderby Island (710 ha). These and other successes have resulted from the synergy of technical developments and increasing confidence in their use. Of particular significance have been the availability of single-dose anticoagulant poisons such as brodifacoum in special bait formulations, and the development of bait stations and aerial application methods for eradicating rodents from islands (Taylor and Thomas, 1989; Veitch, 1994; I.G. McFadden, pers. comm.).

A measure of recent progress in mammal eradications from New Zealand islands is that in the six years after Veitch and Bell (1990) listed all known eradication attempts to 1990, the number of successful eradications of rodents from islands rose from 28 to 57 (Veitch, 1994; C.R. Veitch, pers comm.). The rate of progress in the technical capacity for rodent eradications is graphically illustrated by the rapid recent rise not only in the number of eradications, but also the size of islands now being cleared of these pests (Clout and Saunders, 1995). Only 18 years ago the eradication of rodents from islands was thought to be impossible (Dingwall *et al.,* 1978). Now their removal from islands of up to 250 ha is almost routine and rat eradications are being attempted on islands of up to 2 000 ha in area (C.R. Veitch, pers. comm.). The lessons learned in undertaking these eradications are significant for the potential restoration of islands invaded by introduced rodents elsewhere in the world.

Islands cleared of introduced mammals are now routinely used as important conservation sites for threatened species management. For example, endangered bird species such as takahe (*Porphyrio mantelli*) and kakapo (*Strigops habroptilus*) have been translocated to such islands in attempts to secure their recovery (Clout and Craig, 1995).

Control of mammals at mainland sites

Local control of introduced mammals on the New Zealand mainland has also proved to be both feasible and beneficial to threatened birds in recent years. Temporary mammal control to enhance the breeding success of native birds is now a routine procedure. Examples include stoat control to benefit yellowheads in the Eglinton Valley, Fiordland (O'Donnell *et al.*, 1992), rat control to benefit New Zealand pigeons at Wenderholm, Auckland (Clout *et al.*, 1995), and the control of possums, rats and mustelids to benefit kokako at Kaharoa and Mapara (Saunders, 1990; Hay, 1995).

The largest area currently subject to sustained control of introduced mammals is the central North Island forest reserve of Mapara (1 400 ha), which is being managed to benefit the endangered kokako. Goats, feral pigs, possums, rodents, mustelids and feral cats have been shot, poisoned or trapped since 1989 at Mapara. The number of kokako pairs has subsequently risen by 31%. Productivity has also risen significantly, especially since brodifacoum baits were introduced for possum control in 1993/94. In 1992/93 10 kokako pairs at Mapara raised 15 fledglings, but by 1994/95 productivity had risen dramatically to 55 fledglings from 18 breeding pairs (Hay, 1995; P. Bradfield, pers. comm.). Management experiments like these are vivid illustrations of the pervasive negative effects of introduced species: when key invasives are removed there can be dramatic recovery in the species and the ecosystems which they have been affecting.

Current control of invasive mammals in New Zealand relies heavily on the use of broad spectrum toxins such as sodium monofluoroacetate (compound 1080) and various anticoagulant poisons. Use of these toxins carries the risk of non-target kills and (in the case of anticoagulants) short-term accumulation of toxins. Because New Zealand has no native terrestrial mammals apart from bats, non-target mammal deaths (e.g. deer, carnivores) are normally of little concern in conservation areas. Repeated trials have failed to show negative long term impacts of poison bait distribution on birds or other native wildlife. However, there is growing public unease about the widespread and repeated use of poisons and there is consequently an active search for alternative methods of biological pest control. The biological control of conservation pests is potentially more sustainable than the perpetual use of toxins, but it entails the risk of introducing yet more alien species which might be non-specific in their action, or have other side-effects. Some of the worst conservation pests in New Zealand (e.g. mustelids) were originally introduced as agents of biological control. With recent legislation (see below) such risky introductions would not now be officially sanctioned, although this did not prevent the illegal introduction in 1997 of rabbit calicivirus (RCD) for the biological control of rabbits.

New legal instruments

New Zealand legislation to prevent or reduce the risks from alien species has undergone radical recent change, with the introduction of the Biosecurity Act (1993) and the Hazardous Substances and New Organisms Act (1996).

The Biosecurity Act 1993, has two major components: the prevention of introduction of unwanted organisms not established in New Zealand and the management of unwanted organisms which are established. An "unwanted organism" is any species that the Ministry of Agriculture and Fisheries believes is capable of causing harm to any natural resource. Preventing introductions is attempted by focusing on identified "risk goods" which have to be given specific biosecurity clearance. Maximum penalties for having unauthorised "risk goods" are imprisonment for up to five years and fines of up to NZ$ 200 000. Although the focus is very much on avoiding the introduction of diseases or economic pests, the Act does also seek to avoid ecologically harmful species. Management of established species is achieved through the formal designation of regional councils as pest management agencies, which are required to draw up and enforce regional pest management strategies for listed "harmful organisms". These agencies have the power to use rating provisions to force landowners to control named pests, where necessary.

The Hazardous Substances and New Organisms Act (1996) has among its stated purposes "avoiding or mitigating any harmful effects of...new organisms on the environment; and in particular on all native and valued introduced flora and fauna". It establishes an Environmental Risk Management Agency (ERMA) which has the responsibility for assessing all applications for deliberate introduction of all new organisms, and deciding on the issue of permits. Among the criteria for making decisions on introduction of new organisms are "likely effects on all native and valued introduced fauna and flora and the environment in general". The ERMA is also responsible for assessing all work on genetically modified organisms (also defined in the legislation as "new organisms") in New Zealand.

Conclusions

The global conservation of native biodiversity will in future increasingly involve the management of invasive species. In New Zealand the key threats to native species and terrestrial ecosystems are predation and competition from introduced species in general, and invasive mammals in particular. Experience in managing these threats is relevant not only to the potential restoration of island biotas and ecosystems in New Zealand, but also to similar situations elsewhere. For example, techniques for eradicating mammals from islands have potentially widespread applications in other archipelagos, where vulnerable biotas and ecosystems face similar threats to those encountered in New Zealand. Based on the New Zealand experience, these threats are sometimes manageable but management will have to be conducted carefully, in the right sequence, and in the light of knowledge of the ecological interactions within the system concerned.

In New Zealand, improvements in the control of introduced mammals on islands have raised the potential for ecological restoration at sites on the mainland, applying the knowledge gained on islands. Whilst islands free of introduced mammals will continue to be the primary locations for conserving threatened species for the foreseeable future, ecological restoration programmes on the mainland will make increasingly significant contributions to biodiversity conservation in New Zealand in the future. At present such restoration programmes rely on the perpetual use of toxins such as 1080 and brodifacoum, but biological control will probably play a larger role in future. Extreme caution will be needed to minimise the risks of unexpected invasiveness in any new biological control agents.

Finally, the unpredictable and sometimes severe synergistic impacts of introduced species underline that the most effective way of managing biological invasions is to prevent them from happening in the first place. Strong legislation and quarantine procedures will help with this, but raising awareness of the risks to biodiversity is also vital to ensure the support of the public for action against biological invasions.

Acknowledgements

I thank Ian Atkinson and Dick Veitch for their willing and helpful provision of information used here, and an anonymous referee for constructive comments on my manuscript. I am grateful to the organisers of the Conference on Alien Species for agreeing to a conference session on "impacts on oceanic islands", and for supporting me to present a paper at Trondheim.

References

Anderson, A. (1989) *Prodigious Birds*. Cambridge University Press, Cambridge, UK.

Atkinson, I.A.E. (1990) Ecological restoration on islands: prerequisites for success, in *Ecological Restoration of New Zealand Islands*, (eds D.R. Towns, C.H. Daugherty and I. A. E. Atkinson), Conservation Sciences Publication No. 2, Department of Conservation, Wellington, pp. 73–90.

Atkinson, I.A.E. and Cameron, E.K. (1993) Human influence on the terrestrial biota and biotic communities of New Zealand. *Trends in Ecology and Evolution*, **8**, 447–451.

Atkinson, I.A.E. and Moller, H. (1990) Kiore, in *The Handbook of New Zealand Mammals*, (ed. C.M. King), Oxford University Press, Oxford, UK, pp. 175–192.

Beggs, J.R. and Wilson, P.R. (1991) The kaka *Nestor meridionalis*, a New Zealand parrot endangered by introduced wasps and mammals. *Biological Conservation*, **56**, 23–38.

Brown, K., Innes, J. and Shorten, R. (1993) Evidence that possums prey on and scavenge birds' eggs, birds and mammals. *Notornis*, **40**, 1–9.

Cameron, E.K., de Lange, P.J., Given, D.R., Johnson, P.N. and Ogle, C.C. (1993) *New Zealand Botanical Society Newsletter*, **32**, 14–19.

Campbell, D.J. (1984) The vascular flora of the DSIR study area, lower Orongorongo Valley, Wellington, New Zealand. *NZ Journal of Botany*, **22**, 223–270.

Clout, M.N. and Craig, J.L. (1995) The conservation of critically endangered flightless birds in New Zealand. *Ibis*, **137**, 181–190.

Clout, M.N., Karl, B.J., Pierce, R.J. and Robertson, H.A. (1995) Breeding and survival of New Zealand pigeons (*Hemiphaga novaeseelandiae*). *Ibis*, **137**, 264–271

Clout, M.N. and Saunders, A.J. (1995) Conservation and ecological restoration in New Zealand. *Pacific Conservation Biology*, **2**, 91–98.

Clout, M.N., Denyer, K., James, R.E. and McFadden, I.G. (1995) Breeding success of New Zealand pigeons (*Hemiphaga novaeseelandiae*) in relation to control of introduced mammals. *New Zealand Journal of Ecology*, **19**, 209–212.

Cowan, P.E. (1990a) Brushtail possum, in *The Handbook of New Zealand Mammals*, (ed. C. M. King), Oxford University Press, Oxford, UK, pp. 68–98.

Cowan, P.E. (1990b) Fruits, seeds and flowers in the diet of brushtail possums in the Orongorongo Valley, Wellington, New Zealand, in relation to food-plant availability. *New Zealand Journal of Zoology*, **3**, 399–419.

Dingwall, P.R., Atkinson, I.A.E. and Hay C., eds (1978) *The Ecology and Control of Rodents in New Zealand Nature Reserves*, Department of Lands and Survey Information Series No.4, Wellington, New Zealand.

Freeland, W.J. and Winter, J.W. (1975) Evolutionary consequences of eating: *Trichosurus vulpecula* (Marsupialia) and the genus *Eucalyptus*. *Journal of Chemical Ecology*, **1**, 439–455.

Gardner, R.O. and Early, J.W. (1996) The naturalisation of banyan figs (*Ficus* spp., Moraceae) and their pollinating wasps (Hymenoptera: Agaonidae) in New Zealand. *New Zealand Journal of Botany*, **34**, 103–110.

Harris, R.J., Thomas, C.D. and Moller, H. (1991) The influence of habitat use and foraging on the replacement of one introduced wasp species by another in New Zealand. *Ecological Entomology*, **16**, 441–448.

Hay, J.R. (1995) Mapara revisited. *Conservation Science Newsletter*. April 1995, Department of Conservation, Wellington, pp. 4–5.

IUCN (1996) *1996 IUCN Red List of Threatened Animals*, IUCN, Gland, Switzerland.

King, C. M., ed. (1990) *The Handbook of New Zealand Mammals.* Oxford University Press, Oxford, UK.

Meads, M.J. (1976) Effects of opossum browsing on northern rata trees in the Orongorongo Valley, Wellington, New Zealand. *New Zealand Journal of Zoology,* **3**, 127–139.

Molloy, J. and Davis, A. (1994) *Setting Priorities for the Conservation of New Zealand's Threatened Plants and Animals.* Department of Conservation, Wellington, New Zealand.

Presidente, P.J.A. (1984) Parasites and diseases of brushtail possums (*Trichosurus* spp.): occurrence and significance, in *Possums and Gliders* (eds A. Smith and I. Hume), Surrey Beatty and Sons, Chipping Norton, New South Wales, pp. 171–187.

O'Donnell, C.F.J, Dilks, P.J. and Elliott, G.P. (1992) Control of a stoat population irruption to enhance yellowhead breeding success. *Science and Research Internal Report* No.124, Department of Conservation, Wellington.

Saunders, A.J. (1990) Mapara: island management "mainland" style, in *Ecological Restoration of New Zealand Islands,* (eds D.R. Towns, C.H. Daugherty and I.A.E. Atkinson), Conservation Sciences Publication No. 2, Department of Conservation, Wellington, New Zealand, pp. 147–149.

Taylor, R.H. (1984) Distribution and interactions of introduced rodents and carnivores in New Zealand. *Acta Zoologica Fennica,* **172**, 103–105.

Taylor, R.H. and Thomas, B.W. (1989) Eradication of Norway rats (*Rattus norvegicus*) from Hawea Island, Fiordland, using brodifacoum. *New Zealand Journal of Ecology,* **12**, 23–32.

Thomas, C.D., Moller, H., Plunkett, G.M. and Harris, R.J. (1990) The prevalence of introduced *Vespula vulgaris* wasps in a New Zealand beech forest community. *New Zealand Journal of Ecology,* **13,** 63–72.

Veitch, C.R. and Bell, B.D. (1990) Eradication of introduced mammals from the islands of New Zealand, in *Ecological restoration of New Zealand islands,* (eds D.R. Towns, C.H. Daugherty and I.A.E. Atkinson), Conservation Sciences Publication No. 2, Department of Conservation, Wellington, New Zealand, pp. 137–146.

Veitch, C.R. (1994) Habitat repair: a necessary prerequisite to translocation of threatened birds, in *Reintroduction Biology of Australian and New Zealand Fauna,* (ed. M. Serena), Surrey Beatty and Sons, Chipping Norton, New South Wales, Australia.

24 South Africa's experience regarding alien species: impacts and controls

BRIAN J. HUNTLEY
National Botanical Institute, Cape Town, South Africa

Abstract

The introduction of alien biota to southern Africa, as recorded in cave deposits and archaeological sites, goes back to the Later Stone Age (3 000 BP) for plants and the Early Iron Age (300–600 AD) for domesticated animals. The early introductions included few aliens of invasive potential, and it was not until after the establishment of the first European colonial settlement in 1652 that the major wave of alien species occurred. Currently, over 789 species of naturalised exotic plants are recorded, of which 47 species are of serious concern to conservationists. These are responsible for significant habitat transformation, leading to loss of biodiversity, reduction of ecosystem services and changes in microclimate, soil chemistry and fire regime.

The paper outlines the history of invasion by alien species, and describes South Africa's experience in the identification, evaluation, combat and monitoring of the problem illustrated by two case studies.

The first case study describes the successful eradication of feral domestic cats from the 30 000 ha Marion Island in the sub-Antarctic. From an individual gravid female introduced in 1949, the cat population reached over 3 400 in 1977. By 1980 one species of petrel had become locally extinct. By 1975, the cats were killing 450 000 burrowing petrels per year. Feline panleucopaenia was introduced in 1977, reducing cat numbers to 615 by 1982. Night hunting reduced the population to 89 in 1989. Trapping and hunting eliminated the remainder of the population – no cat has been seen since 1992.

The second case study is the recently launched Water Conservation Programme, an initiative within the Government's Poverty Relief Programme. This unique project has mobilised employment for over 5 000 people in six provinces, in a national network of teams eradicating alien invasive plants from the catchments of the primary urban water resources of an essentially arid country. These case studies will illustrate the research, conservation, political, social and economic information flows and decision paths that were following in dealing with the problem of invasive aliens.

Introduction

The initiation of a short-term project on the Ecology of Biological Invasions by SCOPE (Scientific Committee on Problems of the Environment) in 1982 stimulated South African scientists to undertake a detailed analysis of the history, impacts, management and policies relating to alien species of

O. T. Sandlund et al. (eds.), Invasive Species and Biodiversity Management, 363–375.

animals and plants. The results of the South African contribution to the SCOPE project were published as a comprehensive synthesis volume (Macdonald *et al.*, 1986) and in many other papers (see especially Richardson *et al.*, 1996; for papers on the biological control of weeds in South Africa, and *Biological Conservation,* volume 44, for papers on biological invasions in nature reserves).

This paper will examine South Africa's experience regarding alien species with special reference to two case studies. It is intended to reflect some of the patterns, processes and impacts of alien species introductions, and the management policies and control programmes developed to address the problem especially in relation to the country's rich biodiversity.

South Africa's biodiversity

South Africa occupies 0.8% of the world's land area, situated at the southern tip of Africa at the confluence of the warm Agulhas and cold, upwelling Benguela currents. To the south, in the sub-Antarctic, lie the oceanic islands of Marion and Prince Edward, annexed by South Africa in 1947/48.

Despite its relatively small size, South Africa possesses 8% of the world's diversity of vascular plants, 7% of its birds, and 6% of the world's mammal species. Its flora, embracing the entire Cape Floristic Kingdom, is especially rich, with 18 388 species of indigenous vascular plants (Table 24.1). Detailed accounts of the biodiversity of southern Africa are presented in two synthesis volumes (Huntley, 1989, 1994).

Table 24.1 Species richness of selected non-marine taxa reproducing in South Africa (total area 1 221 000 km²)

Taxon	No of species	Taxon	No of species
Fishes	112	Bryophyta	776
Amphibians	84	Pteridophyta	237
Reptiles	286	Gymnospermae	39
Birds	600	Monocotyledonae	4 377
Mammals	227	Dicotyledonae	13 972
Invertebrates	278 000	Fungi	20 000

The introduction of alien species of animals and plants to South Africa

The history of introduction of alien species of animals and plants to South Africa is detailed in papers by Deacon (1986), Wells *et al.* (1986a, b), Bruton and Van As (1986), Brooke *et al.* (1986) and Shaughnessy (1986). Archaeological analyses of cave deposits indicate that the first alien introductions occurred from the Later Stone Age (ca. 3 000 yrs BP) into the Early Iron Age (ca. 300 AD). These introductions of edible plants and domesticated animals probably had little significant impact on the environment of the time. It was not until after the establishment of the first European colony at the Cape in 1652 that major invasions occurred, and these only since the mid 19th Century.

The motives for introducing alien species were primarily for food, forage and shelter. Van Riebeeck, on arrival in the Cape in 1652, sent an urgent plea to his masters in Amsterdam to "send us anything that will grow" (Fairbridge, 1931). By 1850, overgrazing of the Cape Flats had led to severe wind erosion of the coastal sands, resulting in the importation of many Australian *Acacia* and *Hakea* species, many of which are the most problematic of the invasive species in South Africa today. During the same period, foresters in the Cape promoted the establishment of trees, particularly *Pinus pinaster*, on Table Mountain, by arguing that they would increase the water supply from the mountain catchments (Colonial Botanist, 1865). Cecil Rhodes, in 1897, introduced European starling *Sturnus vulgaris* and several other species to "improve the amenities of the Cape, by diversifying the bird fauna" described as poverty stricken by Shelley in 1875 (Brooke *et al.,* 1986). In addition to the deliberate introductions described above, many species, especially herbaceous plants and invertebrates, were introduced accidentally along with the massive imports of goods during the colonial settlement, and particularly, during the Anglo-Boer War of 1899–1902.

Wells *et al.* (1986b) provide details of 789 species of naturalised exotic plants, which they define as "species of foreign origin, but reproducing as though native". They list a further 91 species of which naturalisation is uncertain, "possible future invasives" and 104 species whose alien origins are in doubt.

Of the 789 invasive alien plant species, 47 were considered of serious concern to conservationists. These "transformers" are responsible for significant habitat transformation, leading to loss of biodiversity, reduction of ecosystem services (such as catchment run-off) and changes in microclimate, soil chemistry and fire regime.

An analysis of the region of origin of 530 of the 789 species (Table 24.2) indicates that 243 or 46% come from Europe and Asia, compared with only 38 or 7% from Australia. The European and Asian species include only eight transformer species, compared with 17 from Australia. Furthermore, the analysis indicates that the vast majority of transformer species are trees and shrubs (Table 24.3). The very high impact of Australian species on South African ecosystems, especially in the Cape fynbos, is ascribed by Macdonald (1985) to Australia's unique evolutionary environment for fire-adapted trees and shrubs suited to temperate climatic conditions on soils of low nutrient availability.

Table 24.2 Regions of origin of alien plant species and of "transformer" species (from Wells *et al.*, 1986b).

Regions of origin	Alien species		Transformer species	
	no of spp.	% of species	no of spp.	% of all species from regions of origin
Europe and Asia	243	46	8	3
South America	139	26	17	12
North America	46	9	4	9
Australia	38	7	17	45
Africa	20	4	1	5
Other	44	7	0	0
Total	530	100	47	9

Table 24.3 Life forms of transformer species (from Wells *et al.*, 1986b).

Life Form	No of species	% of species
Tree/Shrub	41	87.2
Erect herb	2	4.3
Free floating	2	4.3
Attached floating	1	2.1
Shrub/cluster	1	2.1
Total	47	100

In sharp contrast to the substantial impacts that invasive plant species have exerted on South African ecosystems, the impact of alien animals has been minimal, especially when compared with the impact of invasive mammals in Australia and New Zealand. Most alien terrestrial vertebrates are confined to urban or agricultural areas, while the only mammal considered to have a significant potential impact is the Himalayan thar *Hemitragus jemlabicus* restricted in its range to Table Mountain and now under management control. The situation with regard to alien aquatic animals is thought to be less benign, however, with the endangered endemic fish fauna of the western Cape rivers being particularly vulnerable to competition or predation from introduced alien fish (Bruton and Van As, 1986). A similar situation pertains on South Africa's small off-shore and oceanic islands (Cooper and Brooke, 1986) where alien vertebrates exert a profound influence on the native biota.

Case Study I: Feral cats (*Felis domesticus*) on Marion Island

The Prince Edward Islands (Marion and Prince Edward), situated 22 km apart at 47 °S, 38 °E in the southern Indian Ocean, lie 1 800 km from the South African coast, the nearest mainland.

These volcanic islands are extremely rugged and mountainous, Marion rising to 1 230 m, with an oceanic, sub-Antarctic climate (average temperature 5.0 °C, precipitation 2 600 mm). Vegetation comprises peat bogs, mires and feldmark, with areas above 500 m free of vegetation.

A comprehensive description of the ecology of the islands is presented in Van Zinderen Bakker *et al.* (1971) and in the substantial literature resulting from South Africa's participation in the activities of SCAR (Scientific Committee on Antarctic Research). Watkins and Cooper (1986) summarise knowledge on alien invasions on the islands. The indigenous fauna of Marion Island comprises 27 seabird species and three seal species with approximately 38 macroarthropods. Twenty-two species of indigenous vascular plants have been recorded. House mice, *Mus musculus*, have occurred on the island since at least 1802, and in 1949 five domestic cats were introduced by the Meteorological station to control the mice. By 1951 the cats had become feral. By 1964 they occurred throughout the island, and by 1975 the population was estimated to have reached 2 139, increasing at a rate of 23% per annum (Van Aarde, 1979). The 1975 cat population was estimated to be killing 450 000 burrowing petrels and posed a serious threat to remaining populations. Watkins and Cooper (1986) considered that the common diving petrel *Pelecanoides urinatrix* had been exterminated from Marion Island by cats as early as 1965, although the species is not globally extinct.

Concern regarding the impact that the cats were having on the bird populations of sub-Antarctic islands was first raised by the International

Council for Bird Protection in 1962. In 1963 the South African SCAR called for research to establish the effects of cats on Marion Island (Bester, 1993). Following the 1965/66 expedition (Van Zinderen Bakker, 1971) it seemed "to be very difficult, if not impossible, to exterminate these marauders in a terrain which offers so many hiding places". The South African SCAR Committee discussed the urgency of controlling the cat population at successive meetings in the early 1970s, and following wide consultation with wildlife and veterinary specialists, the viral disease, feline panleucopaenia (FPL) was introduced as a primary control measure in 1977. At the time, the cat population was estimated at 3 405, decreasing by 29% per annum to 615 cats in 1982 following the FPL epidemic (Van Rensburg *et al.,* 1987).

The dramatic reduction in cat numbers did not result in the hoped for improvement in the populations of certain bird species, and as the surviving cats developed immunity, the population appeared to stabilize.

Following further debate between researchers, administrators and the SCAR Committee, a second phase of cat control was launched in 1986. Hunting at night, with spotlights and 12-bore shotguns, took place during four eight-month summer seasons (1986/87, 1987/88, 1988,89, 1989/90) and one four-month winter season. From four to eight teams of hunters covered the island below 500 m in a dense grid of hunting sectors, totalling 19 000 ha of the island's 29 000 ha. Despite the extremely rough terrain and inhospitable weather, these dedicated teams killed 458 cats in 1986/87, 206 cats in 1987/88, 143 cats in 1988/89 and 66 in 1989/90.

In the summer of 1988/89, trapping using walk-in cage traps and gin-traps was added to the control programme. In 1988/89 two cats were trapped, and in an intensified effort in 1989/90, 78 cats were trapped. The trapping intensity was increased through 1990/91, 1991/92 and 1992/93, with 109, 8 and 0 cats being trapped in these periods. Hunting continued from 1990/91 through 1992/93 with diminished returns (8, 0, 0 cats shot).

Since July 1991, not a single cat has been seen on Marion Island, despite intensive surveillance.

Case Study II: Woody invasives in the Fynbos Biome

The fynbos biome – the shrubland and heathland vegetation of the nutrient poor soils of South Africa's Western Cape – is unique among fire-prone mediterranean-type ecosystems both in terms of its exceptional floristic diversity and its susceptibility to invasion by woody alien plants. The ecology of fynbos is described in a comprehensive synthesis (Cowling, 1992).

Long recognised as one of the world's six Floral Kingdoms, the global significance of the Cape Floral Region has also been recognised as the world's "hottest" hotspot of plant diversity (Myers, 1990). The Cape flora comprises 8 574 species, 68% of which are endemic (Bond and Goldblatt, 1984). Much of the lowland fynbos has been transformed, particularly in the last 150 years, through agriculture, urbanization and industrialization. Both transformed and untransformed ecosystems have been severely invaded by alien species, such invasions being considered a primary cause of the extinction of 58 species and the inclusion of 3 435 surviving species in the IUCN threatened plant categories of the Red Data List of southern African Plants (Hilton-Taylor, 1996).

Since the initiation of the SCOPE project on ecological invasions, a vast literature has been published on the topic in South Africa, most especially with regard to the fynbos biome (see Macdonald *et al.,* 1986; Richardson *et al.,* 1992; Le Maitre *et al.,* 1996; Van Wilgen *et al.,* 1996). In this brief account, only the nature of the alien plant invasion problem, and management responses, will be discussed.

Although alien plant introductions formed a significant component of the colonization process initiated in 1652, it was only during the 19th Century that aggressive alien species began to make their presence felt.

In the mountain fynbos – the least disturbed by human intervention – species of *Hakea, Pinus* and *Acacia* have spread over wide areas, often in extremely dense stands. *Hakea sericea,* introduced in 1830, had expanded over 4 800 km² by 1984 (Macdonald and Richardson, 1986). The situation in the lowland fynbos is even worse than the mountains. Here *Acacia cyclops* and *A. saligna*, both introduced in the mid 1800s from Australia, are the most important invasives, but other species (*A. longifolia, A. mearnsii, Albizia lophantha* and *Sesbania punicea)* are also of considerable concern. *Acacia* and other thicket forming alien woody plant species were estimated to have invaded some 8 962 km² by 1984 (Macdonald and Richardson, 1986).

The impact of such invasive tree species, in the treeless shrublands and heathlands of the fynbos, has been far more dramatic than woody plant invasions in ecosystems possessing a dominant tree component. The alien trees simply overtop the low fynbos vegetation, resulting in a gradual decrease in indigenous plant diversity, density and cover. Other impacts include increased soil erosion rates, reduction of water yield from catchments, changes in soil nitrogen, phosphorus and organic matter content under *Acacia* species, reduction in the avifaunal diversity at the community level but increases in bird diversity at the biome level (Macdonald and Richardson, 1986).

Management policies relating to exotic plants and animals have changed dramatically over the past 150 years. Although government policy supported the introduction of plants and animals of potential socio-economic utility until the mid 1900s, concern about certain species was expressed as early as 1863 by farmers in the Eastern Cape with reference to the spread of *Hakea sericea* (Phillips, 1938, in Macdonald and Richardson, 1986).

In the 1860s, the Colonial Botanist considered the most important justification for planting trees was the improved hydrological regime that he believed would result. He recommended planting Australian trees during the severe drought of the time because of the desirable effects he believed would follow (Colonial Botanist, 1865). The same motivation was given by foresters to justify the planting of trees, particularly *Pinus pinaster*, on Table Mountain in the 1880s and 1890s (see Shaughnessy, 1986).

A similar policy, of encouraging the use of exotics, especially Australian acacias, for dune stabilization, was promoted by the Forestry Department until at least 1947, when 3 529 ha of the Cape Flats were planted to *A. cyclops* and *A. saligna*.

The afforestation of Table Mountain was further justified in the 1880s on aesthetic grounds, to cover its "bleak and naked appearance", "bare and stoney slopes" which were "an offence and eyesore to Cape Town" (see Shaughnessy, 1986).

This policy of planned afforestation in the 19th Century was followed in the 20th Century by the gradual abandonment of plantations. From these foci, the invasion of alien species radiated widely throughout the fynbos, a sad legacy from an essentially noble cause.

By the 1930s, attitudes towards invasive aliens began to change. From 1936 onwards, those areas beyond demarcated plantations were to be actively cleared of pines. The Weeds Act of 1937 defined policy to deal with noxious exotics, but was ineffective. In 1962 a biological control programme was introduced for *H. sericea*.

The management policy for mountain fynbos had, until 1968, precluded the use of burning. Indeed, from 1652, severe penalties could be imposed for indiscriminate burning, reinforced in 1924 by the Royal Society condemning veld burning. In 1968, following pioneering ecological studies on fynbos, the Department of Forestry accepted prescribed burning as a management policy (Van Wilgen *et al.*, 1994). Gradually a more objective and scientifically based management policy has evolved, benefiting in no small part from the surge of ecological research stimulated by the Fynbos Biome Project of the 1970s and 1980s (Huntley, 1992). The increased interest in biodiversity conservation, especially in the fynbos, and the increase in ecotourism activities, supported the call for action to control invasives. Both government agencies and non-governmental "hack groups" set about removing alien infestations, with reasonable success. Macdonald and Richardson (1986) reported that of 7 592 km² once infested by *Hakea* and *Pinus* species in mountain fynbos, 1 579 km² (21%) had been cleared by the end of 1984. In the lowland fynbos, where control was left largely to volunteer groups, only 870 km² (10%) of the 8 962 km² infested mainly by *Acacia* species had been cleared.

In the early 1990s ecologists and conservationists began to realise that the battle to save the fynbos could not be won without massive financial support, this at a time when the socio-political and economic situation in the country was at its most vulnerable. Successive papers noted the conflict between the urgent priorities for socio-economic reconstruction and development, and for environmental repair. Arguments based on ecological, aesthetic, conservation or spiritual grounds for the maintenance of pristine fynbos were pointless in the prevailing political climate. Fortuitously, two processes then in motion converged to create a new and unprecedented opportunity to address the problem of invasive plants in the fynbos, and indeed throughout South Africa.

The results of catchment studies initiated in the 1940s were being synthesized with new ecological, economic and social information. A series of papers (Van Wilgen *et al.*, 1994, 1996; Le Maitre *et al.*, 1996) demonstrated that the invasion of mountain fynbos catchments by woody weed species reduced the water yield dramatically. This information was not new, but when linked to models of water needs, costs of alien plant eradication, etc, demonstrated that the delivery cost of water to consumers in the rapidly growing urban populations of the Western Cape could be substantially reduced by removing alien vegetation from the catchment areas.

Parallel to these ecological studies, the new Government of National Unity was actively addressing the problem of supplying potable water to the huge influx of people to urban "squatters' camps", plus meeting the longer-term objectives of the Reconstruction and Development Programme (RDP) towards raising the quality of life amongst the previously disadvantaged black majority. In September 1995 the Minister of Water Affairs, Professor Kader Asmal, launched the RDP "Working for Water" programme within the National Water Conservation Campaign. The programme aims, *inter alia*, to clear alien vegetation from catchments in ten different areas of the country (see also Richardson, 1998).

The programme was launched with a grant of R25 million (US$ 5.7 million) from the government's RDP fund, plus a further R7 million contributed by the private sector. During the first six months of operation, over 33 000 ha had been cleared of aliens, over 6 600 jobs created (with over 55% going to women) and over 14 500 person days of training provided. The full R25 million RDP grant had been mobilized.

In early 1996, Minister Asmal announced that the programme would continue with a further R50 million grant from the RDP, for the 1996/97 financial year. Thus in the brief period of two years, more funds and human resources have been allocated to alien plant management than in the entire history of South Africa.

Conclusions

The cat eradication programme at Marion Island can be considered a remarkable achievement, especially in the light of the size of the island, its extremely rugged terrain, its inhospitable weather and the extremely difficult logistic arrangements required to undertake such a project 1 800 km distant from the nearest source of support.

The Marion Island Cat Eradication programme is instructive in a number of respects:

1. The programme was initiated in response to international concern (ICBP, SCAR) developing into national policy and action.
2. The programme was successful because it was based on detailed and ongoing research, not only on the ecology and population biology of the island cat population, but also on immunology, hunting, trapping and poisoning techniques, ethical and environmental constraints, and on the island ecosystem as a whole.
3. Throughout the programme, researchers, managers and administrators maintained constant interaction – policy and practise being developed, tested, revised and implemented in a sustained cycle.
4. Public criticism of cat eradication was noted but was not permitted to halt the SASCAR committee's resolve to complete the project. Specialist technical and ethical advice was sought from a series of working groups and independent evaluators.
5. Absolutely critical to the project's success was the ongoing financial and logistic support of the responsible government ministry.
6. The project cost R3 520 000 (US$ 800 000) to execute over a seven year period (1986/93) (Van Schalkwyk, unpublished). The average cost to kill a cat by hunting in 1986/87 was R534 (US$ 144). In 1991/92 the cost of trapping each of the last eight cats was R36 339 (US$8 259). In 1992/93 the cost to continue the hunting and trapping project was R290 714 (US$ 66 071), without a single cat being sighted!

The success of the project has been demonstrated not only by the elimination of cats from the island, but by subsequent increases in the breeding success of burrowing petrels, (Cooper *et al.*, 1995) and by Australia having approved a cat eradication programme, based on South African experience, for Macquarie Island.

The Working for Water programme has already achieved a high record of success. It includes many of the facets that made the Marion Island cat eradication project a success, plus some that were unique to the programme. These include:

1. A substantial base of information and understanding of the structure and function of South African ecosystems, developed through many decades of collaborative interdisciplinary research.
2. A broad base of public interest in the problem of alien invasives, and experience in controlling such invasions.
3. The existence of an informal network of researchers, conservationists, administrators and other interested and affected parties who could respond to opportunities for action.
4. An emerging capacity in environmental and ecological economics, which could build on many decades of descriptive and experimental ecology.

Most fundamental of all, however, is:

5. The vision and commitment of the political leader responsible for water resource and catchment management, securing the enormous financial and human resources necessary to make a significant impact not only on the alien invasive plant problem, but on achieving employment, quality of life and sustainability goals of the Reconstruction and Development Programme.

The history of alien invasive management and control in South Africa demonstrates both success and failure, elegant science and at times, bungling administration. The literature on the topic is extensive, and this brief account does not do justice to the dedication and productivity of South African researchers and managers in this field. The two case studies offer an introduction to the problem as it has been approached through a long history of changing needs, attitudes, socio-economic and political circumstances. Our experience might be instructive to other countries confronted with similar challenges.

Acknowledgements

I would like to thank the many South African participants in the SCOPE project on the Ecology of Biological Invasions (1982–1986) who stimulated a new awareness of and action on this problem, a momentum which endures to this time. In particular, Brian van Wilgen, Ian Macdonald, John Cooper, Dirk van Schalkwyk, Marten Bester and Guy Preston are thanked for their support in sharing their experience in this field.

References

Bester, M.N. (1993) *The cat eradication campaign on Marion Island: A final report.* Mammal Research Institute, University of Pretoria, 17 pp (unpublished).

Bond, P. and Goldblatt, P. (1984) Plants of the Cape Flora. A descriptive catalogue. *Journal of South African Botany,* **Suppl. Vol. 13**, 1–455.

Brooke, R.K., Lloyd, P.H. and de Villiers, A.L. (1986) Alien and translocated terrestrial vertebrates in South Africa, in *The Ecology and Management of Biological Invasions in Southern Africa* (eds I.A.W. Macdonald, F.J. Kruger and A.A. Ferrar), Oxford University Press, Cape Town, pp. 63–74.

Bruton, M.N. and van As, J. (1986) Faunal invasions of aquatic ecosystems in southern Africa, with suggestions for their management, in *The Ecology and Management of Biological Invasions in Southern Africa,* (eds I.A.W. Macdonald, F.J. Kruger and A.A. Ferrar), Oxford University Press, Cape Town, pp. 47–61.

Colonial Botanist (1865) Report for the year 1864. Cape of Good Hope: Annexures to the Votes and Proceedings of the House of Assembly, in *The Ecology and Management of Biological Invasions in Southern Africa,* (eds I.A.W. Macdonald, F.J. Kruger and A.A. Ferrar), Oxford University Press, Cape Town, G24–1865.

Cooper, J. and Brooke, R.K. (1986) Alien plants and animals on South African continental and oceanic islands: species richness, ecological impacts and management, in *The Ecology and Management of Biological Invasions in Southern Africa* (eds I.A.W. Macdonald, F.J. Kruger and A.A. Ferrar), Oxford University Press, Cape Town, pp. 133–142.

Cooper J., Marais, AvN., Bloomer, J.P. and Bester, M.N. (1995) A success story: breeding of burrowing petrels (Procellariidae) before and after the eradication of feral cats *Felis catus* at sub-Antarctic Marion Island. *Marine Ornithology,* **23**, 33–37.

Cowling, R.M., ed. (1992) *The Ecology of Fynbos.* Oxford University Press, Cape Town.

Deacon, J. (1986) Human settlement in South Africa and archaeological evidence for alien plants and animals, in *The Ecology and Management of Biological Invasions in Southern Africa,* (eds I.A.W. Macdonald, F.J. Kruger and A.A. Ferrar), Oxford University Press, Cape Town, pp 3–21.

Fairbridge, D. (1931) Historic farms of South Africa. Oxford University Press, London. In *The Ecology and Management of Biological Invasions in Southern Africa* (eds I.A.W. Macdonald, F.J. Kruger and A.A. Ferrar), Oxford University Press, Cape Town.

Hilton-Taylor, C. (1996) Red Data List of Southern African Plants. *Strelitzia,* **4**, 117 pp.

Huntley, B.J., ed. (1989) *Biotic Diversity in Southern Africa.* Oxford University Press, Cape Town.

Huntley, B.J. (1992) The Fynbos Biome Project, in *The Ecology of Fynbos* (ed. R.M. Cowling), Oxford University Press, Cape Town, pp 1–5.

Huntley, B.J., ed. (1994) Botanical diversity in southern Africa. *Strelitzia,* **1**,1–412.

Le Maitre, D.C., van Wilgen, B.W., Chapman, R.A. and McKelly, D.H. (1996) Invasive plants and water resources in the Western Cape Province, South Africa: modelling the consequences of a lack of management. *Journal of Applied Ecology,* **33** (in press).

Macdonald, I.A.W. (1985) The Australian contribution to southern Africa's invasive alien flora: an ecological analysis. *Proceedings of the Ecological Society of Australia,* **14**, 225–236.

Macdonald, I.A.W., Kruger, F.J. and Ferrar, A.A., eds (1986) *The Ecology and Management of Biological Invasions in Southern Africa.* Oxford University Press, Cape Town.

Macdonald, I.A.W. and Richardson, D. (1986) Alien species in terrestrial ecosystems of the fynbos biome, in *The Ecology and Management of Biological Invasions in Southern Africa* (eds I.A.W. Macdonald, F.J. Kruger and A.A. Ferrar), Oxford University Press, Cape Town, pp. 77–91.

Myers, N. (1990) The biodiversity challenge expanded: hot-spots analysis. *The Environmentalist,* **10,** 243–255.

Phillips, E.P. (1938) The naturalized species of hakea. Farming in South Africa, 13, 424. In *The Ecology and Management of Biological Invasions in Southern Africa* (eds I.A.W. Macdonald, F.J. Kruger and A.A. Ferrar), Oxford University Press, Cape Town.

Richardson, D.M. (1998) Commercial forestry and agroforestry as sources of invasive alien trees and shrubs, in *Invasive Species and Biodiversity Management,* (eds O.T. Sandlund, P.J. Schei and Å. Viken), Kluwer Academic Publishers, Dordrecht, The Netherlands.

Richardson, D.M., Macdonald, I.A.W., Holmes, P.M. and Cowling, R.M. (1992) Plant and animal invasions, in *The Ecology of Fynbos,* (ed. R.M. Cowling), Oxford University Press. Cape Town, pp. 271–308.

Shaughnessy, G. (1986) A case study of some woody plant introductions to the Cape Town area, in *The Ecology and Management of Biological Invasions in Southern Africa,* (eds I.A.W. Macdonald, F.J. Kruger and A.A. Ferrar), Oxford University Press, Cape Town, pp. 37–43.

Van Aarde, R.J. (1979) Distribution and density of feral house cat *Felis catus* on Marion Island. *South African Journal of Antarctic Research,* **9,** 14–19.

Van Rensburg, P.J.J., Skinner, J.D. and R.J. Van Aarde (1987) Effects of feline panleucopaenia on the population characteristics of feral cats on Marion Island. *Journal of Applied Ecology,* **24,** 63–73.

Van Wilgen, B., Richardson, D.M. and Seydack, A.H.W. (1994) Managing fynbos for biodiversity: constraints and options in a fire-prone environment. *South African Journal of Science,* **90,** 322–329.

Van Wilgen, B.W., Cowling, R.M. and Burgers, C.J. (1996) Valuation of ecosystem services. *Bioscience,* **46,** 184–189.

Van Zinderen Bakker, E.M. (1971) Introduction, in *Marion and Prince Edward Islands,* (eds E.M. Van Zinderen Bakker, J.M. Winterbottom and R.A. Dyer), A.A. Balkema, Cape Town, pp. 1–15.

Van Zinderen Bakker, E.M., Winterbottom, J.M. and Dyer, R.A. (1971) *Marion and Prince Edward Islands.* A.A. Balkema, Cape Town.

Watkins, B.P. and Cooper, J. (1986) Introduction, present status and control of alien species at the Prince Edward Islands, sub-Antarctic. *South African Journal of Antarctic Research,* **16,** 86–94.

Wells, M.J., Poynton, R.J., Balsinhas, A.A., Musil, K.J., Joffe, H., Van Hoepen, E. and Abbot, S.K. (1986a) The history of introduction of invasive alien plants to southern Africa, in *The Ecology and Management of Biological Invasions in Southern Africa,* (eds I.A.W. Macdonald, F.J. Kruger and A.A. Ferrar), Oxford University Press, Cape Town, pp. 21–35.

Wells, M.J., Balsinhas, A.A., Joffe, H., Engelbrecht, V.M., Harding, G. and Stirton, C.H. (1986b) Catalogue of problem plants of southern Africa. *Memoirs of the Botanical Survey of South Africa,* **53**. Government Printer, Pretoria.

25 Managing alien species: the Australian experience

ROGER P. PECH
CSIRO Wildlife and Ecology, Canberra, Australia

Abstract

As an island continent Australia has a viable option of effectively preventing the illegal and accidental importation of unwanted terrestrial species with strict quarantine procedures. An important element of the assessment of legal importations of alien species is the use of ecological criteria to separate beneficial species from those likely to become pests. For existing pest species, the management objective can be eradication but because this has rarely been achieved on a continental scale, a more realistic goal of cost-effective sustained management is often necessary. Sustained management requires an understanding of the relationships between the use of control techniques and the abundance of a pest species, and between pest abundance and the damage it causes.

In Australia there has been widespread public acceptance of the role of introduced pest species as threats to the conservation of native flora and fauna. However public concerns have also been expressed about the ecological consequences of the removal of an alien species once it has become widely established and a conspicuous component of the environment. Part of the solution is to develop management strategies for whole systems, not single species, and to involve the community in the planning and implementation of management programs.

Introduction

Crosby (1986) listed three pre-requisites for the successful creation of what he called "neo-Europes" during the process of global colonisation by Europeans and their flora and fauna. These are (i) the new colony must have a broadly similar climate to Europe, (ii) the prospective colony had to be remote so that its indigenous peoples and native animal and plant species were not adapted to the parasites, competitors and predators imported from Europe and (iii) remoteness would also ensure that Europeans would have a technological edge both in warfare and in exploiting the natural resources of the new land. The Australia of the 18th century satisfied all criteria. The result is that much of the Australian landscape, particularly in the southern temperate areas, is now dominated by alien species. These include a small suite of domesticated animals and plants which have a net benefit, and a

O. T. Sandlund et al. (eds.), Invasive Species and Biodiversity Management, 377–386.

large number of deliberate and accidental introductions with no demonstrable value, many of which have become pests for agriculture and conservation. Though the first category includes species which have become icons of Australian agriculture, these too have been in substantial conflict with native species. For central Australia, Morton (1990) identified competition with sheep and cattle as one of the primary agents leading to the decline of small-to-medium sized mammals. The initial eruption of the sheep population of semi-arid NSW exceeded the resources of the land and although the density of sheep subsequently settled into a series of fluctuations about a lower mean (Caughley, 1987), major environmental impacts had been set in train early in the process of European settlement (Tunbridge, 1991). A similar pattern of initial optimism characterised early land clearing and the expansion of cropping in southern Australia.

Not all more recent introductions of plant and animal species have been beneficial. Of the more than 220 alien plant species declared as noxious weeds in Australia, 46% were introduced intentionally for other purposes, 31% as ornamentals (Panetta, 1993). Lonsdale (1994) reviewed the outcome of the importation of 463 species of grasses and legumes into northern Australia between 1947 and 1985. These species were deliberately introduced to improve cattle production in tropical areas but ultimately 21 were considered useful and of these only 8 were assessed under grazing conditions. Unfortunately 17 of the 21 useful species are listed now as weeds. Thirteen percent of all the introductions are listed as weed species, of which 39 were weeds of conservation. The conclusion was that the programme of plant introductions had increased the need for weed management, that no reliable ecological indicators of weediness have emerged, and that plants which might be useful for cattle production could also become weeds of cropping and conservation.

Current problems with marine invasive species include long-term resident and self-sustaining populations of algae and small benthic organisms. Some of these organisms were introduced in the early stages of coastal shipping and others more recently, several by way of ballast water. Apart from a few high profile species such as the Japanese starfish, the full scope of marine invasions is unknown.

Many arthropods have been accidentally introduced, but their impact has only been assessed in agricultural terms up to now, though some have undoubtedly had significant effects on native flora and fauna. Both accidental and deliberate introductions have generated many of the present day problems in vertebrate pest management. Species such as rabbits, feral horses, feral pigs and foxes, originally introduced for agricultural, sporting or aesthetic reasons, are now considered to have major impacts on agricultural production and conservation values. Fortunately the

Acclimatisation Societies which formed during the second half of the 19th century to promote species introductions have been replaced by stringent controls on the importation of animal species. Assessment procedures now include the risk of a new species becoming established in the wild, the impact it might have and the likelihood of containment and eradication (Bomford, 1991). But difficulties still remain in attempting to predict the characteristics of successful invasive species (Newsome and Noble, 1986).

The National Strategy for the Conservation of Australia's Biological Diversity (Anonymous, 1996) summarises the major conservation issues involving the management of alien species. The strategy includes a programme of research to assess the potential and realised impacts of alien species and further development of biological and other control methods. It also outlines programs for the control or eradication of alien species and procedures to minimise the risk of importing additional pest species into Australia. Under the Commonwealth Endangered Species Protection Act 1992, the Australian Nature Conservation Agency is responsible for the development of Threat Abatement Plans for key threatening processes such as predation by feral cats and foxes, and actively supports related research.

Management strategies for some of Australia's most important vertebrate pest species have been detailed in a series of national guidelines (Braysher, 1993; Dobbie *et al.*, 1993; Williams *et al.*, 1995; Saunders *et al.*, 1995; Parkes *et al.*, 1996; Choquenot *et al.*, 1996). In addition the adoption of "best practice" in pest management is being encouraged by large-scale field projects with the support of the Bureau of Resource Sciences. An important aim of these projects and other recent initiatives in pest management, is to shift the focus from reducing the abundance of pests to reducing the damage they cause and to involve land managers in objectively testing the outcome of pest control. Rather than simply providing sites for research, landholders are an integral part of the projects and help develop management options. The long-term goal is to achieve wider acceptance of cost-effective pest control through practical demonstration and to augment existing conventional methods for providing information. A similar emphasis on integration of approaches and community involvement is part of the National Weeds Strategy.

Objectives for the management of alien species

Much of the management of alien species is linked to the problem of "overpopulation": those species which are too numerous and out of place. Caughley (1981) classified the problems of overpopulation as: (i) the animals threaten human life or livelihood, (ii) the animals depress the densities of favoured species, (iii) the animals are too numerous for their own good, and (iv) the system of plants and animals is off its equilibrium. An important point that emerged from Caughley's review of these problems is that human perceptions are invariably implicit in setting management objectives, particularly in the first three categories, and that the contribution of science is often confused with essentially non-scientific issues in the management process. An example is the conflicting goals of eradication and sustained control.

While both alien and native species can be regarded by humans as pests, it is only for the former category that eradication is sometimes promoted as a management goal. Eradication of species such as the European rabbit and feral goats and cats has been successfully achieved on islands and localised regions of mainland Australia but a review by Bomford and O'Brien (1995) clearly demonstrates that the prospects for eradicating a pest species with a continental distribution is vanishingly small. Despite this landholders have been required by legislation to suppress and destroy pests on their land with an implicit requirement to continue until all are removed.

For most pest species a management objective short of eradication must be accepted. This goal implies a trade-off between the resources expended on control and the extent to which damage caused by the pest can be reduced. The relationships between the abundance of a vertebrate pest and the damage it causes, and between application of a control technique and pest abundance, have been reviewed by Hone (1994, 1996) with particular attention to evidence for non-linear relationships. Non-linearities are important because they are a departure from the view that any effort expended on control will produce a corresponding benefit. In some cases the initial efforts in pest control can produce substantial gains, for example where a high level of damage is caused by a few individuals which can be selectively removed. Alternatively a substantial level of pest control may be required before any significant reduction in damage occurs. This applies if significant damage occurs at very low pest densities such as consumption of seedlings of some native shrub species by rabbits (Lange and Graham, 1983; Cooke, 1987). In addition, use of a control technique may not produce a commensurate decline in pest abundance. In an extreme case, Caughley *et al.* (1992) showed that at some levels of fertility control it is possible to increase the abundance of a pest species and, potentially, the damage it causes.

While the agricultural impacts of some pests, such as feral pigs and several arthropods and weeds, are relatively well quantified, the same is not generally true of impacts on the natural environment. An example is predation by introduced foxes and feral cats on native species. Prey in the critical weight range of 0.1–3 kg are thought to be particularly vulnerable

(Burbidge and McKenzie, 1989) and a growing series of studies have begun to document the importance of predation by foxes and feral cats on native species (for example, Kinnear *et al.*, 1988; Christensen and Burrows, 1994; Friend and Thomas, 1994; Gibson *et al.*, 1994; Priddel and Wheeler, 1994; Southgate, 1994). In some cases, such as the burrowing bettong, *Bettongia lesueur* (Short *et al.*, 1994), it appears that native species are unable to withstand the presence of any foxes or feral cats. In other cases, recent analysis suggests that the persistence of some native species should be considerably enhanced with an intermediate level of predator control and that some species which are predator's primary prey are, paradoxically, good candidates for reintroduction programmes (Sinclair *et al.*, 1998). An example is the woylie or brush-tailed bettong, *Bettongia penicillata*, which persists in areas of forest in south-west Western Australia where there are relatively few resident foxes. The distinction between primary prey and secondary prey is critical for low density prey populations and hence on the level of predator control required need to protect endangered species. Primary prey are characterised by having physical or behavioural protection from predation at low densities whereas secondary prey are essentially "by-catch" and are susceptible to predation at all densities (Pech *et al.*, 1995). In addition, the effects of sub-lethal predation can be important, for example through changes in habitat which affect the degree of exposure to predation. The interaction between predation and shelter is currently the subject of a large-scale manipulative experiment conducted by CSIRO. The aim is to characterise the dynamics of declining populations and hence better define the "declining-population paradigm" proposed by Caughley (1994). This paradigm seeks to deal with the processes which drive species towards extinction rather than the stochastic events which ultimately extinguish small populations.

Techniques for the management of alien species

Pest management can be classified as either "conventional" control, which includes mechanical and chemical techniques, or "biological" control. Conventional techniques are the major component of the national management guidelines which have been produced or are currently being written for key vertebrate pest species (feral horses, rabbits, foxes, feral goats, feral pigs and rodents). Integrated control, combining different approaches, is being emphasised for weeds.

The use of biological controls for managing pest species has a long history in Australia. Biological control of the weed, prickly pear, in the 1930's led to many more programs. There are currently 40 active biological control programs against alien weeds. The spectacular reduction in the abundance of rabbits in the 1950's caused by myxomatosis has tended to mask the more general failure of introduced pathogens to control vertebrate pest populations (Spratt, 1990). However one route to more effective pest management is to enhance the efficacy of an existing agent. A recent example is the release of the arid-adapted flea, *Xenopsylla cunicularis*, to assist the spread of myxoma virus in areas where the European flea does not persist (Williams *et al*., 1995). Also recent advances in genetic engineering have opened new possibilities for biological control agents. A current programme of research into viral-vectored immunocontraception has developed from the idea of engineering myxoma virus so that rabbits which survive the disease are made infertile (Tyndale-Biscoe, 1994). The Cooperative Research Centre for the Biological Control of Vertebrate Pest Populations (VBC) which is conducting this research is also investigating immunocontraceptive agents to be delivered by a virus for house mice (Shellam 1994) and by baits for foxes (Bradley, 1994; Pech *et al*., 1997).

In Australia the efficacy of broadscale fertility control of vertebrate pests is in the process of being tested by the VBC. Clearly, if a sufficiently high level of infertility can be maintained in a population it will decline. Part of the challenge is to develop means of delivering a fertility control agent to a large proportion of a population in a cost-effective way. The ultimate test of fertility control is to determine whether the reduction in pest numbers that can be achieved is sufficient to be useful in meeting management objectives. Preliminary data from large-scale experiments with populations of rabbits sterilised at a range of intensities suggest that moderately high levels of treatment will cause a decline in rabbit abundance over 3 years. However, rabbits appear to be able to compensate through enhanced survival of the offspring of the remaining fertile animals so that the outcome of the control technique is not a commensurate reduction in density. Other compensatory mechanisms, such as changes in the territorial behaviour of breeding and infertile animals, are being studied in foxes.

Consequences of managing alien species

Perhaps surprisingly, new opportunities to control pest species can lead to uncertainty about potential benefits, especially when the pest has become widespread and abundant. For example the introduced plant, *Echium plantagineum*, is considered by farmers to be an invasive, sometimes toxic weed of grazing land but by bee keepers as an essential resource for honey production. Its perceived value is exemplified in the alternative common names, Paterson's Curse and Salvation Jane. The use of biological control agents against *E. plantagineum* was blocked during a protracted, 5-year debate over the claimed detrimental and beneficial consequences of control (Cullen and Delfosse, 1984). The procedure for selecting a target organism for control and for dealing with biological control agents has since been formalised in the Biological Control Act 1984.

A second example of conflicting interests is the recent spread of rabbit calicivirus disease (RCD, also known as rabbit haemorrhagic disease) into mainland Australia in 1995. The agricultural and environmental need to control Australia's most important vertebrate pest is in conflict with the interests of pet owners and commercial enterprises dependent on rabbit products. As well, public concern has been expressed about possible detrimental environmental effects from controlling rabbits, particularly the potential for increased predation on native wildlife by foxes and feral cats following RCD-induced declines in rabbit populations. These issues have been reviewed by Newsome *et al.* (1997). Apart from any short-term impacts of the initial introduction of RCD, the likely longer-term effects of changes in the abundance of rabbits are uncertain. Models which include the interactions of foxes, rabbits, alternative prey and RCD, predict substantial benefits in semi-arid Australia (Pech and Hood, 1998). As well as reducing the mean abundance of rabbits, RCD is likely to decrease the frequency of rabbit plagues both directly and by enhancing the regulatory effect of predators. Recruitment of perennial plant species should improve because there should be fewer occasions when pasture biomass decreases to levels where there is intense competition between rabbits, sheep, kangaroos and other herbivores. The predicted decline in the abundance of introduced predators such as foxes should enhance the prospects for persistence and recovery of small to medium sized native mammals. Other consequences of introducing RCD may include declines in the abundance of some native raptors which have become dependent on rabbits as primary food, and increases in weed species which are currently suppressed by rabbits. An extensive national program has been established to conduct further research on the epidemiology of the virus and to assess its effects on agriculture and the environment.

Summary

Alien species have made significant changes to the Australian landscape over the last 200 years but there remain substantial opportunities to redress some of the imbalances. In terms of pest management, the shift in emphasis from reducing pest numbers to controlling the damage they cause represents a major advance in developing management strategies. In the context of conservation, methods for identifying the contribution of alien species to threatening processes and the early recognition of native species under threat are issues which require more attention in the future. New technologies such as immunocontraception offer the prospect of species-specific, humane methods of controlling vertebrate pests, especially if coupled with cost-effective dissemination techniques such as viral-vectors.

Acknowledgments

Thanks are due to A. Newsome, L. Hinds, B. Walker, M. Braysher, J. Cullen and K. Colgan for useful comments on the manuscript.

References

Anonymous (1996) *The National Strategy for the Conservation of Australia's Biological Diversity.* Commonwealth Department of Environment, Sport and Territories, Canberra.

Braysher, M. (1993) *Managing Vertebrate Pests: Principles and Strategies.* Bureau of Resource Sciences, Australian Government Publishing Service, Canberra.

Bomford, M. (1991) *Importing and Keeping Exotic Vertebrates in Australia. Criteria for the Assessment of Risk*, Bureau of Rural Resources, Bulletin No. 12, Australian Government Publishing Service, Canberra.

Bomford, M. and O'Brien, P. (1995) Eradication or control for vertebrate pests? *Wildlife Society Bulletin,* **23**, 249–255.

Bradley, M.P. (1994) Experimental strategies for the development of an immunocontraceptive vaccine for the European red fox, *Vulpes vulpes*, in *Immunological Control of Fertility: from Gamètes to Gonads,* (ed. M. Bradley), *Reproduction Fertility and Development*, **6,** 307–317.

Burbidge, A.A. and McKenzie, N.L. (1989) Patterns in the modern decline of Western Australia's vertebrate fauna: causes and conservation implications. *Biological Conservation,* **50**, 143–198.

Caughley, G. 1981. Overpopulation, in *Problems in Management of Locally Abundant Wild Mammals,* (eds P.A. Jewell and S. Holt), Academic Press, New York, pp. 7–19.

Caughley, G. (1987) Introduction to the sheep rangelands, in *Kangaroos: Their Ecology and Management in the Sheep Rangelands of Australia,* (eds G. Caughley, N. Shepherd and J. Short), Cambridge University Press, Cambridge, pp. 1–13.

Caughley, G. (1994) Directions in conservation biology. *Journal of Animal Ecology,* **63**, 215–244.

Caughley, G., Pech, R. and Grice, D. (1992) Effect of fertility control on a population's productivity. *Wildlife Research,* **19**, 623–627.

Christensen, P. and Burrows, N. (1994) Project desert dreaming: experimental reintroduction of mammals to the Gibson Desert, Western Australia, in *Reintroduction Biology of Australian and New Zealand Fauna,* (ed. M. Serena), Surrey Beatty & Sons, Chipping Norton, pp. 199–207.

Choquenot, D., McIlroy, J. and Korn, T. (1996) *Managing Vertebrate Pests: Feral Pigs.* Bureau of Resource Sciences, Australian Government Publishing Service, Canberra.

Cooke, B.D. (1987) The effects of rabbit grazing on regeneration of sheoaks, *Allocasuarina verticilliata*, and saltwater ti-trees, *Melaleuca halmaturorum*, in the Coorong National Park, South Australia. *Australian Journal of Ecology*, **13**, 11–20.

Crosby, A.W. (1986) *Ecological Imperialism. The Biological Expansion of Europe, 900–1900*. Cambridge University Press, Cambridge.

Cullen, J.M. and Delfosse, E.S. (1985) *Echium plantagineum* catalyst for conflict and change in Australia, in *Proceedings of the VIth International Symposium on the Biological Control of Weeds*, (ed. E.S. Delfosse), 19–25 August 1984, Vancouver, Canada. Agriculture Canada, pp. 249–292.

Dobbie, D., Berman, D. and Braysher, M. (1993) *Managing Vertebrate Pests: Feral Horses.* Bureau of Resource Sciences, Australian Government Publishing Service, Canberra.

Friend, J.A. and Thomas, N.D. (1994) Reintroduction and the numbat recovery programme, in *Reintroduction Biology of Australian and New Zealand Fauna*, (ed. M. Serena), Surrey Beatty & Sons, Chipping Norton, pp. 189–198.

Gibson, D.F., Johnson, K.A., Langford, D.G., Cole, J.R., Clarke, D.E. and Willowra Community (1994) The rufous hare-wallaby *Lagorchestes hirsutus*: a history of experimental reintroduction in the Tanami Desert, Northern Territory, in *Reintroduction Biology of Australian and New Zealand Fauna*, (ed. M. Serena), Surrey Beatty & Sons, Chipping Norton, pp. 171–176.

Hone, J. (1994) *Analysis of Vertebrate Pest Control.* Cambridge University Press, Cambridge.

Hone, J. (1996) Analysis of vertebrate pest research, in *Proceedings of the 17th Vertebrate Pest Conference*, (eds R.M. Timm and A.C. Crabb), University of California, Davis, CA, pp. 13–17.

Kinnear, J.E., Onus, M.L. and Bromilow, R.N. (1988) Fox control and rock-wallaby population dynamics. *Australian Wildlife Research*, **15**, 435–450.

Lange, R.T. and Graham, C.T. (1983) Rabbits and the failure of regeneration of arid zone *Acacia*. *Australian Journal of Ecology*, **8**, 377–381.

Lonsdale, W.M. (1994) Inviting trouble: introduced pasture species in northern Australia. *Australian Journal of Ecology*, **19**, 345–354.

Morton, S.R. (990). The impact of European settlement on the vertebrate animals of arid Australia. *Proceedings of the Ecological Society of Australia*, **16**, 201–213.

Newsome, A.E. and Noble, I.R. (1986) Ecological and physiological characters of invading species, in *Ecology of Biological Invasions: An Australian Perspective*, (eds R.H. Groves and J.J. Burdon), Australian Academy of Science, Canberra, pp. 1–20.

Newsome, A., Pech, R,. Smyth, R., Banks P., and Dickman, C. (1996) *Potential Impacts on Australian Native Fauna of Rabbit Calicivirus Disease.* Environment Australia, Canberra.

Panetta, F.D. (1993). A system for assessing proposed plant introductions for weed potential. *Plant Protection Quarterly*, **8**, 10–14.

Parkes, J., Henzell, R. and Pickles, G. (1996) *Managing Vertebrate Pests: Feral Goats.* Bureau of Resource Sciences and Australian Nature Conservation Agency, Australian Government Publishing Service, Canberra.

Pech, R.P., Sinclair, A.R.E. and Newsome, A.E. (1995) Predation models for primary and secondary prey species. *Wildlife Research*, **22**, 55–64.

Pech, R.P. and Hood, G. (1998). Foxes, rabbits, alternative prey and rabbit calicivirus disease: ecological consequences of a new biological control agent for an outbreaking species in Australia. *Journal of Applied Ecology*, (in press).

Pech, R., Hood, G.M., McIlroy, J. and Saunders, G. (1997) Can foxes (*Vulpes vulpes*) be controlled by reducing their fertility? *Reproduction, Fertility and Development*, **9**, 41–50.

Priddel, D. and Wheeler, R. (1994) Mortality of captive-raised malleefowl, *Leipoa ocellata*, released into a mallee remnant within the wheat-belt of New South Wales. *Wildlife Research*, **21**, 543–552.

Saunders, G., Coman, B., Kinnear, J. and Braysher, M. (1995) *Managing Vertebrate Pests: Foxes*. Bureau of Resource Sciences and Australian Nature Conservation Agency, Australian Government Publishing Service, Canberra.

Shellam, G.R. (1994) The potential of murine cytomegalovirus as a viral vector for immunocontraception, in *Immunological Control of Fertility: from Gametes to Gonads*, (ed. M. Bradley), *Reproduction Fertility and Development*, **6,** 401–409.

Short, J., Turner, B., Parker S. and Twiss, J. (1994) Reintroduction of endangered mammals to mainland Shark Bay: a progress report, in *Reintroduction Biology of Australian and New Zealand Fauna* (ed. M. Serena), Surrey Beatty & Sons, Chipping Norton, pp. 183–188.

Sinclair, A.R.E., Pech, R.P., Dickman, C., Newsome, A., Hik, D. and Mahon, P. (1998) Predicting the effects of predation on conservation of endangered prey. *Conservation Biology*, **12,** 564-575.

Southgate, R.I. (1994) Why reintroduce the Bilby?, in *Reintroduction Biology of Australian and New Zealand Fauna,* (ed. M. Serena), Surrey Beatty & Sons, Chipping Norton, pp. 165–170.

Spratt, D.M. (1990) The role of helminths in the biological control of mammals. *International Journal for Parasitology,* **20**, 543–550.

Tunbridge, D. (1991) *The Story of the Flinders Ranges Mammals*. Kangaroo Press, Kenthurst.

Tyndale-Biscoe, H. (1994) The CRC for the Biological Control of Vertebrate Pest Populations: fertility control of wildlife for conservation. *Pacific Conservation Biology,* **1**, 160–162.

Williams, K., Parer, I., Coman, B., Burley J. and Braysher, M. (1995) *Managing Vertebrate Pests: Rabbits*. Bureau of Resource Sciences and CSIRO, Australian Government Publishing Service, Canberra.

Part 6

Where do we go from here?

26 International instruments, processes, organizations and non-indigenous species introductions: is a protocol to the Convention on Biological Diversity necessary?[2]

LYLE GLOWKA and CYRILLE de KLEMM
IUCN Environmental Law Centre, Bonn, Germany, and IUCN Commision on Environmental Law, Paris, France

Abstract

References to non-indigenous species can be traced to a number of international instruments and processes at the global and regional levels, with the citation rate increasing markedly from the 1980s until the present. International organizations are also increasingly involved with non-indigenous species issues. But despite the international community's acknowledgment that the intentional and unintentional introduction of non-indigenous species can lead to ecological damage, the consequent loss of biological diversity as well as potentially huge economic and development losses, the extent to which introductions occur appears to be increasing and suggests controls in many countries are far from adequate.

The ratification of the Convention on Biological Diversity by over 170 States and Regional Economic Integration Organizations suggests that the means may exist to develop a comprehensive global approach to the intentional and unintentional introduction of non-indigenous species. The Convention's "country-driven" nature allows Party's to tailor their own approach to the introduction, control or eradication of non-indigenous species through their national biodiversity planning processes. The emphasis on national action and priority-setting is desirable from several standpoints. However, consistency of approach and harmonization of goals – the results of coordination and common priority setting – may be sacrificed without further action by the Convention's Conference of Parties (COP).

This paper will provide a brief survey of the international instruments, processes and organizations which have addressed non-indigenous species to date. A summary of work thus far by the COP will be made. Possible options for a more positive role by the COP will be suggested.

Introduction

References to the introduction of non-indigenous species can be traced to a number of international instruments and processes at the global and regional

[2] Previously published in Environmental Policy and Law, Vol. 26, No. 6, November 1996.

O. T. Sandlund et al. (eds.), Invasive Species and Biodiversity Management, 389–405.

levels, with the citation rate increasing markedly from the 1980s until the present. International organizations are also increasingly involved with non-indigenous species introduction issues. But despite the international community's acknowledgment that such introductions can lead to ecological damage, the consequent loss of biological diversity as well as potentially huge economic and development losses, introductions appear to be increasing suggesting measures in many countries are far from adequate.

Because it has been very widely ratified, the Convention on Biological Diversity (CBD) offers the best opportunity for developing a comprehensive global approach to the intentional and unintentional introduction of non-indigenous species and their eradication or control. The CBD process offers the means to develop consistency of approach and normalization of standards and practices – the results of coordination and common priority setting – through further action by the Conference of Parties (COP).

This paper will provide a brief survey of the international instruments which address non-indigenous species introductions [3] (section II.A) and provide a brief overview of the status of national legislation (section II.B).

Establishing the basis for future action under the Convention on Biological Diversity, including a summary of work on the issue thus far undertaken by the COP and basic principles to guide future work, is discussed (section III).

International instruments which address non-indigenous species introductions and national legislative responses

International instruments

There is currently no binding global treaty dealing solely with the full spectrum of non-indigenous species introductions, although the 1951 International Plant Protection Convention applies to phytosanitary issues.[4]

[3] Introduction is the intentional or accidental dispersal by human agency of a living organism outside its historically known native range (IUCN, 1987). We acknowledge that this definition is broad enough to include genetically modified organisms (GMOs) though this paper will not address GMOs. For a survey of existing instruments on GMOs, see CBD (1995).

Re-introduction (or, more appropriately, re-establishment) and restocking, because of their potential impacts on genetic, taxonomic and ecosystem diversity, are issues related to non-indigenous species introductions which, though beyond the scope of this paper, ideally should be treated together with non-indigenous species introductions. In May 1995, IUCN approved Guidelines on Re-introductions available from the Species Survival Commission.

Instead, the issue is addressed by three categories of instruments which represent a continuum of obligations and guidance (1) global and regional multi-lateral treaties (sometimes called "hard" or binding law) containing provisions on non-indigenous species introduction which, after ratification or accession, States or regional economic integration organizations (REIOs) are then bound to implement; (2) "soft law" instruments reflecting political rather than binding legal commitments, although their implementation by States can contribute to the body of customary international law; and, (3) issue-specific international technical guidance documents promulgated by various international organizations often in the form of recommendations. A number of bi-lateral treaties also exist but are beyond this paper's scope.

Global and regional multi-lateral treaties

There are 22 global and regional multi-lateral environment and conservation treaties referencing non-indigenous species introductions (Glowka, 1996). Twelve have entered into force as of June 1996. Of these, the most notable is the Convention on Biological Diversity (CBD) which, in terrestrial, aquatic and marine areas, requires its parties, as far as possible and as appropriate, "to prevent the introduction of, control or eradicate those aliens species which threaten ecosystems, habitats or species" (article 8(h)).

When they are range States of endangered migratory species listed in annex I, the 38 parties to the 1979 Convention on Migratory Species of Wild Animals, to the extent feasible and appropriate, agree to endeavour to prevent, reduce or control "factors that are endangering or are likely to further endanger the species, including strictly controlling the introduction of, or controlling or eliminating, already introduced exotic species" (article III(4)(c)). Furthermore, agreements for annex II migratory species "where appropriate and feasible (...) should provide for (...) (e) (...) strict control of the introduction of, or control of already introduced exotic species detrimental to the migratory species" (Annex II, article V(4)).

[4] Phytosanitary agreements, the most notable being the 1951 International Plant Protection Convention (as amended in 1979), have as their primary purpose preventing the introduction and spread of plant and plant product "pests" primarily in international trade. The 1951 Convention creates an international phytosanitary regime between its 98 parties premised on parties establishing national plant protection offices which, among other things, issue and review phytosanitary certificates for plants and plant products in international trade and undertake eradication and control efforts. The Convention provides that supplementary agreements can be proposed and negotiated for particular regions, "pests", plants and plant products and methods of international transport.

Negotiated under the auspices of the Migratory Species Convention, the Agreement on the Conservation of African-Eurasian Migratory Waterbirds (Hague 1995), which has yet to enter into force, requires parties to (1) prohibit all deliberate introduction into the environment of non-native waterbird species, (2) take all appropriate measures to prevent unintentional release where prejudicial to the conservation status of wild flora and fauna, as well as (3) take all appropriate measures to prevent already introduced species from becoming a potential threat to indigenous species (article III(2)(g)). The Waterbirds Agreement is accompanied by an action plan for species listed in an accompanying table. Section 2.5 deals with introductions and requires parties (1) to prohibit non-native animal and plant introductions if detrimental to listed species, (2) to take precautions to prevent the accidental escape of captive non-native birds and (3) to take measures to ensure that already introduced non-native species do not pose a potential hazard to listed species.

Focusing specifically on the marine environment, the 1982 United Nations Convention on the Law of the Sea codifies existing customary international law of the sea and creates new obligations for its 85 parties. States are to "take all measures necessary to prevent, reduce and control (...) the intentional or accidental introduction of species, aliens or new, to a particular part of the marine environment, which may cause significant and harmful changes thereto" (article 196(1)).

At the regional level, five protocols to the United Nations Environment Programme (UNEP) Regional Seas Conventions require parties to take measures to limit non-indigenous species introductions **either** within the protocol's geographical region of application, marine protected areas **or** both. These are the Protocol Concerning Mediterranean Specially Protected Areas (Geneva 1982); Protocol Concerning Protected Areas and Wild Fauna (Nairobi 1985); the Protocol for the Conservation and Management of Protected Marine and Coastal Areas of the South–East Pacific (Paipa 1989); the Protocol Concerning Specially Protected Areas and Wildlife to the Convention for the Protection and Development of the Marine Environment of the Wider Caribbean Region (Kingston 1990); and the Protocol Concerning Specially Protected Areas and Biological Diversity in the Mediterranean (Barcelona 1995). Of these, only the 1982 Geneva Protocol Concerning Mediterranean Specially Protected Areas is in force. This protocol will be replaced by the more comprehensive 1995 Barcelona Protocol Concerning Specially Protected Areas and Biological Diversity in the Mediterranean upon its entry into force.

Under the Barcelona protocol, negotiated after the CBD's adoption, parties are to regulate introduction of any non-indigenous species into specially protected areas (article 6(d)), regulate the intentional or accidental introduction into the wild of non-indigenous species and prohibiting those which may have harmful impacts on ecosystems, habitats or species (article 13(2)), as well as endeavour to eradicate species which scientific assessments demonstrate to cause or are likely to cause damage to ecosystems, habitats or species (article 13(2)).

In order to maintain the pristine nature of Antarctica, the Antarctic Treaty System creates a very strict international legal regime dealing with non-indigenous species introductions. The 17 Parties to the Agreed Measures for the Conservation of Antarctic Fauna and Flora (Brussels 1964) are to prohibit introduction into the treaty area of non-indigenous plants and animals (article IX(1-4)). Only organisms listed in an annex can be introduced after a permit issued by a competent national authority. Parties are also to take reasonable precautions to prevent introduction of parasites and diseases.

The 1991 Madrid Protocol to the Antarctic Treaty on Environmental Protection, which has not entered into force, builds on the Agreed Measures. It would prohibit non-native species introductions into the treaty area without a permit, allowing their issuance only for organisms listed in an appendix on the condition that containment measures are taken. Permitted organisms are to be destroyed prior to permit expiration, sled dogs and live poultry are banned from the treaty area and the importation of non-sterile soil is to be avoided (Annex II, article 4). From the Antarctic convergence and south of 60 degrees south latitude, the 28 parties to the Convention on the Conservation of Antarctic Marine Living Resources (Canberra 1980) undertake to "prevent changes or minimise the risk for changes in the marine ecosystem (...) not potentially reversible over two or three decades, taking into account the state of available knowledge (...) (including) the effect of alien species" (article II(3)(c)).

At the regional level, outside Europe, excluding the regional seas protocols, there is poor coverage of non-indigenous species introductions. The 1968 African Convention on the Conservation of Nature and Natural Resources, which is in force for 30 States, makes only a tangential reference to prohibiting non-indigenous introductions in the definitional sections for strict nature reserves and national parks (article III(4)(a)(ii)).

However, pursuant to the Agreement on the Preparation of a Tripartite Environmental Management Programme for Lake Victoria (Dar-es-Salaam 1994), Kenya, Tanzania and Uganda have agreed to implement a 5 year programme to strengthen regional environmental management of Lake Victoria including control of water hyacinth (but not the Nile Perch). Biological control will be used after environmental risks are found acceptable by national authorities. Mechanical control means will also be explored.

The Convention on the Conservation of Nature in the South Pacific (Apia 1976) merely states that its 5 Parties "shall carefully consider the consequences of the deliberate introduction into ecosystems of species which have not previously occurred therein" (article V(4)). The ASEAN Agreement on the Conservation of Nature and Natural Resources (Kuala Lumpur 1985), an otherwise progressive conservation treaty under which parties are to endeavour "to regulate and, where necessary, prohibit the introduction of exotic species" (article 3(3)), has not yet entered into force.

In North America, the North American Agreement on Environmental Cooperation, a side agreement of the North American Free Trade Agreement, created the Commission on Environmental Cooperation, whose Council may develop recommendations regarding exotic species which may be harmful. Canada, Mexico and the United States of America are parties.

Neither is coverage strong in Latin America. The Convention for the Conservation of the Biodiversity and the Protection of Wilderness Areas in Central America (Managua 1992) has yet to enter into force, however, article 24 provides that "[m]echanisms shall be established for the control or eradication of all exotic species which threaten ecosystems, habitats and wild species". It appears the only South American instrument is the Protocol for the Conservation and Management of Protected Marine and Coastal Areas of the South–East Pacific (Paipa 1989), a Regional Seas agreement which has not entered into force.

With four instruments (excluding the 1982 Geneva and 1995 Barcelona Regional Seas Protocols mentioned earlier) Europe has the most extensive coverage of any region in the world. Two instruments prohibit non-indigenous species introductions without consent from a competent authority (Convention Concerning Fishing in the Waters of the Danube (Bucharest 1958); Benelux Convention on Nature Conservation and Landscape Protection (Brussels 1982)) and one requires parties to take strict, though unspecified, control measures (Convention on the Conservation of European Wildlife and Natural Habitats (Bern 1979). These three conventions have entered into force with 6, 3 and 31 parties respectively, while the Protocol for the Implementation of Alpine Convention in the Field of Nature Protection and Landscape (Chambéry 1994), the fourth instrument, has not yet entered into force.

Within the European Union two directives, though technically not international instruments *per se*, are worth mentioning, as their primary purpose is to facilitate the Bern Convention's implementation within the European Community. Council Directives 79/409/EEC of 2 April 1979 (Conservation of Wild Birds) and 92/43/EEC of 21 May 1992 (Conservation of Natural Habitats and of Wild Fauna and Flora) (the "birds" and "habitats" directives) both reference non-indigenous species introductions.

In the former "[m]ember States shall see that any introduction of species of bird which does not occur naturally in the wild state in the European territory of the Member States does not prejudice the local flora and fauna" (article 11). Prior to a proposed introduction, Members must consult with the European

Commission. The purpose of this is not clear, but may allow potentially affected member States to be notified, although whether internal procedural rules within the Commission actually exist to enable this is unknown.

Under the latter, member States "shall ensure that the deliberate introduction into the wild of any species which is not native to their territory is regulated so as not to prejudice natural habitats within their natural range or the wild native fauna and flora and, if they consider it necessary, prohibit such introduction" (article 22(b)). A committee established to assist the European Commission with the directive's implementation is to receive assessment results for information.

As a group, whether in force or not, the global and regional instruments can generally be described as confirming that non-indigenous species introductions have been recognized as a problem worldwide in a variety of contexts: phytosanitary measures, biodiversity, birds and other migratory species, oceans, regional seas, mountainous areas or particular river or lake systems. It is estimated that only 14 States have not become a party to some treaty in force referring to non-indigenous species. When the 1951 International Plant Protection Convention is excluded the number rises to 15 States.

Of course, such a simple statistic glosses over the assortment of instruments to which the remaining 181 States have become parties to and have undertaken obligations to implement. While the absolute number of environment and conservation instruments referencing non-indigenous introductions is indeed impressive, there is an equally striking inconsistency in treatment and approach. For example, some instruments apply broadly to introductions both inside and outside protected areas, while others are limited to introductions in protected areas, seemingly premised on the wishful notion that introductions into surrounding areas will not spread into protected areas; consultation with neighbouring States prior to an intentional introduction is only specifically required by the Benelux Convention; accidental introductions are rarely specifically mentioned; and control or eradication measures are generally called for in the global treaties and but regionally, only in the Bern, Central American and Lake Victoria agreements.

Finally, even the most comprehensive obligation does not go into the specifics of implementation. For example, common protocols for decision-making or evaluating environmental risk or impact are not established, usually leaving it to each party to decide what action is appropriate. The lack of detail is probably due to the nature of treaty making, which is driven by the need to seek consensus sometimes between States with widely disparate interests and implementation capacities, but leaves open the possibility for disparity in approach. With regard to this final point, soft law and technical guidance documents are useful supplementary instruments lending themselves to more normalized national approaches.

Soft law and technical guidance documents

The soft law and technical guidance documents referencing non-indigenous species introductions represent a continuum of increasing specificity and guidance. Soft law documents are in two categories.

The first category comprises instruments which reference non-indigenous species introductions in the context of non-binding, action-oriented environment and development plans. For example, Agenda 21 has a number of references to non-indigenous species introductions, or human activities which can contribute to them, in its chapters on combating deforestation (chap. 11), combating desertification and drought (chap. 12), promoting sustainable agriculture and rural development (chap. 14), biodiversity conservation (chap. 15), environmentally sound management of biotechnology (chap. 16), protection of oceans (chap. 17) and freshwater resources (chap. 18).

In general, non-indigenous species are often inconsistently referred to in Agenda 21 as "exotics", "aliens", "biological control agents", "pests", "foreign" plants and animals, "non-indigenous species" and "noxious aquatic species". Astoundingly, other than acknowledging that inappropriate plant and animal introductions have contributed to biodiversity loss, Chapter 15 (Biological Diversity) makes no other reference to non-indigenous species introductions or their control. Agenda 21's most explicit reference calls on States to assess the need for appropriate rules on ballast water discharges to prevent the spread of non-indigenous species (chap. 17.30).

The few calls for regulatory measures tend to be set-out against activities which generally encourage non-indigenous introductions such as such as aquaculture, afforestation and reforestation and biological control in integrated pest management, implicitly reflecting a tension between environmental protection and human activities contributing to development. However, even with its deficiencies, Agenda 21 is the first international instrument to acknowledge that non-indigenous species introductions are not just an environmental issue, but a development issue for developed and developing countries alike.

The Programme of Action for the Sustainable Development of Small Island Developing States (SIDs) acknowledges that certain non-indigenous species have resulted in a significant loss of biodiversity on islands (para. 41), that protection of marine and terrestrial biodiversity from non-indigenous species should be included in national biodiversity strategies (para. 45(A)(i)), biologically significant sites need special protection (para. 45(B)(i)) and quarantine problems must be addressed nationally, regionally and internationally (paras. 55(A-C)). While it does not get into specifics, the SIDs Programme of Action demonstrates political awareness that non-indigenous species introductions are a major environment and development problem for SIDs.

The second category of soft law documents include recommendations by inter-governmental organizations and codes of conduct adopted by the specialized agencies of the United Nations. The voluntary nature of Recommendation No. R(84)14 of the Committee of Ministers to the Council of Europe Member States Concerning the Introduction of Non-native Species (1984), the International Maritime Organization (IMO) Guidelines for Preventing the Introduction of Unwanted Aquatic Organisms and Pathogens from Ships' Ballast Water and Sediment Discharges (1993) (work has begun on an annex to MARPOL 73/78), the UN Food and Agriculture Organization (FAO) Code of Conduct for the Import and Release of Exotic Biological Control Agents (1995) and the aquaculture provisions of the FAO Code of Conduct for Responsible Fisheries (1995) allow them to be drafted and subsequently adopted by States with more specificity than if they were binding.

These instruments generally set out the responsibilities of various actors – both public and private – involved with a particular activity and recommend practices and procedures to diminish the risk of a particular introduction. The principles they espouse and the detail that they provide make them useful sources upon which to base national legislation in those countries where it is lacking. In some cases such instruments are supplemented by more detailed implementation guidelines as will be the case for the FAO Responsible Fisheries Code.

Because they represent a convergence of international opinion regarding a particular activity of international concern, they establish a body of generally accepted international rules and standards against which to measure State activities and those of other actors such as the private sector. Consequently, they influence State practice and, therefore, international law.

These instruments also have their deficiencies, among them a failure to address control or eradication after release, as well as no references to responsibility or liability for damage resulting from non-indigenous species introductions. Only the FAO Biological Control Agents Code references the necessity of creating emergency procedures for an introduction gone awry.

A slight variation on the code of conduct is the technical guidance document. These documents, such as the International Council for the Exploration of the Seas (ICES)/European Inland Fisheries Advisory Commission (EIFAC) Code of Practice on the Introductions and Transfers of

Marine Organisms (1994), which is applicable to freshwater introductions and transfers as well, and the IUCN Position Statement on Translocation of Living Organisms (1987) are technically not soft law. However, like the codes of conduct adopted by the UN agencies, they too help establish the corpus of generally accepted international rules and standards by providing recommendations to States and other actors on their activities.

The ICES/EIFAC document is especially useful for suggesting information to be included in a pre-introduction prospectus to regulators, including a detailed analysis of potential aquatic ecosystem impacts. It is innovative in suggesting that prior to introduction, ICES/EIFAC members submit the prospectus to their respective organizations for an opinion.

The IUCN document is a holistic document looking at organism translocation in general. Its broadest principles are that (1) intentional introductions should occur only if clear and well-defined benefits exist and if no suitable native organism can be used; (2) accidental introductions should be "discouraged" whenever possible, especially in sensitive isolated ecosystems or habitats, by identifying and controlling human mediated pathways; (3) non-indigenous species causing negative impacts should be removed if possible, recognizing the need for prioritization, if necessary, especially if impacts occur in sensitive or unique ecosystems or habitats; (4) pre-existing scientific and regulatory structures should be relied upon and enhanced; (5) national legislation should require a permit for intentional introductions, regulate activities contributing to accidental introductions, establish criminal penalties and civil liability for deliberate introductions without a permit and negligence leading to accidental introductions; and (6) neighbouring States should be notified and consulted when transboundary movement of an introduction is probable.

The ICES and EIFAC are developing updated implementation guidance. IUCN's Invasive Species Specialist Group is developing guidelines on intentional introductions and control/eradication.

National legislative initiatives on non-indigenous species introductions

A global survey of national legislation on non-indigenous species introductions has not yet been undertaken. However, under the auspices of the Bern Convention's Group of Experts on Legal Aspects of Introduction and Re-introduction of Wildlife Species, a study was commissioned to review the status of national legislation on introductions (including sanitary regulations and genetically modified organisms), reintroductions (or re-establishment) and restocking in European States party to the Bern Convention, and States outside of Europe invited to join the Convention (de Klemm, 1995).

The analysis revealed considerable differences between countries: some countries have detailed rules while others have only adopted very general provisions or provisions that cover only a particular domain such as freshwater species. A handful of States do not appear to have any legislation at all.

A number of observations were made in the report. Among them legislation rarely addresses intentional introductions holistically (e.g. all environments; all organisms likely to be introduced) and is usually sector-oriented; sectoralism, typically of historical and administrative origins, impedes complete legal coverage and normalization; marine introductions and biological control agents are largely ignored, while non-native plant introductions are rarely regulated; few States regulate the introduction of organisms inter-regionally within their borders; sub-species and races are rarely addressed; "introduction" and "non-native" species is rarely defined; sanitary and phytosanitary controls are not targeted at preventing introductions in the broadest sense; provisions to limit range or eradicate introduced species are rare; accidental or unintentional introductions are rarely addressed; and offenses are generally considered minor despite damage done and civil liability for deliberate or negligent introductions is rarely mentioned.

Among the study's proposals, it was concluded that within Europe "a minimum number of rules, accepted and applied by everyone, aiming at anticipating and repairing the damage caused by inappropriate introductions. Such rules should be based essentially on principles that are now universally recognised, especially the principles of prevention and precaution, and the polluter-pays principle" (de Klemm, 1995). Though primarily limited to Europe, the study is instructive for the inferences which can drawn from it regarding the possible state of national legislation in the rest of the world. A global legislative survey is surely needed.

Establishing the basis for future action under the Convention on Biological Diversity

The Convention on Biological Diversity comprises a comprehensive set of far-reaching obligations for each Party to address the conservation of biological diversity at the genetic, taxonomic and ecosystem levels. A unique characteristic of the Convention is that its provisions are mostly expressed as overall goals and policies, rather than precisely defined obligations.

Action at the national level is emphasized and the CBD's "country-driven" nature allows each Party to tailor its own approach to biodiversity conservation through national biodiversity planning processes (article 6).

Through such processes the basis for implementing article 8(h) would be established.

The emphasis on national action and priority-setting is desirable from several stand points particularly flexibility of approach. However, consistency of approach and normalization of standards and practices – the results of coordination and common priority setting – may be sacrificed for article 8(h) without further action by the CBD's Conference of Parties (COP).

The COP could assist parties by fashioning common approaches in the areas of legislation, institutions, environmental impact (including risk) assessment and eradication and control of non-indigenous species introductions thereby enhancing the article's implementation and, ideally, facilitating international cooperation. This could be taken one step further by encouraging development of integrated approaches to intentional non-indigenous species introductions, the assessment and release of genetically modified organisms and the control of organisms of sanitary or phytosanitary concern at the national level. Common eradication and control practices could be explored.

Emphasizing integrated approaches which develop a country's "biosecurity" – to borrow the term and approach used by New Zealand – may actually increase efficiency and cost effectiveness, especially in developing countries, while building broad scientific, administrative and regulatory capacities. It could, in turn, facilitate the implementation of articles 8(h) and 8(g) (LMOs). A synergistic effect on the implementation of existing phytosanitary and conservation conventions and perhaps related trade agreements, as well as the future biosafety protocol on living modified organisms, may also result.

There are several tools at the COP's disposal, usable either together or separately, to facilitate normalized treatment of and harmonized approaches to non-indigenous species introductions. These are COP decisions and the development of annexes or protocols.

COP decisions

The COP's decisions are the most immediate way for it to facilitate normalized treatment of non-indigenous species introductions by parties. At its second meeting, the COP addressed the issue when it examined marine and coastal biological diversity conservation and sustainable use, a thematic area on its medium term work programme. It "supported" some of the recommendations made to it by the Conventions's Subsidiary Body for Scientific, Technical and Technological Advice (SBSTTA) (COP, 1995).

SBSTTA highlighted non-indigenous species invasions as one of the most important present and potential threats to marine and coastal biodiversity (SBSTTA, 1995). Its recommendations focused separately on mariculture and non-indigenous species in general, although a number overlapped. SBSTTA recommended *inter alia* (1) treating use of non-indigenous species in mariculture as an introduction into the wild because of the high risk of escape; (2) adhering, as a "minimum requirement", to the ICES Code and that of the International Epizootic Organization in mariculture operations; (3) undertaking rigorous pre-introduction environmental impact assessment, including risk assessment, based on the precautionary principle and assessing whether post-introduction monitoring can take place; (4) undertaking post-introduction monitoring; (5) assessing and giving preference to indigenous or local species alternatives; (5) addressing article 8(h)'s implementation in a party's national plan including implementation of international protocols and guidelines; (6) notifying neighbouring states on shared watercourses prior to introduction; (7) determining whether any adverse effects can be reversed within two human generations; (8) assessing the environmental impact of canal construction which may link coastal water bodies should be assessed; (9) conducting public education on the possible dangers releasing ornamental species and sport fishery species; (10) undertaking research on non-indigenous species impacts on *in-situ* conservation; (11) supplying information assessing the effectiveness of prevention, eradication and control technologies via the CBD's Clearinghouse Mechanism; (12) supporting and providing input into IMO's work on ballast water guidelines as well as review; and (13) contacting relevant international bodies and instruments to ensure adequate controls on alien or living modified organisms are addressed.

The first treatment of non-indigenous species introductions by the SBSTTA and COP is noteworthy for the relative comprehensiveness of recommendations parties received. However, problems in substance and approach can be discerned even at this early stage.

Substantive problems relate not to what was recommended but to that which was not. For example, failure to recommend that Parties create (a) a national legislative and institutional framework, including the use of incentive measures, for addressing non-indigenous species introductions, their eradication or control and (b) establish liability for subsequent damage, can only be explained by the lack of a minimum set of principles for negotiators to refer to as they worked. Such an oversight at so early a stage in the Convention's implementation foreshadows the potential for other substantive oversights as the COP and SBSTTA address non-indigenous species introductions in other contexts such as terrestrial and aquatic areas.

A fragmented approach to the issue, that is, addressing non-indigenous species introductions in the context of other thematic areas on the COP's medium term work programme might actually contribute to additional substantive oversights and inconsistencies as a new set of experts will be on the scene. Furthermore, such an approach may actually encourage the maintenance or proliferation of fragmented approaches not only at the national level, but globally as well especially if the COP provides inputs to other relevant international bodies and instruments as recommended by SBSTTA in the marine and coastal area.

Consequently, it may be advantageous to consider the desirability of placing non-indigenous species introductions, their control and eradication on the COP's medium term work programme as a separate, comprehensive thematic area. This could in turn generate a comprehensive set of principles and criteria for Parties to use within their own national contexts, and which could provide the basis for the COP's future inputs to other international bodies and instruments.

Annexes and protocols

As the COP addresses non-indigenous species introductions, it may be worthwhile for it to consider the level of normalization it would like to achieve on the issue. Decisions by the COP are not binding on individual parties, especially since the CBD does not provide for such decisions. At most they can only be considered "soft law". Consequently, they may not deliver a desirable level of normalization to achieve a particular goal. Annexes or protocols are the only means available to concretize a set of obligations, normalize practices and facilitate harmonized approaches.

Articles 23(4)(c) and (f) empower the COP to consider and adopt protocols and additional annexes to the CBD. Annexes form an integral part of the Convention. As a result, the COP has oversight over an annex's implementation and the Convention's financial mechanism could, in theory, support its implementation. However, creating new substantive obligations or rules in an annex may not be possible as an annex's content is restricted to procedural, scientific, technical and administrative matters (article 30(1)). Annexes would enter into force for all Convention parties, except those which have deposited a declaration of objection, one year later, after COP adoption (article 30(c)).

Protocols are separate legal instruments and parties to the Convention are not obliged to become protocol parties. The Convention does not specify the subject matter of protocols, implying that any matter covered by it may be covered by a protocol. A protocol typically contains additional substantive obligations or rules and, if necessary to promote particular goals, may actually exceed the scope of the Convention. Protocols may, if necessary, provide for their own international machinery such as a conference of parties, secretariat and financial mechanism separate from the underlying convention. A protocol under the CBD would enter into force only after a specified number of COP parties ratified it.

Negotiating an annex or protocol may take an equal amount of effort, but the primary advantage of an annex is the simplified procedure for its entry into force. This is especially attractive considering the slim likelihood that all CBD parties would become parties to a protocol, in addition to the possibility for extended delays in a protocol's entry into force, if at all.

A basic set of principles to guide future work under the Convention

It is ultimately up to the COP to decide whether and how to establish the basis for future action on non-indigenous species introductions. Primary consideration should be given to the level of normalization desired, coverage, expediency, ease of amendment and cost in terms of implementation and administration. In any event, existing international instruments and scholarly work could supply the substantive principles and criteria upon which the COP could base its future work. A non-exhaustive list of themes is provided in Table 26.1.

Conclusion

The Convention on Biological Diversity provides the best opportunity for developing a comprehensive global approach to the introduction of non-indigenous species and their eradication or control because of its global nature and wide ratification. The COP has already recognized the significance of non-indigenous species introductions to marine and coastal biodiversity, but risks promoting a fragmented approach to the overall issue unless it is formally included in the COP's medium term programme of work as a stand alone item. Placing the issue on the medium term work programme is a necessary first step and would provide the basis for initiating any process for future work on the subject.

Table 26.1 Basic themes for principles and criteria to guide future action on non-indigenous species introductions.

Principles and criteria might refer to:

- applying the preventative and precautionary approaches to all non-indigenous species introductions;
- establishing the "polluter or originator pays" principle for harmful non-indigenous species introductions;
- establishing clear State responsibilities with regard to neighbouring States including notification, consultation and liability;
- clarifying use of terms especially for "non-indigenous species" and "introduction";
- addressing non-indigenous species introductions and their eradication or control in national biodiversity strategies and action plans;
- creating a legal and institutional framework to address non-indigenous species introductions, including the use of non-indigenous biological control agents, while exploring possibilities for integrated approaches in the areas of GMOs and organisms of sanitary and phytosanitary concern;
- designating a single national focal point or creating a coordinating mechanism to clearly establish institutional authority for introductions and, where necessary, their eradication or control in terrestrial, aquatic and marine areas;
- creating a general prohibition on intentional introductions, whether by importation, or inter-regionally within a country, without authorization from a competent authority;
- shifting the burden of establishing no harmful impact to the originator of the proposed introduction;
- requiring pre-introduction environmental and risk assessment, including an alternatives analysis, as a minimum pre-requisite to obtaining a permit;
- monitoring after introduction;
- ensuring that organisms authorized for release are free from pathogens and other organisms which could affect biodiversity;
- identifying and controlling pathways of accidental introductions, including restricting the import and sale of non-indigenous pets, ornamental plants, birds and fish to those which cannot survive in the wild;
- assessing and, where necessary, prohibiting development projects which could lead to introductions;
- building alliances with relevant businesses, industry and other organizations;
- avoiding the inadvertent protection of non-indigenous species under negative lists of protected species;
- developing control and eradication plans for already introduced species which are harmful, prioritizing where necessary, and ensuring that the means ultimately chosen are first assessed for their environmental impact and risk;
- establishing an early warning system to detect introductions;
- establishing fast response or emergency procedures early after an introduction is detected;
- providing criminal penalties and civil liability for unauthorized intentional introductions and liability for negligence resulting in harmful accidental introductions;

Table 26.1 continued

- ensuring statutes of limitation reflect the long lead time it may take for harm to be detected;
- conducting public education and awareness campaigns;
- undertaking research and training; and
- providing adequate financial resources, eliminating perverse incentives and establishing incentive measures to prevent non-indigenous species introductions and ensure there eradication or control.

Acknowledgments

The views expressed here do not necessarily reflect the views of IUCN or its members. Many thanks to Ms Ulrike Deuschel in the Environmental Law Centre's, as well as Ms Anni Lukacs and Mr Torsten Wäsch who operate the IUCN Environmental Law Information System.

References

CBD (1995) *Secretariat Note Prepared for the Open-ended Ad Hoc Group of Experts on Biosafety.* (CBD/Biosafety Expert Group/Inf. 2), Convention on Biological Diversity Secretariat, Montreal.

COP (1995) Medium-term Programme of Work of the Conference of the Parties, Decision II/10, in *Report of the Second Meeting of the Conference of the Parties to the Convention on Biological Diversity,* Conference of the Parties to the Convention on Biological Diversity, (U.N. Doc. UNEP/CBD/COP/1/17), pp. 60–65.

Glowka, L. (1996) *Non-Indigenous Species Introductions: References in International Instruments.* IUCN, Bonn (Draft).

IUCN (1987) *Translocation of Living Organisms.* IUCN Position Statement. IUCN, Gland.

Jenkins, P. (1996) Free Trade and Exotic Species Introductions. *Conservation Biology,* **10**, 300–302.

Klemm, Cyrille de (1995) *Introductions of Non-native Organisms Into the Natural Environment.* Report to the Group of Experts on Legal Aspects of Introduction and Re-introduction of Wildlife Species (T-PVS (95) 17). Council of Europe, Strasbourg.

OTA (1993) *Harmful Non-indigenous Species in the United States.* Office of Technology Assessment, Washington, DC.

SBSTTA (1995) Scientific, Technical and Technological Aspects of the Conservation and Sustainable Use of Coastal and Marine Biological Diversity, Recommendation I/8, in *Report of the First Meeting of the Subsidiary Body on Scientific, Technical and Technological Advice,* Subsidiary Body of the Convention on Biological Diversity, (U.N. Doc UNEP/CBD/COP/2/5), pp. 34–43.

27 A global strategy for dealing with alien invasive species

HAROLD A. MOONEY
Stanford University, Stanford, California, USA

The need for a global strategy

A mandate of the Biodiversity Convention (Article 8h) is to "prevent the introduction of, control or eradicate those alien species which threaten ecosystems, habitats or species". Alien, (non-indigenous, introduced species, exotic) are those species that occur in places different from their current area of natural distribution. It is generally those alien species that are aggressive, or threatening, that are of concern – these are normally called invasive species, or harmful non-indigenous species. I will use the shorthand of invasive species here since it is principally aliens that fall into this category.

This is a very sweeping charge, and one that we cannot fulfill with our current knowledge base and management tools. We are going to have to develop a whole new way of approaching this problem if we are going to achieve this grand, and crucially important, objective.

Why, in fact, has such a difficult challenge been put forth? There are a number of reasons which include:

1) Alien species are often aggressive (invasive) when successfully established, since they have escaped, at least temporarily, their population regulators, such as consumers, parasites, etc.
2) There are many alien species that are threatening biotic systems and causing species extinctions. These aliens may be either purposeful, or inadvertent introductions.
3) Invaders can disrupt ecosystem processes that provide free services to society.
4) Invasives can cause great direct economic loss.
5) There are no broadly agreed-upon principles or guidelines for either allowing or excluding the introduction of biological material internationally.
6) Our predictive capacity on which species will become a successful invader is poor; however, we can predict which of those that do become established might do the most harm.

O. T. Sandlund et al. (eds.), Invasive Species and Biodiversity Management, 407–418.

The problem of invasives is great and probably will be become more severe since:

1) The numbers of invasive species is already very large in many parts of the world.
2) The growing human population will result in more disruption of ecosystems and hence systems that are more invasion prone.
3) Global commerce is rapidly increasing – hence the opportunity of inadvertent introductions.
4) Global change will increase the success of many short-lived, invasive-type species.

What this means is that at the same time there will be a need for greater ecosystem management due to an increasing human intervention into natural systems, our capacity to provide it will become diminished.

The components of a Global Invasive Species Programme (GISP)

There is no question that understanding and dealing with the invasive problem is an enormous challenge. We need to develop a new strategy for dealing with the invasive alien species problem. To meet this need SCOPE (Scientific Committee on Problems of the Environment) is calling for a focused, coordinated, and broadly-based approach to the invasive species problem that would engage the large, and global community, concerned with these issues. What follows are the elements needed for developing such a strategy:

Assess our current knowledge base

An update of our scientific understanding of the predictability of invasive capacity and potential ecosystem impact of introduced species. Since the 1989 global overview (Drake *et al.*, 1989) of the invasive problem there has been a considerable amount of new research in this area that needs to be assessed and incorporated into policy options. The U.S. Congressional report in 1993 does an excellent job of updating our knowledge but it is focused principally on the United States (OTA, 1993). Specifically we need to address the current human dimensions of the alien invasive species, including invasions through time, ethical and cultural concerns relating to invasives, the drivers of invasions, including how globalization trends in all sectors are affecting the mode and tempo of invasions. We need to reassess our current capacity to predict the population dynamics of invaders, including transport, establishment, and spread. We need to examine carefully the current vectors for invaders and the role of modern commerce in serving as new vector modes as well as how new international trade agreements are affecting the invasive problem. Finally, in reviewing our knowledge base we need to examine how current and projected land-use practices and atmospheric and climate global change are affecting invasive success.

Evaluating best practices and developing new tools

Early warning systems

The distribution of an organism is controlled to some degree by climate. It follows that species invading a system which has a comparable climate to that from which it originated has a greater likelihood of success than if it encounters a dissimilar climate. Thus a comparative analysis of successful invaders in the worlds similar climatic regions can give some indication of problems that can be avoided. For example, the star thistle, which is a native plant of the Mediterranean region has become a devastating weed in comparable climatic regions in California, yet is still at small population sizes in Chile, which shares a similar climate. It follows that close attention should be given to the Chilean populations. There is a clear need for comparative listings of problem plants in regions of comparable climates that could serve as early warnings of potential problems. Specifically we need to examine current global data bases on the status and movement of invasive species and their use as early warning. Further we need to develop the structures for more comprehensive data systems as well as the means to convey this information to all nations.

Rapid assessment of the status of invaders

We need to develop techniques to make rapid assessments of the degree of the success and movement of invaders and of their potential ecosystem impacts. These techniques have to be suitable across taxonomic groups and comparable through time. At present our knowledge about the status of invaders is generally of two states only – it is present (which is derived from floral and faunal lists) or it is firmly established and doing devastating damage (often learned from the popular press). Information is needed between these two extremes. In order to acquire this information a rapid sampling approach is needed that would produce a quantitative assessment of the status of invading species, that could be repeated at intervals, to provide a clear focus on emerging problems, helping to alleviate the crisis management approach to invaders. Efforts are needed to utilize developing technologies for tracking invasives including remote sensing and GIS. Further, we need to develop global maps of the distributions of the most abundant and most devastating invasive species, as well as the most sensitive ecosystem types.

Costing the full economic impact of invasives and of their control

A global assessment of the reduction in biodiversity that is incurred through the impacts of invasives is needed as well as the development of guidelines for the protection of native biota from invasions. The reduction in biodiversity that is being widely chronicled is due to many factors including habitat destruction, fragmentation, over-harvesting, etc. It is however not readily appreciated how much biodiversity loss is due directly to the action of invasive species. A more careful analysis of these losses is needed.

A full analysis of the economic and environmental costs and benefits of the control (chemical, biological, genetic) of invasives in relation to ecosystem services (e.g. clean and abundant water, clean air, sediment control) needs to be made. Decision making in relation to control of invasives is often made with incomplete information, or is based mainly on simple economics. In recent years we have learned a great deal about the "free services" that are provided by the components of ecosystems that are of considerable value and that need to be incorporated into full assessments of the impact of invaders as well as in their control.

Similarly we need to need a full costing of the impact of invasives and of their control on human health and the crops and timber that supports society.

Risk analysis of the consequences of the release of exotics

An analysis of the comparative treatment and concern over the release of genetically-modified organisms (GMOs) versus the release of biocontrol agents and introductions is needed as is the development of a more uniform policy of dealing with alien species and GMOs. In the past very strict procedures have been imposed for testing of GMOs before they can be "released" for general use. These policies have been prudent and have greatly reduced the possibilities for inadvertent consequences of a release. Similarly, care is taken before the introduction of a potential biocontrol organism to test for its selectivity and effectiveness. Oddly enough, no such care is generally taken prior to the purposeful introduction of horticultural material. What is needed is a comparative risk analysis of the rationale and effectiveness of the various approaches to the environmental release of exotic material, independent of their origins and purposes.

International agreement – laws and other instruments for dealing with alien invasives

An assessment is needed of the effectiveness of international agreements for controlling invasives, including those related to international trade agreements. At present there are three international agreements related to the control of invasive species. One of these is the International Plant Protection Convention (IPPC) that deals with agriculture pests only. It attempts to "strengthen international efforts to prevent the introduction and spread of pests of plants and plant products". The second is The Convention on Biological Diversity which calls for each contracting party to: "prevent the introduction of, control or eradicate those alien species which threaten ecosystems, habitats or species". Finally, the Convention on the Law of the Sea, would "prevent, reduce and control...the intentional or accidental introduction of species, alien or new, to a particular part of the marine environment, which may cause significant and harmful changes thereto". Unfortunately, not all nations are signatories to these agreements, or have ratified them, and further there are few provisions for enforcing the intent of these agreements. A comparative analysis of the effectiveness of these agreements needs to be made as well as consideration of the development of an agreement specifically targeted at the international transport of biotic material.

An analysis needs to be made of the current country mechanisms for dealing with invasives and their effectiveness. Importantly, new mechanisms for dealing with the transport and release of aliens needs to be developed including liability, insurance and bonding. Further codes of conduct and responsibilities for the various sectors of society need be proposed.

Increase education about invasives

Surprisingly, considering the enormous economic and ecological damage that can result from biotic introductions, there is a relatively low level effort made for educating the public about these consequences. A great deal can be done in education at all levels, from grade schools to the public at large. To begin with we need an evaluation of the success of those few programs we do have to educate the public on the issues related to invasive species. We need to educate tourists, but especially operators of international commerce carriers of all sorts, and those in the horticultural and pet trade. An investment in education could result in a large economic payback if even a fraction of the import of invasive species were curtailed.

We have a vast collective information base about the success and failures of various methods of eradication and control of invasive species. This information is not readily available globally. A clearing house mechanism needs to be developed where such information is readily available.

We need a mechanism to continuously update the priority research needs that will provide the tools for successfully dealing with alien invasives.

Building a Global Strategy

There has been agreement by the initial program partners (SCOPE, IUCN, UNEP, CABI) to work over the next three years to produce a global strategy (Global Invasive Species Programme (GISP)) for dealing with invasives, an action plan for policy makers for fulfilling the strategy, and a broadly popular document explaining the needs and requirements for the strategy. This program is a component of DIVERSITAS, a larger effort dedicated to the science of biodiversity. The initial Scientific Steering Committee of GISP is composed of the following people: D. Andow, M. Clout, L. Glowka, R. Hobbs, A. Holt, M. Lonsdale, R. Mack, J. McNeely, C. Perrings, M. Rejmánek, D. Richardson, P.J. Schei, J. Waage, M. Williamson and H.A. Mooney as Chair.

An outline of many of the issues that will be addressed by GISP is given below, as well as a flow diagram of the interactions among project components (Figure 27.1). The actual work to accomplish these objectives will take the dedication of a great number of concerned and expert natural and social scientists, lawyers, resource managers, educators and media persons.

I. The knowledge base – a global perspective

1) Human dimensions
    a) Historical perspective on invasions
    b) Socio-economic issues
        i) The players
        ii) Cultural concerns

   iii) Markets
   iv) Human settlement patterns
 c) Ethical issues and value systems
   i) Inter-generational obligations
   ii) Scale of concern
 d) Strategic directions
   i) Institutional responses
   ii) Building public awareness
   iii) The role of protected areas

2) Ecology of invasives
 a) Introduction
   i) A brief history of invasions
   ii) Invasive organisms: an overview of taxa
 b) Dynamics of expansion
   i) Natural modes of dispersal
   ii) Human-mediated dispersal
   iii) Are there distinct phases in the invasion process?
 c) What makes an organism invasive?
   i) Life cycles and life histories of invasive and non-invasive species
   (incl. hypotheses and testing based on existing databases)
   ii) Environmental requirements, phenotypic plasticity and fitness homeostasis
   iii) Population dynamics/demography
   iv) Genetic polymorphism, hybridization (genetic pollution) including pesticide resistance and specific problems with GMOs)
   v) Invaders vs. rare/endangered species: a mirror image?
 d) What makes an ecosystem invasible?
   i) Pools of potential invaders and major environmental gradients
   ii) Competition
   iii) Food-web interactions
   iv) Interactions among invaders
   v) Disturbance, fragmentation, stress, end eutrophication (including atmospheric fertilization)
   vi) Succession
 e) Predictions
   i) Modeling of dispersal in homogeneous environments
   ii) Modeling of dispersal in heterogeneous environments (incl. relevant epidemiological models)

iii) Predicting impacts of invaders on other organisms and ecosystems
iv) Screening procedures and expert systems

3) International pathways of harmful invasive species and their changing nature
   a) Quantitative assessments of pathways for possible case study countries
   b) Overall assessment of current international pathways for terrestrial invasive species
      i) Intentional introductions
      ii) Unintentional introductions
      iii) Unassisted spread from neighboring country where introduced
      iv) Escaped from cultivation or captivity, feralized, discarded
      v) Human disease vectors in relation to pathways of terrestrial invasive species
   c) Assessment of current international pathways for aquatic (marine and freshwater) invasive species
      i) Vessels
      ii) Aquaculture, other fisheries and aquarium industries
      iii) Other commercial, government and private activities
      iv) Scientific research
      v) Canals
      vi) Unassisted spread from neighboring country where introduced
      vii) Human disease vectors in relation to pathways of aquatic invasive species
   d) Forecasting the changing pathways of harmful introductions
      i) Socioeconomic and political trends affecting future pathways
      ii) Technological factors affecting future pathways
      iii) Projected future international pathways of greatest concern

4) Invasives and global change
   a) Introduction
      i) Changing trade pathways
   b) Impact of invasives
      i) Land use change overview
      ii) Changing agricultural practices
      iii) Climate change:
         Aquatic systems
         Terrestrial systems

Human diseases
iv) Atmospheric change:
CO_2 effects – plants
CO_2 effects – animals
Ozone effects
Nitrogen deposition
v) Biotic change
Herbicide resistance
vi) Country studies of potential impacts
vii) Modeling predictions of global change impacts
viii) New experimental approaches for exploring global change impacts on invasives

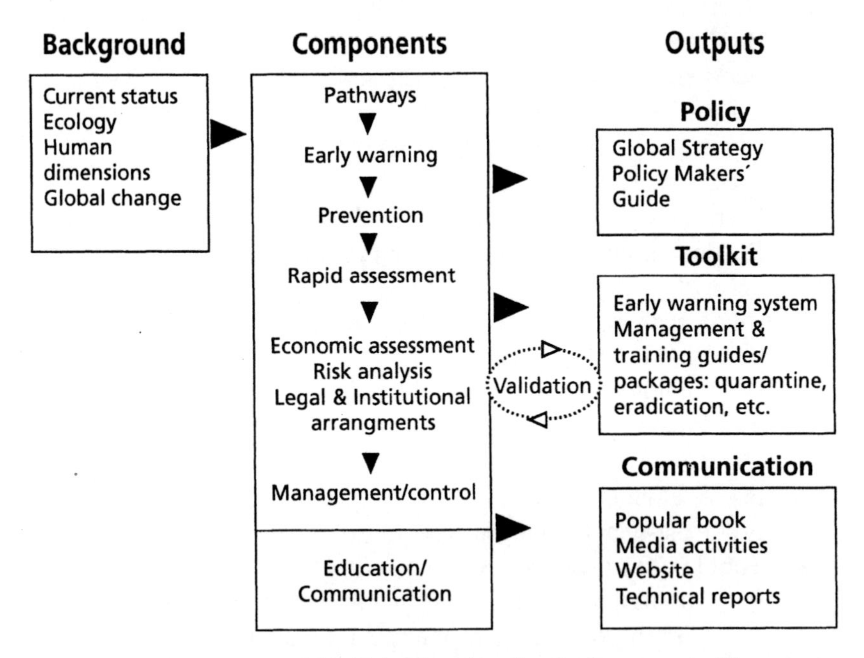

Figure 27.1 The elements of the Global Invasives Species Programme. The program will review our current knowledge of invasives, develop new tools to deal with them, and will communicate these results through a wide range of products.

II. The new tools needed

5) Early warning systems
   a) History of early warning systems
   b) Development of new systems

- c) Description and operation of current systems
  - i) Access
  - ii) Global use and interpretation
  - iii) Local use and interpretation
  - iv) Update process

6) Assessing the status of invaders
   - a) Worldwide assessment of status of invasions
   - b) Scales of detection currently available
   - c) Methods by which invasions can be rapidly and routinely reassessed
   - d) Methods for monitoring alien species not yet deemed invaders
   - e) Training program approaches for development officers

7) Economics of invasives
   - a) Overview
   - b) Biodiversity
   - c) Ecosystem functioning: the costs and benefits of invasives
   - d) Human health and epidemiology
   - e) Agriculture, forestry and fisheries
   - f) Trade, transport and travel
   - g) Distributional consequences of invasives
   - h) Recommendations: regulations and incentives

8) Control of invasives
   - a) Selection of target species
     - i) Ecology of invasive species problems and their management
     - ii) Present status of invasive species management in different countries
     - iii) Protocols for selecting target systems and species
     - iv) Cost-benefit analyses for control
   - b) Tool kits for invasive species management
     - i) Population ecology of invasive species management
     - ii) Current tools for invasive species management: physical, chemical, biological
     - iii) Risks and environmental impact of management tools
     - iv) New technologies in development
   - c) Approaches to management of major pest taxa
     - i) Management options: exclusion, eradication, control
     - ii) Management of vertebrates
     - iii) Management of terrestrial invertebrates
     - iv) Management of weeds

        v) New targets (marine invertebrates, diseases)
    d) Establishing management programs
        i) Organization of local and national invasive species management
        ii) International coordination of management programs

9) Comparative risk analysis approaches
    a) Inadvertent introductions
    b) Purposeful introductions
    c) Biocontrol agents
    d) Release of GMOs for control of invasives

10) Legal and institutional aspects of invasive alien species: legal component of the global invasive species strategy
    a) Knowledge base on existing international instruments, institutions and processes related to invasive species
    b) Knowledge base on existing legal instruments and institutional approaches at the national and sub-national level
    c) Comparison of the national and sub-national legal instruments and institutional approaches of selected case study countries
    d) Synthesis: Deficiencies in approaches at all levels
    e) Proposed new legal/institutional means for prevention/control/eradication of invasives

11) New approaches to education about invasives
    a) Overview of existing education programs at local to international levels
    b) Desired outcomes of education and messages to promote these outcomes
    c) Target audiences and how to reach them
    d) Assessing effectiveness of education efforts

12) Capacity building and enhancement
    a) Technical knowledge exchange (clearing house mechanisms on methods of eradication and control). Web page development and printed materials made easily available
    b) Develop the means of exchanging expertise and the development of short courses on prevention and control of invasives

Acknowledgments

The initial development of the above strategy was through the SCOPE executive committee and by important input from Peter Jenkins, Calestous Juma, Veronique Plocq, Cyriaque Sendashonga and Wendy Strahm. Subsequent input came as a result of discussions on the floor of the Trondheim Conference on Alien Species hosted by the Norwegian Ministry of the Environment, and input from M. Clout, J. Illueca, Pierre Lasserre, Jeff McNeely, Peter J. Schei, and J. Waage. More recently representatives of the partner organizations that compose the Scientific Steering Committee of GISP (SCOPE, IUCN, UNEP, CABI) have provided guidance. Initial support for this program is being provided by GEF, UNEP, ICSU and NASA.

References

Drake, J.A., Mooney, H.A., di Castri, F., Groves, R.H., Kruger, F.J., Rejmánek, M. and Williamson, M., eds (1989) *Biological Invasions. A Global Perspective.* John Wiley & Sons, Chichester, UK.

OTA (1993) *Harmful Non-Indigenous Species in the United States.* Office of Technology Assessment, U.S. Government Printing Office, Washington, DC.

INDEX

Population and Community Biology Series

1. R.M. Anderson (ed.): *Population Dynamics of Infectious Diseases: Theory and Applications*. 1982. Hb.
2. S.L. Pimm: *Food Webs*. 1982. Hb/Pb.
3. R.J. Taylor: *Predation*. 1984. Hb/Pb.
4. B.F. J. Manley: *The Statistics of Natural Selection*. 1985. Hb/Pb.
5. P.G.N. Digby and R.A. Kempton: *Multivariate Analysis of Ecological Communities*. 1987. Hb/Pb/Reprint.
6. P.A. Keddy: *Competition*. 1989. Hb/Pb/Reprint.
7. B.F.J. Manley: *Stage-Structured Populations*. 1990. Hb.
8. S.S. Bell, E.D. McCoy and H.R. Mushinsky (eds.): *Habitat Structure: The Physical Arrangement of Objects in Space*. 1991. Hb.
9. D.L. DeAngelis: *Dynamics of Nutrient Cycling and Food Webs*. 1992. Pb.
10. T. Royama: *Analytical Population Dynamics*. 1992. Hb/Pb.
11. D.C. Glenn-Lewin, R.K. Peet and T.T. Veblen (eds.): *Plant Succession: Theory and Prediction*. 1992.
12. M.A. Burgman, S. Ferson and H.R. Akçakaya: *Risk Assessment in Conservation Biology*. 1993. Hb/Reprint.
13. K.J. Gaston: *Rarity*. 1994. Hb/Pb.
14. W.J. Bond and B.W. Van Wilgen: *Fire and Plants*. 1996. Hb.
15. M. Williamson: *Biological Invasions*. 1996. Hb/Pb.
16. P.J. den Boer and J. Reddingius: *Regulation and Stabilization: Paradigms in Population Ecology*. 1996. Hb.
17. W.E. Kunin and K.J. Gaston (eds.): *The Biology of Rarity: Causes and Consequences of Rare-Common Differences*. 1997. Hb.
18. Shripad Tujlapurkar and H. Caswell (eds.): *Structured-Population Models in Marine, Terrestrial, and Freshwater Systems*. 1997. Pb.
19. J.P. Grover: *Resource Competition*. 1997. Hb.
20. J. Fryxell and P. Lundberg: *Behaviour and Trophic Dynamics*. 1998. Pb.
21. T. Czárán: *Spatiotemporal Models of Population and Community Dynamics*. 1998. Pb.
22. M. Scheffer: *Dynamics of Shallow Lake Communities*. 1998. Hb.
23. R.H. Karlson: *Dynamics of Coral Communities*. 1999 ISBN 0-7923-5534-2
24. O.T. Sandlund, P.J. Schei and A. Viken (eds.): *Invasive Species and Biodiversity Management*. 1999 ISBN 0412-84080-4
25. J.J. Worrall (Ed.): *Structure and Dynamics of Fungal Populations*. 1999 ISBN 0412-80430-1

KLUWER ACADEMIC PUBLISHERS – DORDRECHT / BOSTON / LONDON

Printed in the United Kingdom
by Lightning Source UK Ltd.
9521600001B

9 780412 840807